Lecture Notes in Biomathematics

Managing Editor: S. Levin

53

T. M.

Evolutionary Dynamics of Genetic Diversity

Proceedings of a Symposium held in Manchester, England, March 29 – 30, 1983

Edited by G. S. Mani

Springer-Verlag
Berlin Heidelberg New York Tokyo 1984

Editor

G. S. Mani
Department of Physics, Schuster Laboratory
The University, Manchester M13 3PL, England

AMS-MOS Subject Classifications (1980): 92-02, 92A10, 92A15

ISBN 3-540-12903-0 Springer-Verlag Berlin Heidelberg New York Tokyo
ISBN 0-387-12903-0 Springer-Verlag New York Heidelberg Berlin Tokyo

Printed in Germany

Printing and binding: Beltz Offsetdruck, Hemsbach/Bergstr.
2146/3140-543210

TABLE OF CONTENTS

TABLE OF CONTENTS CONT'D.

PROLOGUE

This volume is a result of a symposium on "The Basis of Genetic Diversity" held in Manchester on the 29th and 30th of March, 1983 under the auspices of the Genetical Society. The need for a symposium on this topic arose out of the long and often confusing discussions on the nature and the origin of polymorphism between Laurence Cook and me over innumerable cups of coffee. We both felt the need for a meeting of the proponents of different points of view to discuss the problem of genetic diversity and hopefully arrive at a concensus of understanding. We are very grateful to the Genetical Society and in particular to Dr. B. Burnet, Secretary of the Society, for financial assistance and encouragement without which the symposium would not have been possible.

There were two sessions devoted to invited talks and one session to contributed papers. There were seven invited talks and these were given by L. M. Cook (Manchester), E. Nevo (Israel), M. Nei (U.S.A.), G. S. Mani (Manchester), B. Clarke (Nottingham), C. Wills (U.S.A.) and R. Harris (Manchester). Of the seven speakers, Prof. B. Clarke and Prof. R. Harris did not submit a manuscript for inclusion in this volume. Prof. Clarke's talk was entitled "Diversity Due to Frequency-dependent Selection" and Prof. Harris gave a talk on "HLA and Interacting Polymorphism on Disease Resistance".

This volume is divided into five chapters, each chapter being the contribution from one of the five speakers. In the first chapter Laurence Cook traces the tortuous history of polymorphism and discusses what the problem is concerning the maintenance of polymorphism. He stresses the need for introducing ecological parameters and constraints in the general framework of population genetics to enable a better understanding of genetic diversity.

In the second chapter on "The Evolutionary Significance of Genetic Diversity", Nevo presents a very exhaustive compilation of all the available data on allozyme polymorphism. Nevo also presents the results of an extensive analysis of the data to determine the correlation between the observed genetic diversity and the ecological, the demographic and the life-history parameters. He shows that if one considers the enzyme polymorphism on a global scale, then one is led to the indisputable fact that the data can only be explained by including the various extrinsic factors mentioned above. He thus demonstrates the shortcoming of the "neutral mutation-drift" theory in as much as it disregards these various factors. The inclusion of these factors would imply the existence of selective features at the enzyme level. Part of these correlations obtained by Nevo, it may be argued, arises through random sampling and through fine sample sizes (see, for example, R. K. Selander and H. Ochman's discussion of enzyme polymorphism in molluscs, Isozymes, Vol. 10 (1983), pp. 93-123). Theoretical models behave according to Wright's "shifting balance theory", moving from one "adaptive" peak to another depending on the available data and on the current fashion in theoretical biology. Thus, the interpretation of these results could change with time. On the other hand, the extensive data and reference compilation by Nevo has a more permanent character and I hope would be of use to both experimentalists and theorists for years to come.

In the third chapter, Nei discusses the data on variability both at the enzyme and at the DNA level. He compares these results with the predictions of the neutral model and the various models based on balancing selection. He demonstrates that both at the enzyme level and at the DNA level the data are consistent with the neutral theory and cannot be explained through classical selection theories based on balancing and on overdominant selection. He also makes some critical comments on the ecological-genetic theory (see chapter four). I discuss some of these objections in the Appendix to chapter four.

The fourth chapter describes an ecological genetic model for enzyme polymorphism. The model shows that ecological constraints such as resource availability, interactions between populations and environmental fluctuations could lead to relatively weak selection at the genetic level. It is shown that the model yields predictions that are compatible with the enzyme data. Further, since the model incorporates ecological and environmental features, it can in principle explain the types of correlations that Nevo obtains.

The last chapter by Wills on "The Possibility of Stress-triggered Evolution" may appear to be outside the problem discussed in the earlier chapters. In this chapter Wills argues persuasively for the possibility that discontinuous and perhaps rapid evolution could proceed through the process of "stress-triggered" evolution. Such evolution is the result of the interaction between internal co-evolved mutagenic agents such as viruses, plasmids and transposable elements which trigger genomic changes in the presence of specific environmental stresses. Such co-evolution processes, if they exist and are responsible for triggering evolutionary changes, would proceed through a mechanism not very dissimilar to the ecological-genetic model applied, for example, to the co-evolution of the host-parasite system and to the variability created through the interaction of such systems with the environment. Wills suggests some experimental ideas to verify the mechanism for discontinuous evolution that he proposes.

Finally I would like to express my deep gratitude to Dr. L. M. Cook who was the chief motivator behind this symposium and whose enthusiasm and encouragement was largely responsible for the success of the symposium. I would like to take this opportunity to thank Mrs. Barbara Barlow for the secretarial assistance both in regard to the symposium and to this volume. But for her help and patience, this volume would have taken much longer to produce.

G. S. Mani

THE PROBLEM

L. M. Cook,
Department of Zoology,
University of Manchester,
Manchester.
M13 9PL

The maintenance of genetic diversity is probably the most fundamental issue in population genetics. Only one generally accepted way exists for generating diversity - mutation - and there is only one force shaping the overall course of evolution - natural selection. There is no doubt, however, that the picture revealed by study of natural populations is quite incompatible with the suggestion that if most genetic loci are selected the diversity is simply the result of the balance of mutational input and selective elimination. There is far too much polymorphism for this to be true. There is also a great ability to respond in an adaptive manner to new environmental conditions, almost as if a high level of polymorphism is maintained for the purpose of making such responses possible. The central problem is to determine the purposeless combination of factors that brings this situation about. The literature of population genetics has circled round this issue for at least fifty years, and it is perhaps not too frivolous to compare the situation with one which arises in a science fiction novel called the Hitch Hiker's Guide to the Galaxy (Adams 1979). An immensely powerful computer called Deep Thought is set the task of finding the answer to the Ultimate Question of Life, the Universe and Everything. After an enormously long time it comes up with the answer, which turns out to be 42. "I checked it very thoroughly" said the computer, "and that quite definitely is the answer. I think the problem, to be quite honest with you, is that you've never actually known what the question is."

In relation to genetic diversity we now have a fair picture of how much polymorphism there is (the answer is 24 per cent, or thereabouts, for diverse sets of data such as random samples of enzyme loci (Nevo, Chapter 2) or visual shell characters in mollusc species (Clarke *et al*. 1978)). We also know the levels of heterozygosity and with less certainty, the average number of alleles per polymorphic locus. One of the main ways of trying to determine what this implies has been to ask the question: are the alleles concerned largely under the influence of natural selection or are they neutral?

Anyone who holds that most polymorphisms are maintained by balancing selection would be prepared to refer back to the Darwinian argument that evolution proceeds as a result of natural selection based on a struggle for existence and survival of the fittest. Most selectionist arguments in population genetics, however, are based on formal representations such as

$$\Delta q = \frac{q(1-q)}{k.\bar{w}} \cdot \frac{d\bar{w}}{dq}$$

This approach has yielded a very rich scenario of possibilities to be investigated. It has, nevertheless, an undefined relation, and possibly no relation at all, to the verbal statement of Darwinian natural selection, because it by-passes the fact that the struggle for existence is a struggle, and therefore density-regulated. This issue was discussed in the 1950's (Cain and Sheppard 1956 give references), but with few exceptions such as the work of Roughgarden (1979), it seems to have died of exhaustion rather than to have been properly explored.

The classical selectionist position is that polymorphisms are maintained by balancing selection with constant fitnesses resulting in heterozygote advantage. One might reasonably ask how it came about that so many allele combinations, including multiple alleles, resulted in heterozygote advantage. An answer is available (the Sheppard, Caspari theory, Sheppard, 1975) which suggests that heterozygote advantage arises by modification of phenotypic expression as a result of selection of modifiers. Experimental evidence shows that modification of expression can be achieved over a very few generations (Ford, 1971), and the argument is an extension of Fisher's theory of the evolution of dominance. Yet it is almost impossible to see how either could come about as a result of change in frequency of alleles with modifying effect situated at unlinked loci unless the genome contains a ready supply of selectively neutral genic variability (Ewens, 1979).

To illustrate the bones of the problem, consider the argument in the following terms. A mutant showing incomplete dominance appears in a population where its net effect is advantageous. In order to become established as a polymorphism the advantageous effects of the gene have to become more dominant and the disadvantageous ones more recessive. A second, unlinked locus has a modifying effect on it. The relation betwe the two may be represented as follows.

			Locus Studied			
	Modifying locus		++	+a	aa	Genotypes
Genotypes	Frequency	Fitness	u^2	$2u(1-u)$	$(1-u)^2$	Frequency
++	q^2	$1-t$	$1-s$	$1-\frac{1}{2}s$	1	Fitness
+m & mm	$1-q^2$	1	$1-s$	1	1	

The problem is for the locus studied to move from the fitnesses shown in the top row to those in the second row. This can only occur if the modifier frequency q decreases. Change in q may be described approximately as,

$$\Delta q = -\frac{tq^2(1-q)}{1-tq^2} - \frac{su(1-u)q^2(1-q)}{1-suq^2-su^2(1-q^2)}$$

If s and t are small, this means that the term $su(1-u)$ must overcome any selection on the modifier locus, measured by the selective coefficient t, tending to move the frequency in the other direction. The number of generations during which $u(1-u)$ is reasonably large is inversely related to s, so that the total pressure applied to the modifying locus by an advantageous gene can never be large, and if this terminology is realistic, an effect is only conceivable if t is near zero. If heterozygote advantage is to arise an equivalent type of modification has to act on the disadvantageous properties of the gene, too.

One alternative is to consider the situation in which the modifier is closely linked to the locus it modifies (O'Donald and Barrett, 1973, Charlesworth and Charlesworth, 1976), but we then have to ask how commonly close linkage would be expected. Another possibility is that a genetic mechanism exists for modifying dominance, perhaps through the gene regulation system, but that it is not adequately represented by the type of formulation outlined above, being much more efficient. O'Donald (1968) obtained modification of dominance in a computer simulation by representing the modifying agency as genetic variance in the expression of fitness i.e. by using the methods of quantitative genetics. We know too little about gene regulation in higher organisms to chose between these types of representation. They illustrate the general problem of dealing with fitness conferred by interacting loci, using the existing analytical methods and the assumptions on which they are based.

Quantitative genetics is certainly seen by selectionists as part of thei: own programme. Genetic variability is maintained by stabilizing selection acting on the phenotype, and since selection is stabilizing for much of the time, variability is available to allow appreciable and rapid response to directional selection when it occurs. The formulation of the models of genetic systems giving rise to the phenotypes, however, suggests that to a large extent there is interchangeability between alleles, so that different sets may produce the same phenotype. In othe words, they are selectively neutral.

If evolution of heterozygote advantage raises difficulties it is worth investigating the possibility that heterozygote advantage would arise by chance if fitnesses were assigned to genotypes at random. Depending on the model the probability of chance occurrence may be quite high (e.g Avery, 1977). One simple formulation of the problem could run as follow Starting as in Fisher's demonstration of the importance of mutants of small effect, we may ask what the probable consequences are of a rare mutation when it arises in a new environment. The mutant has an effect which is random with respect to the environment. If we compare the mutant homozygote with the "wild type" homozygote the probability distribution of its relative fitness is likely to be normally distributed. The chance that a given mutant is advantageous is calculable if we know the mean for such mutants and the standard deviation of the distribution Now, the heterozygote is functionally related to the homozygotes, so that it must have a fitness which is on average related in some way to that of the homozygote. Suppose that x_1 and x_2 are the selective coefficients of the homozygote and the heterozygote, and that m_1 and m_2 are the means for all such homozygotes and heterozygotes. Because the heterozygote has an allele performing each of the alternative functions controlled by the locus, it is reasonable to suggest that $m_2 = \frac{1}{2}m_1$ and that $s_2 = \frac{1}{2}s_1$, where s_i are the respective standard deviations. For stable polymorphism to arise by chance it is necessary that $x_2 < x_1$ and $x_2 < 0$. A more restricting but simpler condition is that $x_2 < 0 < x_1$. We therefore need to know the probability that this will occur, which is given by the product $\int_0^{+\infty} y_{x_1}\,dx \int_{-\infty}^{0} y_{x_2}\,dx$, the y's being the respective probability distributions. If mutants are on average neutral, although individual fitnesses may be widely distributed about neutrality then so long as the probability distributions are symmetrical the chance of heterozygote advantage at any locus is $\frac{1}{4}$.

If the average fitness is not zero then the answer depends on the values given to the s_i. It would be reasonable to suggest that, from the

Poisson distribution, $s_1 = \sqrt{m_1}$, so that

$$y_1 = \frac{1}{\sqrt{2\pi m_1}} \exp\left[\frac{-(x_1-m_1)}{2m_1}\right]$$

$$y_2 = \frac{1}{\sqrt{\frac{1}{2}\pi m_1}} \exp\left[\frac{-(2x_2-m_1)}{m_1}\right]$$

On these assumptions we may guess at the probability that a particular mutant will arrive by chance at a state of heterozygote advantage. This probability is quite high even when mutants are on average disadvantageous; with a disadvantage of 20 per cent there is still a chance of about 22 per cent that the mutant will form a polymorphism (Figure 1).

These assumptions are unrealistic, however, in several respects. In the first place they assume no correlation between the fitnesses of the heterozygote and the homozygote. With complete correlation heterozygote advantage would be impossible. Curves indicating the probability when there is partial correlation are shown in Figure 1, obtained by comparing the circular distribution of y_1 and y_2 when $r = 0$ with elliptical distributions in which the axes are $1+r$ and $1-r$. With a correlation in fitness between genotypes of about 0.3 and with an average disadvantage of new mutants of as much as 20 per cent there is still a 12 per cent probability that heterozygote advantage will arise by chance. This type of argument would not, however, account for the fact that the majority of rare recurrent mutations affecting the phenotype are recessive.

A different selectionist position holds that fitnesses are not constant but are frequency-dependent, and that stable equilibria arise from the frequency dependence. There is an increasing amount of evidence for frequency dependence (Clarke, 1979) whereas there is an embarrassing shortage of examples of heterozygote advantage. Frequency dependent selection is capable of maintaining diversity when applied to phenotypes controlled by dominant alleles and in haploid and inbreeding species. With respect to genetic load there is a noteworthy difference between frequency-dependent and frequency-independent selection. Genetic load is a concomitant of polymorphism in frequency-independent systems as they are usually represented, and it becomes the more weighty the more polymorphisms are present. As frequency-dependent systems are represented,

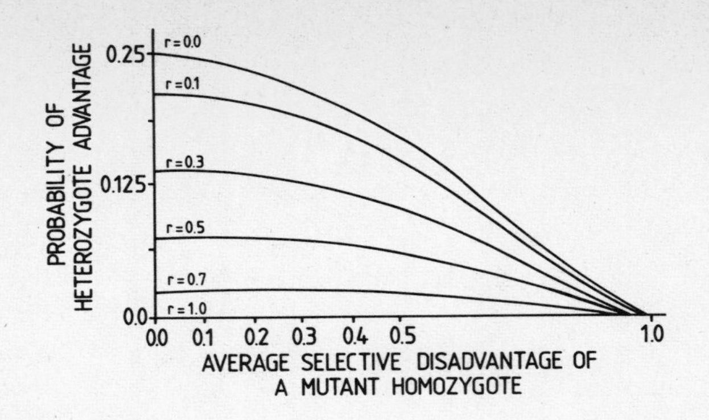

Figure 1. The probability that heterozygote advantage will arise by chance in a large population, given the parameters discussed in the text. It is intended to show that the probability is not necessarily negligible even when new mutations are assumed on average to be disadvantageous.

Table 1. Methods by which Polymorphism may be Maintained

Average level of selection	Environment: Spatially homogeneous	Environment: Spatially heterogeneous
$S \leqslant 1/2N$ "neutralist"	(a) mutation balance (b) mutation, selection balance	(h) mutation/migration balanced with selection
$S >> 1/2N$ "selectionist"	(c) selective balance (d) genotype frequency dependent selection (e) cyclic selection (f) sex, self incompatibility meiotic drive and other systems (g) advantage to heteromeric molecules	(i) selection, migration balance of various kinds

On the left are systems which do not require spatial heterogencity in the environment, on the right those which do. If the average level of selective difference is low then categories (a), (b) or (h) are the most important. For selective differences which are large category (i) is favoured by many modern workers. There is no evidence favouring (g) as a general mechanism maintaining polymorphism. Selective balance (c) may arise through adjustment of gene expression. If this is achieved by selection of alleles at other polymorphic loci then categories (a), (b) or (h) are implicated to maintain the variation at the modifying loci. (modified from Bishop and Cook, 1981.)

the load disappears at equilibrium. Nevertheless, concentration on the distinction between the two types of selection does not have as much heuristic value as these considerations seem to suggest.

Within the selectionist domain one of the important questions is whether or not the polymorphisms usually require environmental heterogeneity for their maintenance (Table 1). Heterozygote advantage can arise within a single habitat as a result of a balance of different types of selection acting on the homozygotes at some stage in the life cycle. Frequency-dependent selection (e.g. on a prey species as a result of predator behaviour) may do the same, and there are useful distinctions to be made between the two processes. A third class of models to be investigated involves selection acting in different directions on genotypes in different habitats, with movement between them. These models always require a limitation of the numbers of individuals of one morph which a segment of the environment may contain. This may be formulated as frequency-dependence. Often, but not necessarily, they result in net heterozygote advantage as well. Whether selection is frequency-dependent or independent is not as important as whether or not polymorphism depends on a varied environment.

With regard to the question of load, models of frequency-dependent selection predict only the load generated as a result of drift from the equilibrium frequency, whereas for frequency-independent selection there is also an appreciable load imposed at equilibrium. On average, this is half the average selective coefficient per locus. Ecological reality suggests, however, that most frequency-dependent selection is likely to be accompanied by some sort of load factor at equilibrium. Polymorphism resulting from frequency-dependent predation depends on the occurrence of predation, and that is not likely to disappear because the equilibrium frequency has been reached. If the selection is based on favoured niches then the load will arise as a consequence of imperfect ability to locate the favoured niche. The difference between the two types of selection in this respect is likely to be less great than it appears at first sight.

Does the load matter? The neutralist school would claim that it does, so that a scenario in which most polymorphic loci are subject to substantial selection is untenable. A variety of arguments have been put forward, however, to suggest that this is not so (discussed by Wallace, 1970, Wills, 1981). These involve truncation selection models, in which loci are selected in groups rather than individually, or "soft" selection models involving density-dependent modulation of the selection applied. The favoured phenotypes may be characterized by possession of particular genes, or gene combinations, or by a high average heterozygosity, and they may be subject to intraspecific competition. The supposed connectedness of phenotype to genotype is relaxed, as it is in quantitative models.

According to the neutralist position evolution takes place via selection on the phenotype. Because of the loose connectedness between phenotype and genotype this leaves a large number of mutations out of the selective net and functionally equivalent to each other. Mutations are effectively neutral to the product if the selective coefficient and the population size is less than 0.5. The selected part of the genome will be largely monomorphic; by picking on the polymorphic sector we are electing to study neutral mutations in a process of random drift.

A number of predictions follow, made possible because the mathematics of neutral systems is simple compared with that involving selection. The average rate of allele substitution should be constant in time. Comparing mutant with mutant, the rate of substitution should be inversely related to the functional constraints on the molecule. Thus, synonymous codons should be substituted at a higher rate than non-synonymous ones, mutations in introns at a higher rate than those in coding parts of the gene. The expected distributions of heterozygosity, allele frequency etc. are predictable as functions of mutation rate and population size.

Proponents of the neutralist position have emphasised its predictive power. It is claimed to be testable, and susceptible to refutation, where selectionist theories fail to make quantitative predictions (Kimura, 1979).

The fit to data has been tested. A number of predictions have been supported. Base substitution does appear to be higher in the third position (Brown et al. 1982). The expected relation between average heterozygosity and the variance in heterozygosity is obtained (Fuerst et al. 1977). Others fail the test. For example, the increase in heterozygosity with increase in population size does not follow the expected trend. Some enzyme loci which might have been expected to have neutral alleles have been shown experimentally to be subject to selection, most notable the alcohol dehydrogenase systems (e.g. Clarke, 1979, Kreitman, 1983). A large category produce results open to a variety of interpretations. The difference in frequency of silent-base and coding-base substitutions may be taken to show how important selective constraints on bases 1 and 2 actually are, or the silent bases may be subject to variable selection. The supposed constancy of amino-acid substitution rate may be disputed (e.g. Fitch 1976). Differences in frequency distributions between groups of organisms may indicate the effect of ecological factors on the loci concerned (Nevo, 1978 and Chapter 2). The neutralist model has not emerged unscathed from the testing process. It is necessary to postulate weakly deleterious mutations in quantities which, because of the definition of effective neutrality, are functions of population size. Worse, from the point of view of distinguishing it from weak selectionist theories, it is necessary to postulate population bottlenecks which have altered gene distributions at some unstudied time in the past.

These considerations suggest that the framework for discussion needs to be reassessed. The question why there is so much genetic diversity is a general one, and any answer will have to be general too. It is likely to come from examination of large data sets, predictions from computer simulations etc., and not from detailed case histories taken species by species. That point was realized at an early stage by the neutralists. The difficulty with the stepwise approach is its assymetry; a demonstration of selection shows that under some circumstances the locus is not neutral but failure to demonstrate selection does not imply neutrality. On the other hand, a general model may predict neutrality. An example is the case of the evolution of dominance, where the assumption that modifiers are unlinked polymorphic loci sets an upper limit to the primary selection which can act on them.

The approach should always be ecological genetic, however, in the sense that the investigator should always pay attention to how the environment shapes the phenotype. When this is done it is seen that selection may generally be expected to be density regulated. When environmental conditions are ameliorated selection is relaxed, and variant forms which previously were selected become functionally equivalent. Population size increases, and selection will again come into action. All alleles have the same status in this view, fluctuating from neutrality to selective constraint depending on environmental conditions and on gene combinations at other loci. The difference between selected and neutral genes is one of degree, and few mutants are likely to escape selection indefinitely.

The distinction between frequency-dependent and frequency-independent selection also becomes blurred, since most fitness values are to some degree density or frequency-dependent. Similarly, the problem of genetic load disappears. Species survive, flourish or fail in an ecological realm, and estimates of genetic load do no more than record how their genotypes equip them to do so.

Finally, we arrive at a rather uncomfortable conclusion. Many of the difficulties seem to result from trying to extract explicit solutions from the formal representations of population genetics. From the start it has been unclear what relation the representation of mean fitness has to real ecology. It is also unclear how one generalizes from single locus to multilocus systems, and whether the algebra of gene interaction is a reasonable way of describing genome organization. Representations such as those used above have only a limited value. They may show what is possible but cannot be used to test what is probable. For that, a larger number of parameters have to be considered together. The neutralist school avoided the problems by simplifying the algebra to the point where solutions can be obtained. The models then become unrealistic.

What are now required are simulations of evolving populations which are sufficiently complex to have some hope of providing insights about the real world. These must certainly include selection and finite, fluctuating population size. It will then be possible to match the predictions of the model to real data and to make critical comparisons between the various hypotheses. We would see whether the results obtained by the neutralists are model-dependent, or whether they would arise in selectionist models as well, and the robustness of the results under various sets of assumptions could be compared. The simulation approach also allows the problem of maintenance of diversity to be extended from the level of genetic polymorphism to that of the coexistence of interacting species. The result will be a much more comprehensive representation of evolution than those which we use at present.

References

Adams, D., 1979 The hitch hiker's guide to the galaxy. Pan Books, London

Avery, P. J. 1977 The effect of random selection coefficients on populations of finite size - some particular models Genet. Res. Camb. 29, 97 - 112

Bishop, J. A. and Cook, L. M. (ed.) 1981 Genetic consequences of man made change. Academic Press, London. Ch. 1.

Brown, W. M. Prager, C. M., Wang, A. and Wilson, A. C. 1982 Mitochondrial DNA sequences in Primates: tempo and mode of evolution. J. mol. evol., 18, 225-239

Cain, A. J. and Sheppard, P. M. 1956 Adaptive and selective value. Amer. Natur., 90, 202-203

Charlesworth, D. and Charlesworth, B. 1976 Theoretical genetics of Batesian mimicry. II. Evolution of supergenes. J. theor. Biol., 55, 305-324

Clarke, B. C. 1979 The evolution of genetic diversity. Proc. Roy. Soc. Lond. B, 205, 453-474

Clarke, B., Arthur, W., Horsley, D. T. and Parkin, D. T. 1978 Genetic variation and natural selection in pulmonate molluscs. in Fretter, V and Peaks, J. F. (ed.) Pulmonates. Vol. 2A. Systematics, evolution and ecology. Academic Press, London. pp 219-270

Ewens, W. J. 1979 Mathematical population genetics Springer, Berlin.

Fitch, W. M. 1976 Molecular evolutionary clocks. in Ayala, F. J. (ed.) Molecular evolution. Sinauer, Sunderland, Mass.

Ford, E. B. 1971 Ecological genetics. Chapman and Hall, London. 3rd ed.

Fuerst, P. A., Chakraborty, R. and Nei, M., 1977 Statistical studies on protein polymorphism in natural populations. I. Distribution of single locus heterozygosity. Genetics, 86, 455 - 483

Kimura, M. 1979 The neutral theory of molecular evolution. Sci. Am. 241 (5), 94 - 104

Kreitman, M. 1983 Nucleotide polymorphism at the alcohol de-hydrogenase locus of Drosophila melanogoster. Nature, Lond., 304, 412 - 417

Nevo, E. 1978 Genetic variation in natural populations: patterns and theory. Theoret. pop. Biol., 13, 121 - 177

O'Donald, P. 1968 Models of the evolution of dominance. Proc. Roy. Soc. Lond. B., 171, 127 - 143

O'Donald, P. and Barrett, J. A. 1973 Evolution of dominance in polymorphic Batesian mimicry. Theoret. pop. Biol., 4, 173 - 192

Roughgarden, J. 1979 The theory of population genetics and evolutionary ecology: an introduction. Macmillan, New York.

Sheppard, P. M. 1975 Natural selection and heredity. Hutchinson, London. 4th ed.

Wallace, B. 1970 Genetic load, its biological and conceptual aspects. Prentice-Hall, Englewood Cliffs, N.J.

Wills, C. 1981 Genetic variability. Clarendon, Oxford.

THE EVOLUTIONARY SIGNIFICANCE OF GENETIC DIVERSITY: ECOLOGICAL, DEMOGRAPHIC AND LIFE HISTORY CORRELATES

Eviatar Nevo, Avigdor Beiles, and Rachel Ben-Shlomo

Institute of Evolution

University of Haifa, Mount Carmel, Haifa, Israel

THE PROBLEM

The evolutionary significance of genetic diversity of proteins in nature remains controversial despite the numerous protein studies conducted electrophoretically during the last two decades. Ironically, the discovery of extensive protein polymorphisms in nature (reviewed by Lewontin, 1974; Powell, 1975; Selander, 1976; Nevo 1978, 1983b; Hamrick et al., 1979; Nelson and Hedgecock, 1980), did not resolve the disagreement between the dichotomous explanatory models of selection (e.g., Ayala, 1977; Milkman, 1978; Clarke, 1979; Wills, 1981) versus neutrality (Kimura, 1968; Kimura and Ohta, 1971; Nei, 1975; and modifications in Kimura, 1979a,b). The more general problem of the relative importance of the evolutionary forces interacting in genetic population differentiation at the molecular levels of proteins and DNA, i.e., mutation, migration, natural selection and genetic drift, remains now as enigmatic as ever.

Many attempts have been made to resolve the enigma of protein diversity. These include the theoretical testing of neutral models (Nei, 1975 and later works; Ewens, 1979 and references therein); genetic-environmental correlations (e.g., Bryant, 1974; Hedrick et al., 1976; Hamrick et al., 1979; Nevo, 1978, 1983a); biochemical kinetics *in vitro* and/or *in vivo* (reviewed in Nevo, 1983 b, and exemplified by Koehn, 1978);

physiological function (Johnson, 1979); and biological testing of allozymes (Nevo et al., 1983). Each of the above mentioned methodologies has its pros and cons, and only their combined multidisciplinary efforts might ultimately illuminate the obscure interactions of the operating forces, and the adaptive significance of an enzyme polymorphism. Nevertheless, it is important to recall that the selectionist - neutralist debate is essentially quantitative. Selectionists do not ignore genetic drift but believe that selection is of prime importance. Likewise, neutralists believe that adaptive evolution occurs by positive Darwinian selection, but only a small portion of gene substitutions is sufficient for this (Nei, 1975), while extensive evolutionary change at the molecular level is propelled primarily by random drift (Kimura, 1979b).

Genetic-environmental correlations demonstrate inferentially the adaptive significance of enzyme polymorphisms. Hence, they need additional evidence to support a causal mechanism and evaluate the selective and nonselective causes through fitness tests (see critical discussion in Nevo et al.,1983 and references therein). However, the correlative methodology has the obvious merit that it may easily cover many loci, populations and species. The strength of the methodology increases manyfold if the correlations are made, as in the present study, with ecological, demographic and life history characteristics. All these biological parameters were specified by the individual research workers for each species they studied genetically, thereby avoiding any specific bias of the present authors.

The objectives of this review are (a) to estimate the levels of genetic diversity as reflected in protein variation of higher taxa, involving the analysis of 1111 species studied primarily for allozymic variation, where in most cases each was tested for at least 14 gene loci; (b) to estimate the mean levels of genetic diversity of categories within 21 biotic variables: 7 ecological, 5 demographic, and 9 life history and other biological characteristics; (c) to associate the levels of genetic diversity with the ecological, demographic and life history factors in an attempt to estimate their relative roles and importance in genetic differentiation of populations and species; and lastly, (d) to compare and contrast the number of explanatory models concerned with the causes of genetic diversity in nature, based on the largest number of species analyzed to date.

MATERIALS AND METHODS

General

The data-set presented in Appendix I, includes taxonomic and genetic indices of polymorphism and heterozygosity for 1111 species, and ecological, demographic and life history characteristics for 814 species, of animals and plants. The genetic indices were derived from routine electrophoretic studies of allozymic variation in natural populations in which in most cases 14 or more loci (mean 23) and a minimum of 10 individuals per species (mean 199) were studied. This set of data is the largest assembled so far, and represents the most recent review of proteins, primarily allozymes, in natural populations and species.

Taxonomic classification

The 1111 species listed in Appendix I have been analyzed separately in three major taxa (vertebrates, invertebrates, and plants) which were subdivided into ten higher taxa (mammals, birds, reptiles, amphibians, fishes, crustaceans, insects excluding *Drosophila*, *Drosophila*, molluscs, and dicotyledons). Other higher taxa were excluded from the analysis due to small sample sizes. Appendix I lists for each species, the number of individuals, populations, number of gene loci tested, genetic indices and its biotic profile (i.e., ecological, demographic and life history characteristics) along with the reference.

This review compares and contrasts genetic variation among species living under varied conditions, based on routine electrophoretic studies. Hence, the comparisons are valid notwithstanding the well known underestimates of P and H derived from electrophoretic studies (Lewontin, 1974). Likewise, the levels of genetic diversity reported here are commensurate, regardless of the different levels of P and H obtained by either bidimensional (Brown and Langley, 1979), or sequential (Ramshaw et al., 1979) electrophoresis.

Genetic data

The genetic data involve the proportion of polymorphic loci per population (P, criterion 1%) and observed proportion of loci heterozygous per individual (H). Since genetic estimates vary in the literature we transformed all estimates of P to the 1% criterion of polymorphism. Likewise, all heterozygosity estimates appearing here are observed heterozygosity (see procedures of empirical transformations to P-1% and He to H in Appendix III). To obtain the best species estimates of P and H in cases where more than one report was available, we weighted (by the number of individuals, multiplied by the number of loci analyzed) the original estimates. In cases of species where genetic indices were reported separately for mainland and island populations, the minority of either mainland or island populations was excluded from the species means. Thus, discrepancies between the the genetic indices reported here and the original material result from the aforementioned transformation and exclusions. Lastly, in all the analyses no selection of species was made unless otherwise specified.

Biotic data

In order to correlate the genetic with the biotic structures we sent a questionnaire to the individual researchers inquiring about ecological, demographic and life history characteristics (Appendix II). This procedure was an attempt to obtain the best available biotic characterization of species and to avoid our personal biases. We are cognizant of the fact that while some of the variables in this questionnaire are totally objective (e.g., life zone), others (e.g., species size) are partly or totally subjective; still others which obviously vary with the taxonomic categories (e.g., body size), we standardized. A notorious and widely criticized ecological variable is habitat range, dichotomizing species into generalists and specialists. We agree that a clear cut distinction between specialists and generalists varies with (i) the taxonomic category, (ii) individual researcher, (iii) niche structure and characterization, (iv) habitat classification, (v) life cycle stage, (vi) intraspecific variation such as central versus marginal populations and (vii) combinations of all these factors as well as others. In any event, it should be recalled that generalism, as well as specialism, always remain relative terms and hence are potentially hazardous, inspite of their straightforward biological appeal. For remarks on other biotic factors see Appendix II.

Statistical analysis

The data were analyzed by uni- and multivariate analyses. The means and standard deviations of _P_ and _H_ were computed for different taxa, and for each category of the 21 biotic variables (7 ecological, 5 demographic, and 9 life history and other biological variables). Heterogeneity was tested by single classification analysis of variance, and group differences were tested by multiple comparisons analysis, Modified Least Significant Differences, LSDMOD (Hull and Nie, 1981). The pairwise comparisons between frequency distributions of _P_ and _H_ were tested by Kolmogorov-Smirnov test. Pearsonian correlations among all genetic and biotic variables were conducted after reranking categories in some of the variables in an attempt to approach a sequential order. This reranking permits us to use the variables in future analyses. The multivariate pattern of the data was analyzed by means of Smallest Space Analysis, SSA, which is a nonmetric technique designed to plot the intercorrelations of multivariate data in a space of minimal dimensionality (Guttman, 1968; Levy, 1981). Likewise, we have run stepwise multiple regression analysis, MR, to determine whether and what kinds of biotic factors influence, or are associated with genetic variables. The SSA computer program appears in Roskam and Lingoes (1970), and all other statistical techniques mentioned above in Hull and Nie (1981). Lastly, in all multivariate analyses only species characterized by a complete relevant biotic profile were included (listwise deletion).

THE EVIDENCE

Patterns Among Taxonomic Groupings (Table 1; Fig. 1)

Table 1a gives the means of $\underline{P}$ and $\underline{H}$ with their standard deviations, and the correlation between them $r(\underline{P}, \underline{H})$. Table 1b gives the results of all the multiple tests for major groupings. Mean genetic indices over all 1111 species were $\underline{P} = 0.284$ ($N = 1042$) and $\underline{H} = 0.073$ ($N = 968$), and their overall correlation $r(\underline{P},\underline{H})$ was 0.79[3] ($p < 0.001$). These estimates are similar to those reported in Nevo (1978) for 243 species.

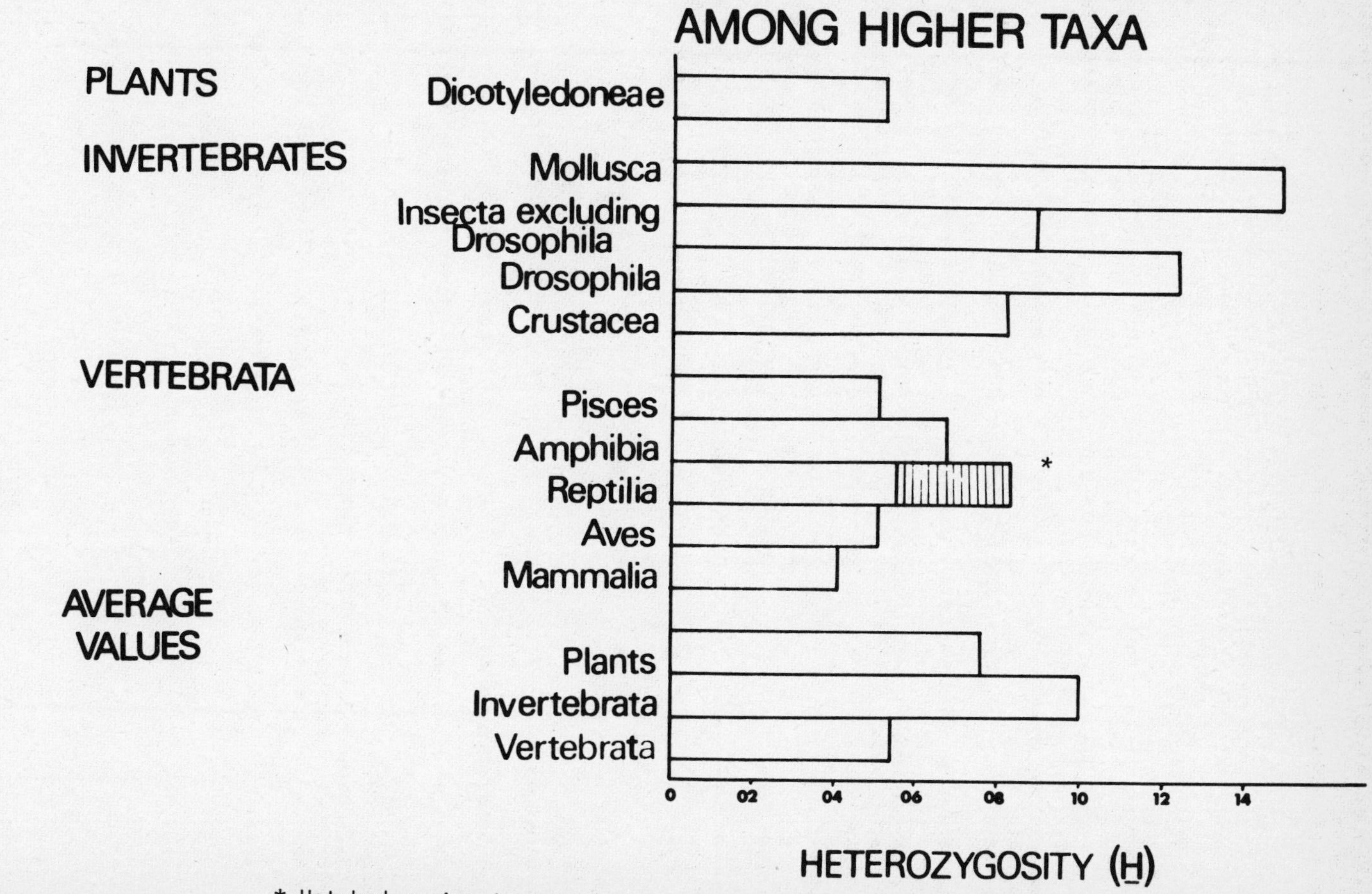

* Hatched region is due to parthenogenetic species

Fig. 1

Table 1a. Heterozygosity ($\underline{H}$) and polymorphism ($\underline{P}$) and their coefficient of correlation (r) over all species listed in appendix I, classified in seventeen higher and in three major taxa respectively.

	H			P			
	No. of species	Mean	S.d.	No. of species	Mean	S.d.	r ($\underline{H}$,$\underline{P}$)
I. Seventeen taxa							
Mammalia	184	0.041	0.035	181	0.191	0.137	0.821 ***
Aves	46	0.051	0.029	56	0.302	0.143	0.497 ***
Reptilia	$ 75	0.083	0.119	84	0.256	0.148	0.814 ***
exc.parthenogentic	70	0.055	0.047	84	0.256	0.148	
Amphibia	61	0.067	0.058	73	0.254	0.151	0.735 ***
Pisces	183	0.051	0.035	200	0.209	0.137	0.845 ***
Echinodermata	15	0.126	0.083	17	0.505	0.181	0.836 ***
Drosophila	34	0.123	0.053	39	0.480	0.143	0.552 ***
Insecta exc.*Dros.*	122	0.089	0.060	130	0.351	0.187	0.753 ***
Crustacea	122	0.082	0.082	119	0.313	0.224	0.879 ***
Chelizerata	6	0.080	0.033	6	0.269	0.098	0.876 *
Mollusca	46	0.148	0.170	44	0.468	0.287	0.764 ***
Brachiopoda	3	0.137	0.087	3	0.526	0.247	0.984 ns
Vermes	6	0.072	0.079	6	0.289	0.222	0.949 **
Coelenterata	5	0.140	0.042	5	0.481	0.191	0.840 ns
Monocotyledoneae	7	0.116	0.091	12	0.378	0.275	0.985 **
Dicotyledoneae	40	0.052	0.049	56	0.235	0.204	0.751 ***
Gymnospermeae	7	0.146	0.065	5	0.734	0.186	-0.948 ns
II. Three major taxa							
Vertebrata	551	0.054	0.059	596	0.226	0.146	0.792 ***
Invertebrata	361	0.100	0.091	371	0.375	0.219	0.769 ***
Plants	56	0.075	0.069	75	0.295	0.251	0.842 ***
Significance (for part II)		***			***		
Total	968	0.073	0.076	1042	0.284	0.197	0.793 ***

Abbreviations:
Ins.Ex.Dr. = Insecta excluding *Drosophila*.
Dicotyled. = Dicotyledoneae
$ Reptilia exc. parthenogenetic = describes genetic indices after exclusion of the parthenogenetic reptilian species.
Significance: * = $p < 0.05$; ** = $p < 0.01$; *** = $p < 0.001$; ns = $p > 0.05$

Table 1b. Multiple comparisons of heterozygosity (H) and polymorphism (P) among the ten higher and three major taxonomic groupings (LSD-modified test).

I. Ten higher taxa

Taxon	1		2		3		4		5		6		7		8		9	
	H	P	H	P	H	P	H	P	H	P	H	P	H	P	H	P	H	P
1.Mammalia	--	--																
2.Aves	ns	**	--	--														
3.Reptilia	**	ns	ns	ns	--	--												
4.Amphibia	ns	ns	ns	ns	ns	ns	--	--										
5.Pisces	ns	ns	ns	*	ns	ns	ns	ns	--	--								
6.Crustacea	***	***	ns	ns	ns	ns	ns	ns	**	***	--	--						
7.Ins.ex.Dr.	***	***	ns	ns	ns	**	ns	**	***	***	ns	ns	--	--				
8.Drosophila	***	***	***	***	ns	***	**	***	***	***	ns	***	ns	**	--	--		
9.Mollusca	***	***	***	***	***	***	***	***	***	***	***	***	***	**	ns	ns	--	--
10.Dicotyled.	ns	ns	ns	ns	ns	ns	ns	ns	ns	ns	ns	ns	ns	**	***	***	***	***

II. Three major taxa (I = Vertebrata; II = Invertebrata; III = Plants)

	H	P
I - II	***	***
I - III	ns	**
II - III	ns	**

The following results were indicated:

1. The three major taxa differed significantly in their P and H estimates (ANOVA ; $p < 0.001$). Furthermore, all pairwise combinations varied significantly in P (LSDMOD; $p < 0.01$) but only vertebrates varied significantly in H ($p < 0.001$).

2. Both P and H mean values decreased in the following order: invertebrates > plants > vertebrates.

3. Within vertebrates the lowest mean values of heterozygosity and polymorphism were those of mammals. Within invertebrates *Drosophila* and molluscs displayed the highest mean values of the genetic indices, whereas crustaceans and insects excluding *Drosophila*, exhibited the lowest values. The lowest values of

invertebrates are similar but still higher than those of amphibians (but equal to reptiles when the parthenogenetic species are included). The analysis in this and later sections excludes all the higher taxa with small sample size of species, including echinoderms, chelicerates, brachiopods, worms, coelenterates, monocotyledons and gymnosperms.

Relative frequency distribution of genetic indices (Table 2; Fig. 2)

To compare variability patterns we also analayzed the frequency distributions of polymorphism and heterozygosity among the higher taxa. The results are shown in Fig. 2 A-D, and their pairwise comparison by Kolmogorov-Smirnov tests are given in Table 2. The following patterns were found:

1. The distribution patterns obtained for both genetic indices showed striking heterogeneity among higher taxa. The major extremes in $\underline{H}$ were generally exhibited by mammals or vertebrates, as opposed to molluscs or most invertebrates. The mammals, and probably all vertebrates, showed a distinct J - type distribution. By contrast, particularly in molluscs, but similarly in all invertebrates, a shallow distribution approaching evenness was indicated. This heterogeneity is not only qualitative, but also quantitative (Table 2).

2. Vertebrates differed significantly from invertebrates and less so from plants in both genetic indices. Likewise, invertebrates differed from plants primarily in $\underline{P}$.

3. Within vertebrates, mammals displayed a distinct skewed J - distribution in $\underline{H}$, thereby differing significantly from all other classes. The same trend seemed to occur in $\underline{P}$, but less distinctly.

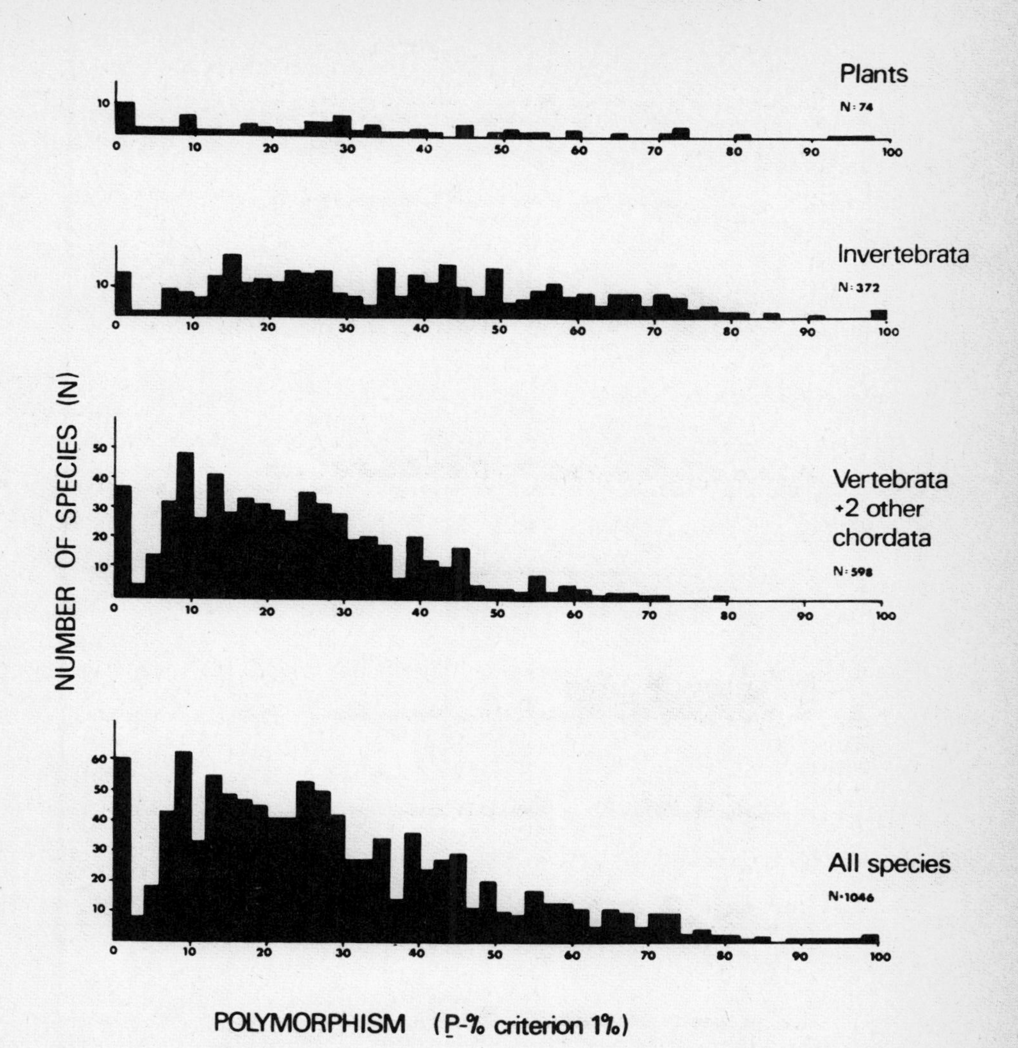

FREQUENCY DISTRIBUTION OF POLYMORPHISM (P)
Plants
N: 74
Invertebrata
N: 372
Vertebrata
+2 other
chordata
N: 598
All species
N-1046
NUMBER OF SPECIES (N)
POLYMORPHISM (P-% criterion 1%)

Fig. 2a

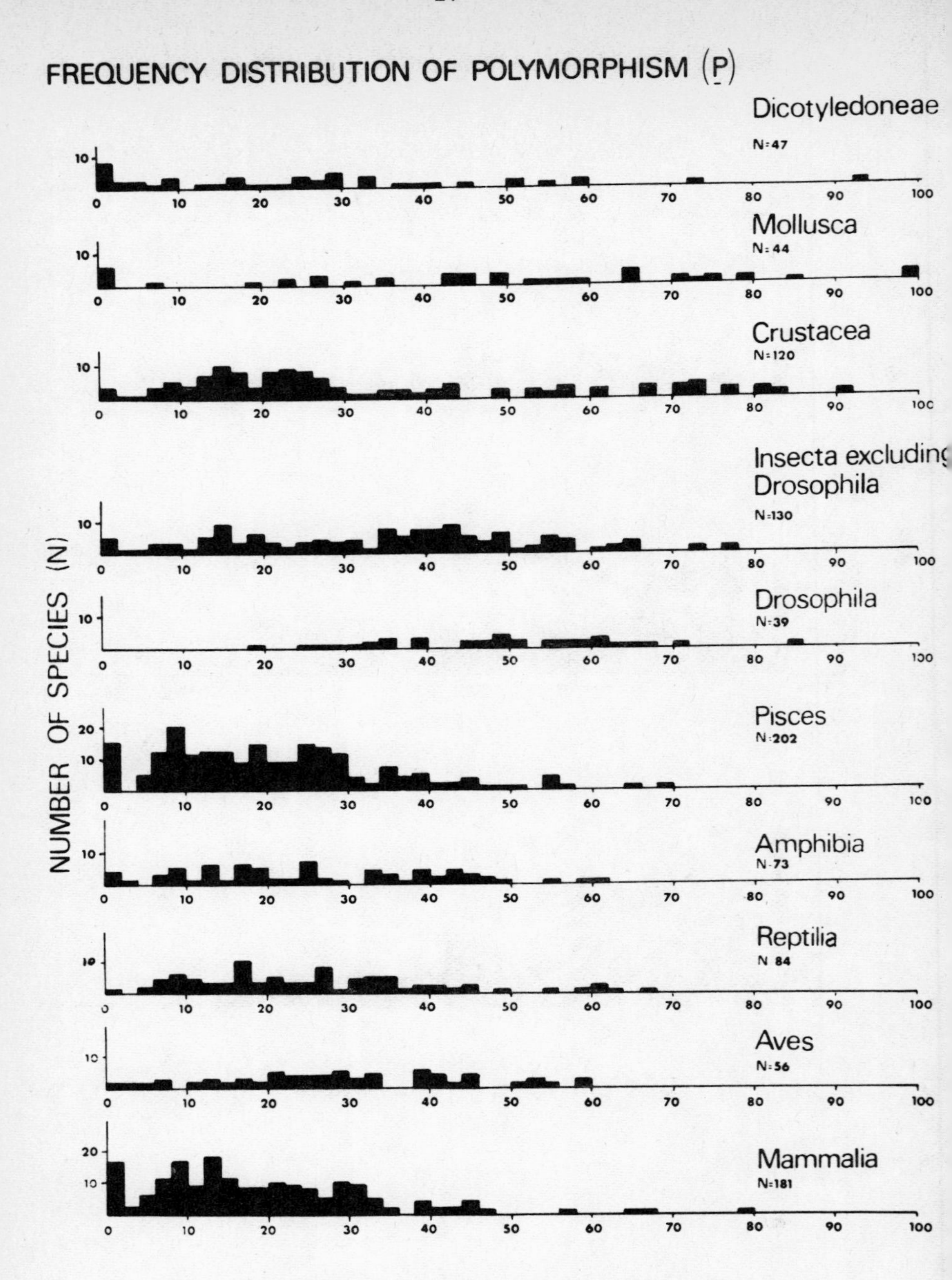
FREQUENCY DISTRIBUTION OF POLYMORPHISM (P)
Dicotyledoneae
N=47
Mollusca
N=44
Crustacea
N=120
Insecta excluding Drosophila
N=130
Drosophila
N=39
Pisces
N=202
Amphibia
N=73
Reptilia
N=84
Aves
N=56
Mammalia
N=181
NUMBER OF SPECIES (N)
0
10
20
30
40
50
60
70
80
90
100
POLYMORPHISM (P-% criterion 1%)

Fig. 2b

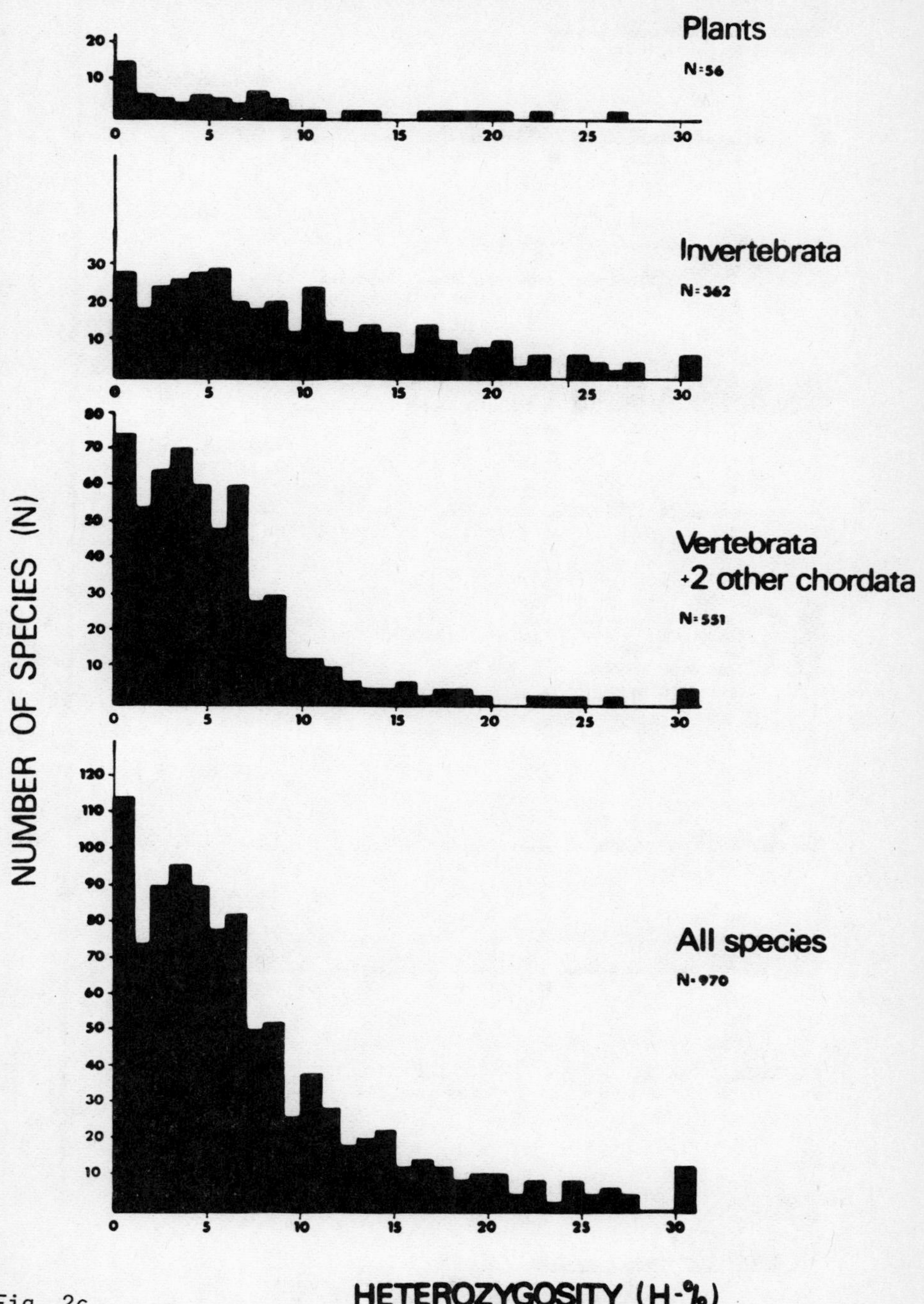
FREQUENCY DISTRIBUTION OF HETEROZYGOSITY (H)
Plants
N=56
Invertebrata
N=362
Vertebrata
+2 other chordata
N=551
All species
N=970
NUMBER OF SPECIES (N)
HETEROZYGOSITY (H-%)

Fig. 2c

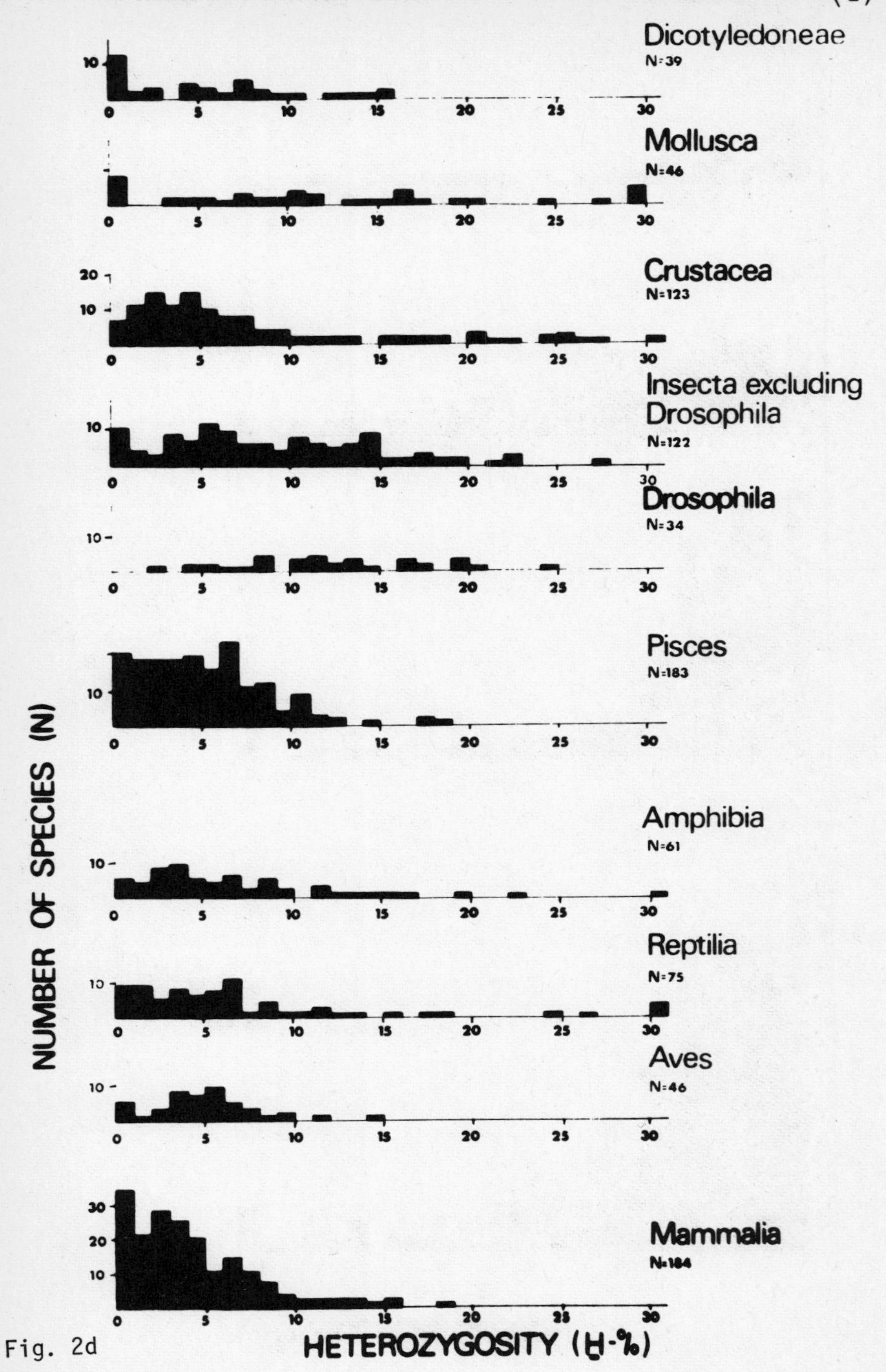
FREQUENCY DISTRIBUTION OF HETEROZYGOSITY (H)
Dicotyledoneae
N=39
Mollusca
N=46
Crustacea
N=123
Insecta excluding Drosophila
N=122
Drosophila
N=34
Pisces
N=183
Amphibia
N=61
Reptilia
N=75
Aves
N=46
Mammalia
N=184
NUMBER OF SPECIES (N)
HETEROZYGOSITY (H-%)
0
5
10
15
20
25
30

Fig. 2d

Table 2. Pairwise comparisons of the relative cumulative frequency distribution of heterozygosity (H) and polymorphism (P) among the ten higher and three major taxonomic groupings (Kolmogorov- Smirnov test).

I. Ten higher taxa

Taxon	1		2		3		4		5		6		7		8		9	
	H	P	H	P	H	P	H	P	H	P	H	P	H	P	H	P	H	P
1. Mammalia	--	--																
2. Aves	**	***	--	--														
3. Reptilia	*	**	ns	ns	--	--												
4. Amphibia	*	**	ns	ns	ns	ns	--	--										
5. Pisces	**	ns	ns	***	ns	ns	ns	*	--	--								
6. Crustacea	***	***	ns	*	ns	ns	ns	ns	**	**	--	--						
7. Ins. ex. Dr.	***	***	***	ns	***	***	**	**	***	***	**	***	--	--				
8. Drosophila	***	***	***	***	***	***	***	***	***	***	***	***	**	**	--	--		
9. Mollusca	***	***	***	***	***	***	***	***	***	***	**	***	ns	**	ns	ns	--	--
10. Dicotyled.	ns	ns	ns	*	ns	ns	*	ns	ns	ns	*	*	**	***	***	***	***	***

II. Three major taxa (I = Vertebrata; II = Invertebrata; III = Plants)

	H	P
I - II	***	***
I - III	**	*
II - III	ns	**

Abbreviations:
Ins.ex.Dr. = Insecta excluding Drosophila.
Dicotyled. = Dicotyledoneae
Significance: * = $p < 0.05$; ** = $p < 0.01$; *** = $p < 0.001$; ns = $p > 0.05$

4. Within invertebrates, genetic variation in crustaceans was skewed towards the lower values, whereas the distribution flattened out in this order: insects excluding Drosophila, Drosophila, molluscs. These patterns differed significantly in both P and H in most comparisons.

5. Notably, the distributional patterns of P (which is not dependent on the mating system) in the two sedentary groups, dicotyledons and molluscs, were similar.

LEVELS OF HETEROZYGOSITY OF BIOTIC FACTORS

ECOLOGIC VARIABLES

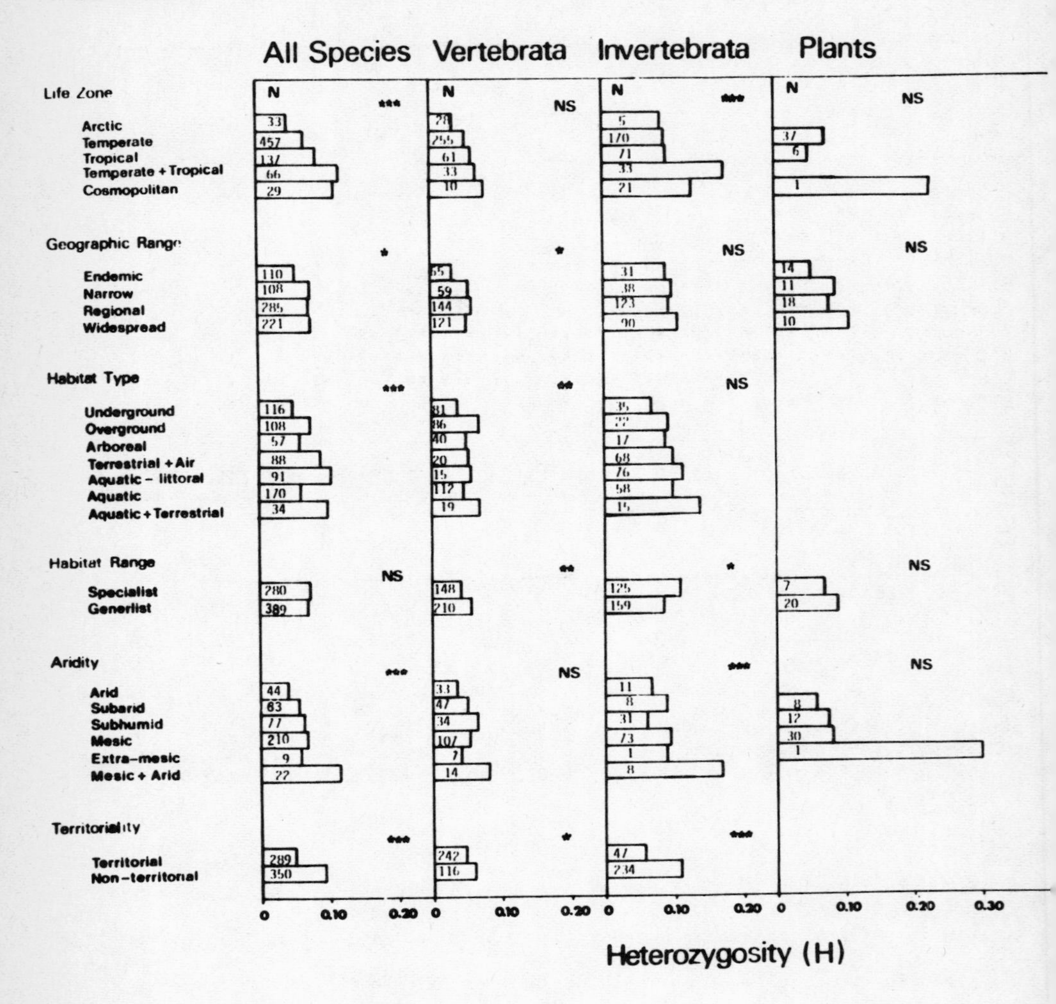

N = number of species analyzed
Significance levels as in Tables

Fig. 3a

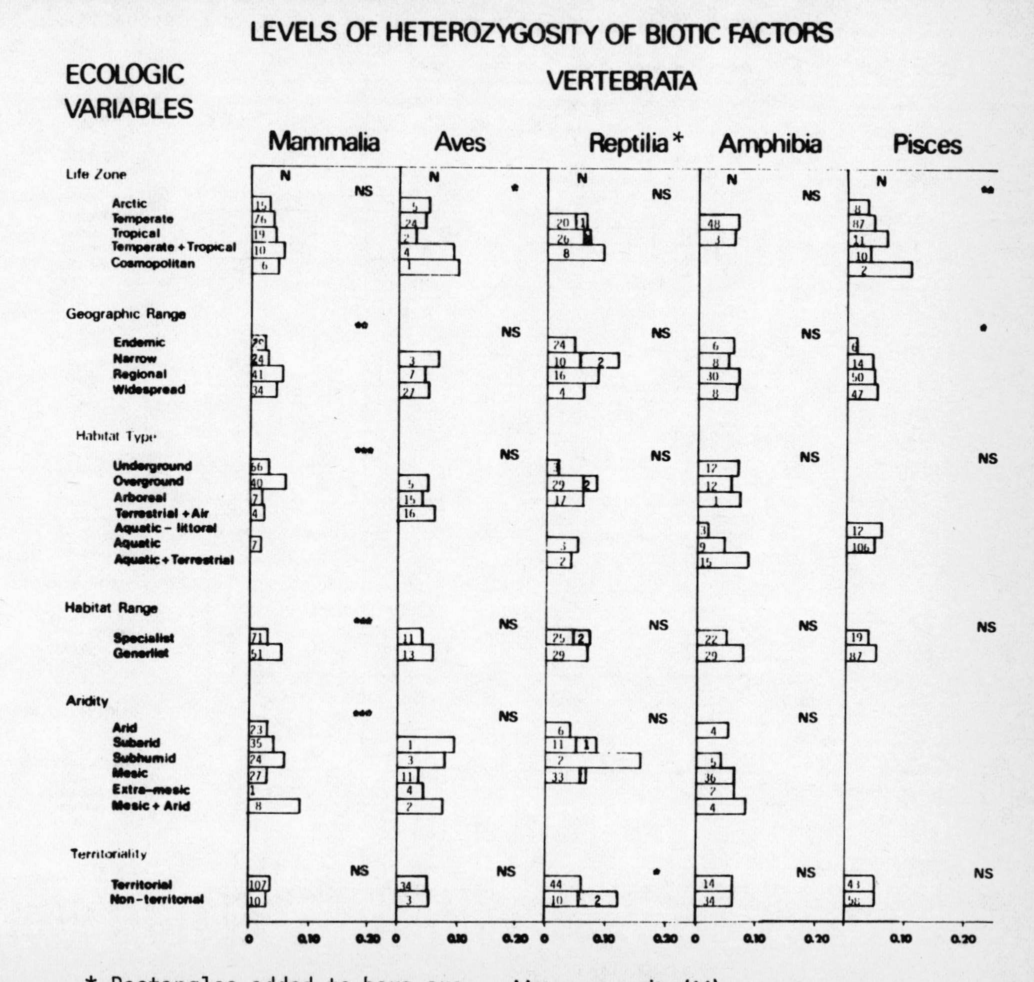

Fig. 3b * Rectangles added to bars are due to parthenogenetic species

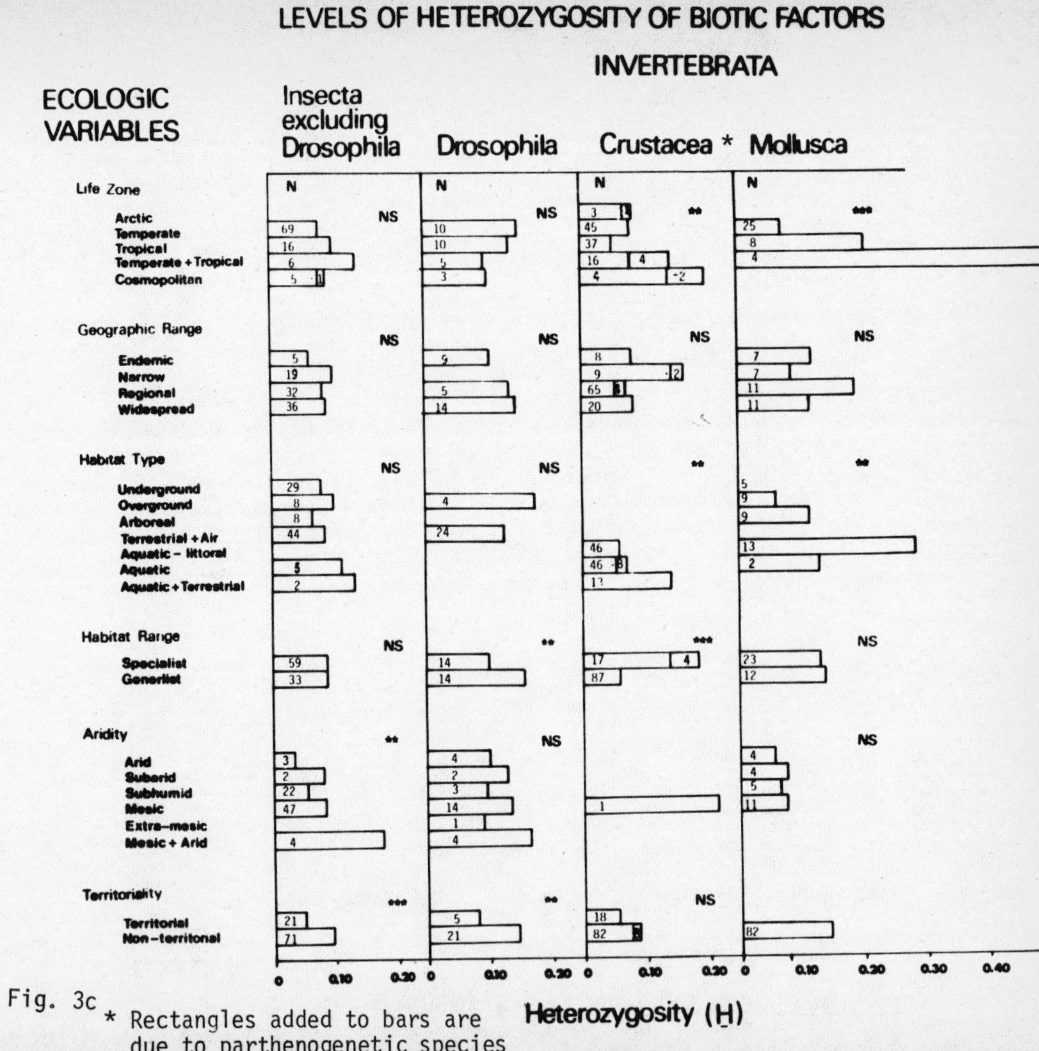

Fig. 3c
* Rectangles added to bars are due to parthenogenetic species

Patterns Among Ecological Parameters (Fig. 3)

The following analysis presents the results only for H as a genetic measure of diversity. The reasons are: (a) H was, in general, highly correlated with P ($r = 0.794$; $p < 0.001$); (b) H was much less dependent on either sample size or the criterion defining P (i.e., percent polymorphism).

Table 3a. Heterozygosity (H) within different life zones.

Taxon	Arctic	Temperate	Tropical	Temperate and Tropical	Cosmopolitan	Significance
All species						
Mean	0.045	0.065	0.076	0.119	0.114	***
S.d.	0.036	0.057	0.078	0.142	0.082	
N	33	457	137	66	29	
Vertebrata						
Mean	0.037	0.050	0.059	0.063	0.076	ns
S.d.	0.031	0.046	0.054	0.051	0.063	
N	28	255	61	33	10	
Invertebrata						
Mean	0.085	0.087	0.092	0.175	0.128	***
S.d.	0.040	0.062	0.092	0.179	0.085	
N	5	166	71	33	18	
Plants						
Mean	--	0.073	0.051	--	0.226	ns
S.d.	--	0.061	0.081	--	--	
N	0	36	5	0	1	
Mammalia						
Mean	0.034	0.041	0.042	0.057	0.048	ns
S.d.	0.034	0.034	0.040	0.034	0.048	
N	15	76	19	10	6	
Aves						
Mean	0.050	0.044	0.027	0.092	0.098	*
S.d.	0.031	0.019	0.028	0.061	--	
N	5	24	2	4	1	
Reptilia						
Mean	--	0.064	0.071	0.094	--	ns
S.d.	--	0.096	0.071	0.072	--	
N	0	20	26	8	0	
Amphibia						
Mean	--	0.067	0.061	0.0	--	ns
S.d.	--	0.061	0.026	--	--	
N	0	48	3	1	0	

Table 3a (contin.)

Taxon	Arctic	Temperate	Tropical	Temperate and Tropical	Cosmopolitan	Significance
Pisces						
Mean	0.037	0.047	0.067	0.038	0.109	**
S.d.	0.026	0.029	0.025	0.020	0.101	
N	8	87	11	10	2	
Crustacea						
Mean	0.083	0.078	0.051	0.143	0.142	**
S.d.	0.028	0.071	0.047	0.140	0.087	
N	3	45	37	16	4	
Insecta exc. <u>Dr</u>.						
Mean	--	0.078	0.102	0.135	0.077	ns
S.d.	--	0.048	0.067	0.080	0.114	
N	0	69	16	6	5	
<u>Drosophila</u>						
Mean	--	0.149	0.136	0.094	0.102	ns
S.d.	--	0.059	0.056	0.025	0.042	
N	0	10	10	5	3	
Mollusca						
Mean	--	0.071	0.205	0.501	--	***
S.d.	--	0.059	0.184	0.254	--	
N	0	25	8	4	0	
Dicotyledoneae						
Mean	--	0.052	0.016	--	--	ns
S.d.	--	0.046	0.022	--	--	
N	0	23	4	0	0	

Abbreviations: N = sample size;
S.d. = standard deviation;
Insecta exc.<u>Dr</u>. = insecta excluding <u>Drosophila</u>.
Significance: * = $p < 0.05$; ** = $p < 0.01$; *** = $p < 0.001$; ns = $p > 0.05$

Table 3b. Multiple comparisons of heterozygosity (<u>H</u>) among different life zones.

Life zone	1	2	3	4	5
1.Arctic	--				
2.Temperate	ns	--			
3.Tropical	ns	ns	--		
4.Temperate + Tropical	**	**	**	--	
5.Cosmopolitan	**	**	ns	ns	--

Abbreviations:
Significance: * = $p < 0.05$; ** = $p < 0.01$; ns = $p > 0.05$ (LSD-modified test).

Table 3c. Pairwise comparisons of the relative cumulative frequency distribution of heterozygosity (H) among different life zones (Kolmogorov - Smirnov test).

Life zone	1	2	3	4	5
1. Arctic	--				
2. Temperate	ns	--			
3. Tropical	ns	ns	--		
4. Temperate + Tropical	**	**	ns	--	
5. Cosmopolitan	***	***	**	ns	--

1. Life zone (Table 3)

For each of the higher taxonomic categories Table 3 gives the mean of *H* with its standard deviation for different life zones accompanied by statistical testing of differences. The following results were indicated:

a. The life zones differed significantly in *H* (ANOVA; $p < 0.001$).

b. *H* decreased in the following order: cosmopolitan, temperate + tropical > tropical > temperate > arctic.

c. Multiple comparisons indicate that each of the two life zones, arctic and temperate, differed significantly in *H* ($p < 0.01$) from the compound life zones and wide distribution (i.e., temperate + tropical, and cosmopolitan). Likewise, the temperate + tropical species harboured significantly higher *H* than tropical species ($p < 0.01$). This pattern, though not always significant, was true for both vertebrates and invertebrates.

d. The same relationships shown for mean *H* values in a-c also held for the pairwise group comparisons of the relative frequency distribution of *H* among the life zones (Table 3c).

e. Within vertebrates, all five classes displayed the above mentioned general trends. In contrast, although the invertebrates as a whole showed a dichotomous pattern, the individual subgroups (see Table 3a) differed distinctly from the overall pattern, possibly due to the small sample size.

The pattern obtained here appears to be, in principle, in line with that described in Nevo (1978). However, instead of the high and significant genetic variation in tropical as compared to temperate ones the present analysis, involving many more species, indicated that the rise in genic diversity is between either temperate or tropical to temperate + tropical species. Note, that the category of temperate + tropical, as well as that of cosmopolitan species embraced species widespread over either life zones or continents.

Table 4. Heterozygosity ($\underline{H}$) within different geographical ranges.

Taxon	Endemic and Relict	Narrow	Regional	Widespread	Significance
All species					
Mean	0.053	0.075	0.074	0.078	*
S.d.	0.057	0.070	0.077	0.073	
N	110	108	285	221	
Vertebrata					
Mean	0.035	0.055	0.058	0.053	*
S.d.	0.036	0.071	0.048	0.037	
N	65	59	144	121	
Invertebrata					
Mean	0.090	0.104	0.093	0.108	ns
S.d.	0.073	0.065	0.101	0.092	
N	31	38	123	90	
Plants					
Mean	0.052	0.087	0.078	0.107	ns
S.d.	0.058	0.052	0.057	0.104	
N	14	11	18	10	
Mammalia					
Mean	0.025	0.031	0.055	0.045	**
S.d.	0.024	0.022	0.041	0.039	
N	29	24	41	34	
Aves					
Mean	--	0.068	0.042	0.050	ns
S.d.	--	0.073	0.029	0.027	
N	0	3	7	27	

Table 4 (contin.)

Taxon	Endemic	Narrow	Regional	Widespread	Significance
Reptilia					
Mean	0.044	0.120	0.085	0.063	ns
S.d.	0.031	0.147	0.070	0.041	
N	24	10	16	4	
Amphibia					
Mean	0.061	0.054	0.069	0.066	ns
S.d.	0.080	0.040	0.062	0.056	
N	6	8	30	8	
Pisces					
Mean	0.016	0.046	0.047	0.056	*
S.d.	0.011	0.014	0.031	0.033	
N	6	14	50	47	
Crustacea					
Mean	0.082	0.145	0.071	0.084	ns
S.d.	0.103	0.090	0.083	0.074	
N	8	9	65	20	
Insecta exc.Dr.					
Mean	0.060	0.098	0.083	0.089	ns
S.d.	0.045	0.041	0.056	0.073	
N	5	19	32	36	
Drosophila					
Mean	0.103	--	0.136	0.145	ns
S.d.	0.057	--	0.063	0.046	
N	9	0	5	14	
Mollusca					
Mean	0.120	0.085	0.190	0.114	ns
S.d.	0.070	0.078	0.221	0.182	
N	7	7	11	11	
Dicotyledoneae					
Mean	0.052	0.070	0.059	0.013	ns
S.d.	0.058	0.041	0.034	0.010	
N	14	9	8	5	

Abbreviations: N = sample size;
S.d. = standard deviation;
Insecta exc.Dr. = insecta excluding Drosophila.
Significance: * = $p < 0.05$; ** = $p < 0.01$; *** = $p < 0.001$; ns = $p > 0.05$.

2. Geographical range (Table 4)

a. The four geographical ranges differed significantly in H (ANOVA; $p < 0.05$).

b. H increased in the following order: endemic < narrow, regional, widespread.

c. The endemic species are significantly lower in genetic variation as compared with the combined group of narrow + regional + widespread (ANOVA; $p < 0.01$).

Notably, the regional category (in mammals, amphibians and particularly in molluscs) and the narrow category (in insects excluding *Drosophila*, reptiles and particularly in crustaceans) harboured the highest level of *H* (compare this pattern with that reported by Hamrick et al., 1979, where the widespread category displayed less genetic diversity than the regional species).

Table 5. Heterozygosity (*H*) and polymorphism (*P*) of mainland and island populations within species.

	Heterozygosity (*H*)						Polymorphism (*P*)					
	Vertebrates			Invertebrates			Vertebrates			Invertebrates		
	N	Mean	S.d.	N	Mean	S.d.	N	Mean	S.d.	N	Mean	S.d.
Mainland	11	0.073	0.048	8	0.139	0.048	13	0.248	0.112	8	0.526	0.13
Island	11	0.057	0.052	8	0.125	0.057	13	0.230	0.164	8	0.466	0.11
Total	11	0.065	0.050	8	0.132	0.051	13	0.239	0.138	8	0.496	0.12

3. Mainland versus island distribution within species (Table 5)

Table 5 compares the means of genetic indices between mainland and island populations in each of 19 species for *H* and 21 species for *P*. In general, mainland populations harboured slightly, though nonsignificantly, higher levels of *P* and *H*.

Table 6a. Heterozygosity (H) within different habitat types.

Taxon	Under ground	Over ground	Arboreal	Air and Terrestrial	Aquatic $	Aquatic littoral	Aquatic and Terrestrial	Aquatic and Air @	Significance
All species									
Mean	0.049	0.075	0.062	0.088	0.062	0.104	0.103	0.093	***
S.d.	0.046	0.068	0.047	0.059	0.066	0.123	0.091	0.074	
N	116	151	57	88	184	95	31	3	
Vertebrata									
Mean	0.040	0.069	0.049	0.052	0.046	0.057	0.076	0.008	**
S.d.	0.037	0.070	0.035	0.028	0.033	0.035	0.072	--	
N	81	86	40	20	125	16	18	1	
Invertebrata									
Mean	0.069	0.096	0.092	0.098	0.099	0.113	0.141	0.136	ns
S.d.	0.057	0.066	0.057	0.061	0.099	0.132	0.103	0.008	
N	35	22	17	68	58	79	13	2	
Plants									
Mean	--	0.075	--	--	0.0	--	--	--	ns
S.d.	--	0.065	--	--	--	--	--	--	
N	0	43	0	0	1	0	0	0	
Mammalia									
Mean	0.035	0.062	0.021	0.023	0.018	--	0.0	--	***
S.d.	0.026	0.044	0.042	0.011	0.015	--	--	--	
N	66	40	7	4	7	0	1	0	
Aves									
Mean	--	0.051	0.044	0.059	--	--	--	0.008	ns
S.d.	--	0.019	0.038	0.027	--	--	--	--	
N	0	5	15	16	0	0	0	1	
Reptilia									
Mean	0.020	0.085	0.064	--	0.058	--	0.043	--	ns
S.d.	0.019	0.106	0.019	--	0.057	--	0.014	--	
N	3	29	17	0	3	0	2	0	
Amphibia									
Mean	0.072	0.061	0.072	--	0.045	0.017	0.085	--	ns
S.d.	0.069	0.039	--	--	0.042	0.010	0.075	--	
N	12	12	1	0	9	3	15	0	
Pisces									
Mean	--	--	--	--	0.047	0.059	--	--	ns
S.d.	--	--	--	--	0.032	0.018	--	--	
N	0	0	0	0	106	12	0	0	
Crustacea									
Mean	--	--	--	--	0.086	0.060	0.141	--	**
S.d.	--	--	--	--	0.103	0.041	0.103	--	
N	0	0	0	0	43	49	13	0	
Insecta exc.Dr.									
Mean	0.080	0.101	0.067	0.085	0.112	--	--	0.136	ns
S.d.	0.054	0.074	0.042	0.062	0.067	--	--	0.008	
N	29	8	8	44	5	0	0	2	

Table 6 (contin.)

Taxon	Under ground	Over ground	Arboreal	Air and Terrestrial	Aquatic	Aquatic littoral	Aquatic and Terrestrial	Aquatic and Air	Sig
Drosophila									
Mean	--	0.173	--	0.122	--	--	--	--	ns
S.d.	--	0.055	--	0.051	--	--	--	--	
N	0	4	0	24	0	0	0	0	
Mollusca									
Mean	0.0	0.057	0.114	--	0.127	0.285	--	--	**
S.d.	0.0	0.030	0.061	--	0.049	0.244	--	--	
N	5	9	9	0	2	13	0	0	
Dicotyledoneae									
Mean	--	0.047	--	--	0.0	--	--	--	ns
S.d.	--	0.044	--	--	--	--	--	--	
N	0	29	0	0	1	0	0	0	

Abbreviations: N = sample size; S.d. = standard deviation;
Insecta exc.Dr. = insecta excluding *Drosophila*.
Significance: * = $p < 0.05$; ** = $p < 0.01$; *** = $p < 0.001$; ns = $p > 0.05$.
$ Aquatic, without the littoral species.
@ In all analyses species from the "air + aquatic" habitat type were included in the "aquatic + terrestrial" habitat type.

Table 6b. Multiple comparisons of heterozygosity (H) among different habitat types,in all species.

Habitat type	1	2	3	4	5	6	7
1.Under ground	--						
2.Over ground	ns	--					
3.Arboreal	ns	ns	--				
4.Air + terrestrial	**	ns	ns	--			
5.Aquatic $	ns	ns	ns	ns	--		
6.Aquatic littoral	**	ns	*	ns	**	--	
7.Aquatic + terrestrial @	**	ns	ns	ns	ns	ns	--

Table 6c. Multiple comparisons of heterozygosity (H) among different habitat types,in vertebrate species.

Habitat type	1	2	3	4	5	6	7
1.Under ground	--						
2.Over ground	**	--					
3.Arboreal	ns	ns	--				
4.Air + terrestrial	ns	ns	ns	--			
5.Aquatic $	ns	**	ns	ns	--		
6.Aquatic littoral	ns	ns	ns	ns	ns	--	
7.Aquatic + terrestrial @	ns	ns	ns	ns	ns	ns	--

4. Habitat type (Table 6)

a. The habitats differed significantly in H (ANOVA; $p < 0.001$).

b. H increased in the following order: underground < arboreal, aquatic < overground < air + terrestrial < terrestrial + aquatic, and littoral-aquatic species. Notably, in both vertebrates and invertebrates, underground species displayed the lowest, and aquatic + terrestrial species the highest level of genetic variation.

c. Multiple comparisons indicated that underground species harboured significantly lower genetic variation than species living in all compound habitats ($p < 0.01$). In addition, underground vertebrate species harboured significantly lower genetic variation than overground species ($p < 0.01$).

d. Aquatic vertebrates had significantly ($p < 0.01$) lower H than overground terrestrial species, and fresh water vertebrates harboured lower genetic variation than marine species, although not significantly.

e. Aquatic, littoral and largely sedentary molluscs harboured dramatically ($p < 0.001$) more genetic variation than their terrestrial counterparts, and more than aquatic mobile crustaceans. Notably, *Balanus* and *Chthamalus*, the sessile crustaceans, had higher genetic variation than mobile crustaceans.

Particularly in vertebrates, but also in some invertebrate taxa, the greater the buffering effects on the habitat the lower the genetic variation. Thus, genetic variation is lower in underground, arboreal and aquatic pelagic habitats compared with overground aquatic-littoral and aquatic + terrestrial habitats.

Table 7. Heterozygosity (H) within different habitat ranges.

Taxon	Specialist	Generalist	Significance
All species			
Mean	0.075	0.072	ns
S.d.	0.086	0.064	
N	280	389	
Vertebrata			
Mean	0.043	0.059	**
S.d.	0.055	0.044	
N	148	210	
Invertebrata			
Mean	0.112	0.088	*
S.d.	0.101	0.080	
N	125	159	
Plants			
Mean	0.070	0.091	ns
S.d.	0.053	0.069	
N	7	20	
Mammalia			
Mean	0.032	0.054	***
S.d.	0.024	0.046	
N	71	51	
Aves			
Mean	0.040	0.057	ns
S.d.	0.031	0.040	
N	11	13	
Reptilia			
Mean	0.073	0.070	ns
S.d.	0.111	0.040	
N	25	29	
Amphibia			
Mean	0.050	0.078	ns
S.d.	0.052	0.063	
N	22	29	
Pisces			
Mean	0.040	0.051	ns
S.d.	0.025	0.032	
N	19	87	
Crustacea			
Mean	0.186	0.060	***
S.d.	0.136	0.049	
N	17	87	
Insecta exc. *Dr*.			
Mean	0.087	0.086	ns
S.d.	0.049	0.076	
N	59	33	

Table 7 (contin.)

Taxon	Specialist	Generalist	Significance
Drosophila			
Mean	0.102	0.157	**
S.d.	0.047	0.046	
N	14	14	
Mollusca			
Mean	0.129	0.137	ns
S.d.	0.168	0.171	
N	23	12	
Dicotyledoneae			
Mean	0.071	0.056	ns
S.d.	0.058	0.043	
N	6	10	

Abbreviations: N = sample size;
S.d. = standard deviation;
Insecta exc.Dr. = insecta excluding Drosophila.
Significance: * = $p < 0.05$; ** = $p < 0.01$;
*** = $p < 0.001$; ns = $p > 0.05$.

5. Habitat range (Table 7)

a. Habitat specialists and generalists did not differ in the level of genetic variation when compared over 669 species.

b. In the vertebrates, generalists harboured significantly ($p < 0.01$) more genetic variation than specialists. This pattern was substantiated in all 5 vertebrate classes. However, when the two species of parthenogenetic reptiles which had remarkably high $\underline{H}$ values were included, generalist and specialist species of reptiles displayed a similar level of $\underline{H}$. But even reptiles display the vertebrate pattern of generalist-specialist when $\underline{P}$ is the measure ($\underline{P}$ in specialists = 0.22, versus generalists 0.29, ANOVA; $p = 0.06$).

c. In the invertebrates, Drosophila and molluscs displayed the vertebrate pattern. Insects excluding Drosophila exhibited similar levels of $\underline{H}$ in both specialists and generalists. However, a reanalysis of this group (defining specialist by different combinations of isolated, endemic, underground and arid categories; and generalists by the counterpart categories) gives the vertebrate pattern, although specialists were small in numbers.

d. A drastically contrasting pattern where specialists harboured significantly ($p < 0.001$) more genetic variation than generalists was found in crustaceans. This contrast contributes strikingly, first to the invertebrate pattern at large, and second, to the overall pattern. A critical recent discussion on the genetics of the crustaceans is given by Hedgecock et al., (1982). They found a negative correlation between environmental heterogeneity and trophic generalism. This correlation may account for the contrasting pattern found in crustaceans.

Table 8a. Heterozygosity (H) within different aridity categories.

Taxon	Arid	Sub-arid	Sub-humid	Mesic	Extra mesic	Mesic + arid	Significance
All species							
Mean	0.044	0.058	0.066	0.072	0.060	0.115	***
S.d.	0.037	0.064	0.053	0.065	0.047	0.061	
N	44	63	77	210	9	22	
Vertebrata							
Mean	0.036	0.052	0.065	0.054	0.043	0.083	ns
S.d.	0.030	0.066	0.055	0.057	0.034	0.038	
N	33	47	34	107	7	14	
Invertebrata							
Mean	0.069	0.090	0.063	0.095	0.090	0.169	***
S.d.	0.045	0.048	0.042	0.068	--	0.055	
N	11	8	31	73	1	8	
Plants							
Mean	--	0.060	0.077	0.082	0.150	--	ns
S.d.	--	0.063	0.073	0.068	--	--	
N	0	8	12	30	1	0	
Mammalia							
Mean	0.031	0.041	0.060	0.031	0.0	0.085	***
S.d.	0.022	0.030	0.046	0.032	--	0.030	
N	23	35	24	27	1	8	

Table 8a (contin.)

Taxon	Arid	Sub-arid	Sub-humid	Mesic	Extra mesic	Mesic + arid	Signif.
Aves							
Mean	--	0.094	0.081	0.036	0.042	0.075	ns
S.d.	--	--	0.061	0.024	0.030	0.057	
N	0	1	3	11	4	2	
Reptilia							
Mean	0.043	0.084	0.160	0.067	--	--	ns
S.d.	0.050	0.124	0.127	0.064	--	--	
N	6	11	2	33	0	0	
Amphibia							
Mean	0.054	--	0.044	0.066	0.066	0.085	ns
S.d.	0.033	--	0.030	0.066	0.035	0.053	
N	4	0	5	36	2	4	
Pisces							
Mean	--	--	--	--	--	--	--
S.d.	--	--	--	--	--	--	
N	0	0	0	0	0	0	
Crustacea							
Mean	--	--	--	0.217	--	--	--
S.d.	--	--	--	--	--	--	
N	0	0	0	1	0	0	
Insecta exc.Dr.							
Mean	0.036	0.082	0.057	0.085	--	0.174	**
S.d.	0.052	0.036	0.044	0.060	--	0.061	
N	3	2	22	47	0	4	
Drosophila							
Mean	0.109	0.128	0.095	0.136	0.090	0.164	ns
S.d.	0.023	0.090	0.046	0.057	--	0.058	
N	4	2	3	14	1	4	
Mollusca							
Mean	0.054	0.074	0.063	0.073	--	--	ns
S.d.	0.031	0.030	0.030	0.084	--	--	
N	4	4	5	11	0	0	
Dicotyledoneae							
Mean	--	0.042	0.064	0.050	--	--	ns
S.d.	--	0.039	0.054	0.042	--	--	
N	0	7	7	20	0	0	

Abbreviations: N = sample size;
S.d. = standard deviation;
Insecta exc.Dr. = insecta excluding *Drosophila*.
Significance: * = $p < 0.05$; ** = $p < 0.01$; *** = $p < 0.001$; ns = $p > 0.05$.

Table 8b. Multiple comparisons of heterozygosity (H) among different aridity categories.

Aridity category	1	2	3	4	5	6
1.Arid	--					
2.Sub arid	ns	--				
3.Sub humid	ns	ns	--			
4.Mesic	ns	ns	ns	--		
5.Extra mesic	ns	ns	ns	ns	--	
6.Mesic + arid	**	**	*	*	ns	--

Abbreviations:
Significance: * = $p < 0.05$; ** = $p < 0.01$; ns = $p > 0.05$; (LSD-modified test).

6. Aridity (Table 8)

a. The climatic regions differed significantly in H in the overall (ANOVA; $p < 0.001$) data set, as well as in vertebrates and invertebrates separately.

b. In general, the arid and extra-mesic ecological extremes displayed low levels of genetic variation, whereas the species ranging in a broader climatic spectrum, i.e., in mesic + arid environments showed the highest level of H.

c. Multiple comparisons indicated that the mesic + arid category differed significantly from all other aridity categories, except for the small extra-mesic habitat.

Notably, species living under an arid climatic extreme exhibit lower, whereas species living under intermediate climates, or over broader climatic amplitude displayed higher genetic variation. This pattern conforms with the niche width variation hypothesis.

Table 9. Heterozygosity (H) of territorial and nonterritorial species.

Taxon	Territorial	Non territorial	Significance
All species			
Mean	0.049	0.093	***
S.d.	0.039	0.092	
N	289	350	
Vertebrata			
Mean	0.047	0.060	*
S.d.	0.038	0.059	
N	242	116	
Invertebrata			
Mean	0.058	0.110	***
S.d.	0.048	0.101	
N	47	234	
Mammalia			
Mean	0.038	0.030	ns
S.d.	0.030	0.035	
N	107	10	
Aves			
Mean	0.050	0.051	ns
S.d.	0.031	0.045	
N	34	3	
Reptilia			
Mean	0.061	0.121	*
S.d.	0.053	0.145	
N	44	10	
Amphibia			
Mean	0.061	0.062	ns
S.d.	0.046	0.049	
N	14	34	
Pisces			
Mean	0.050	0.051	ns
S.d.	0.031	0.033	
N	43	58	
Crustacea			
Mean	0.054	0.089	ns
S.d.	0.027	0.092	
N	18	82	
Insecta exc.*Dr*.			
Mean	0.049	0.097	***
S.d.	0.061	0.054	
N	21	71	

Table 9 (contin.)

Taxon	Territorial	Non territorial	Significance
Drosophila			
Mean	0.080	0.147	**
S.d.	0.034	0.049	
N	5	21	
Mollusca			
Mean	--	0.144	--
S.d.	—	0.182	
N	0	37	

Abbreviations: N = sample size;
S.d. = standard deviation;
Insecta exc.*Dr*. = insecta excluding *Drosophila*.
Significance: * = $p < 0.05$; ** = $p < 0.01$;
*** = $p < 0.001$; ns = $p > 0.05$.

7. Territoriality (Table 9)

a. Nonterritorial species were significantly more variable genetically than territorial ones (ANOVA; $p < 0.001$).

b. The above pattern was particularly strong in invertebrates (insects and crustaceans) and in vertebrates (reptiles). Mammals displayed a nonsignificant opposite trend.

The above pattern may be related to the seemingly larger ecological amplitude and competitive interactions experienced by nonterritorial as compared with territorial species. Analysis of each of the species size groups (small, medium and large) indicates the same pattern. Therefore, we assume that the pattern is genuine rather than dependent on species size.

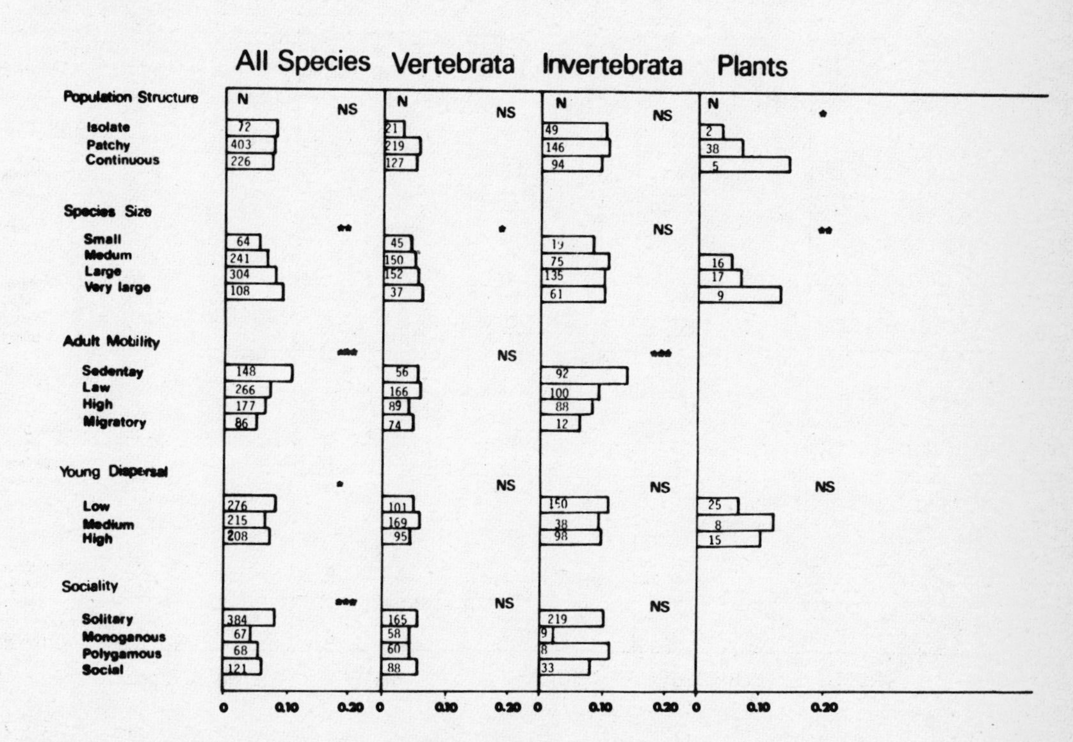

Fig. 4a

N = number of species analyzed
Significance levels as in Tables

LEVELS OF HETEROZYGOSITY OF BIOTIC FACTORS

VERTEBRATA

DEMOGRAPHIC VARIABLES

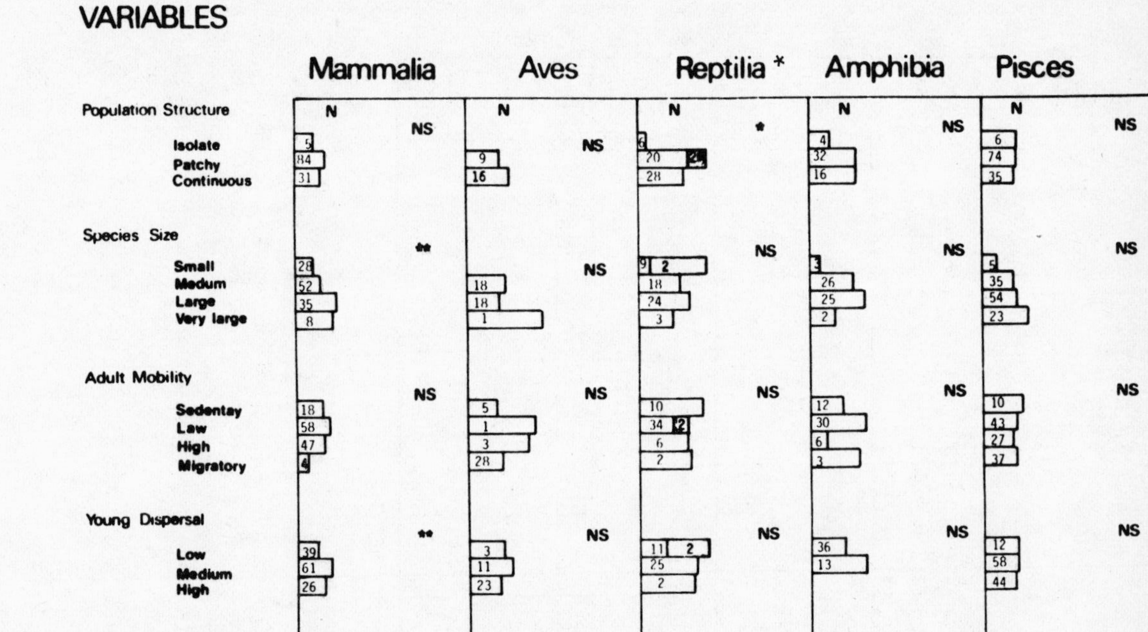

Fig. 4b

* Rectangles added to bars are due to parthenogenetic species

LEVELS OF HETEROZYGOSITY OF BIOTIC FACTORS

INVERTEBRATA

DEMOGRAPHIC VARIABLES

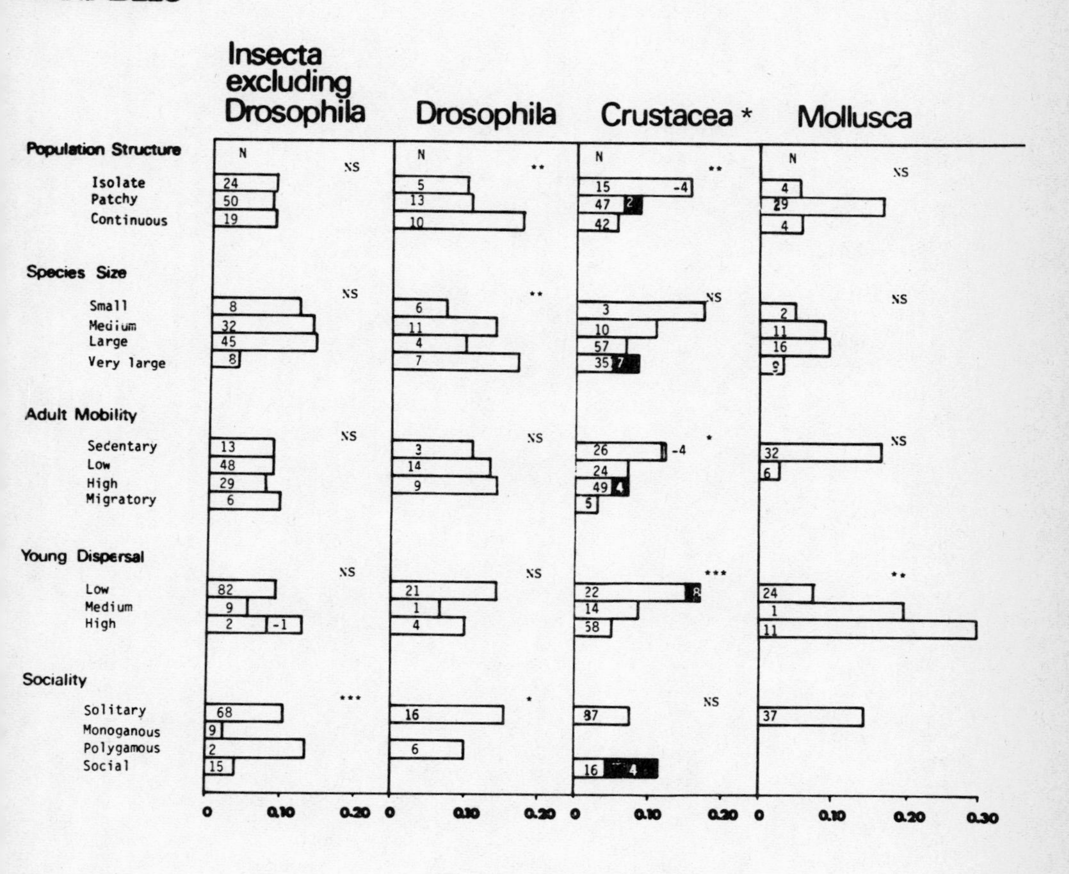

Heterozygosity (H)

Fig. 4c

* Rectangles added to bars are due to parthenogenetic species

Table 10. Heterozygosity ($\underline{H}$) within different population structure categories.

Taxon	Isolate	Patchy	Continuous	Significance
All species				
Mean	0.080	0.074	0.072	ns
S.d.	0.072	0.085	0.060	
N	72	403	226	
Vertebrata				
Mean	0.029	0.054	0.053	ns
S.d.	0.039	0.055	0.037	
N	21	219	127	
Invertebrata				
Mean	0.103	0.105	0.094	ns
S.d.	0.072	0.114	0.072	
N	49	146	94	
Plants				
Mean	0.036	0.064	0.142	*
S.d.	0.038	0.060	0.079	
N	2	38	5	
Mammalia				
Mean	0.026	0.044	0.037	ns
S.d.	0.026	0.035	0.041	
N	5	84	31	
Aves				
Mean	--	0.043	0.056	ns
S.d.	--	0.038	0.037	
N	0	9	16	
Reptilia				
Mean	0.010	0.102	0.063	*
S.d.	0.011	0.119	0.032	
N	6	20	28	
Amphibia				
Mean	0.028	0.068	0.068	ns
S.d.	0.011	0.069	0.040	
N	4	32	16	
Pisces				
Mean	0.051	0.049	0.047	ns
S.d.	0.066	0.030	0.027	
N	6	74	35	
Crustacea				
Mean	0.135	0.087	0.054	**
S.d.	0.100	0.094	0.051	
N	15	47	42	
Insecta exc. *Dr*.				
Mean	0.090	0.084	0.089	ns
S.d.	0.054	0.057	0.075	
N	24	50	19	

Table 10 (contin.)

Taxon	Isolate	Patchy	Continuous	Significance
Drosophila				
Mean	0.104	0.105	0.175	**
S.d.	0.056	0.043	0.036	
N	5	13	10	
Mollusca				
Mean	0.054	0.171	0.057	ns
S.d.	0.031	0.198	0.032	
N	4	29	4	
Dicotyledoneae				
Mean	0.009	0.047	--	ns
S.d.	—	0.044	--	
N	1	29	0	

Abbreviations: N = sample size;
S.d. = standard deviation;
Insecta exc.Dr. = insecta excluding Drosophila.
Significance: * = $p < 0.05$; ** = $p < 0.01$; *** = $p < 0.001$; ns = $p > 0.05$.

Patterns Among Demographic Parameters (Fig. 4)

1. Population structure (Table 10)

a. The different population structures did not differ in their level of H when compared with the set of 701 species analyzed for this character. It appears that in general, isolated vertebrate species harboured lower genetic variation as compared to species whose population structure is either patchy or continuous. This pattern was particularly emphasized by the level of polymorphism (0.149, 0.218, 0.239 for isolated, patchy and continuous population structure, respectively. ANOVA; $p = 0.01$). Invertebrates displayed heterogeneous patterns.

b. A contrasting pattern was revealed by crustaceans where isolated species exhibited significantly ($p < 0.01$) higher H than others. Insects excluding Drosophila did not show at the present level of analysis differentiation according to population structure. Finally, Drosophila exhibited significantly ($p < 0.01$) higher H in species with continuous population structure.

The aforementioned repetitive pattern in most vertebrate classes may suggest a genuine reduction of H in isolated species of vertebrates.

Table 11a. Heterozygosity (H) within different species size categories.

Taxon	Small	Medium	Large	Very large	Significance
All species					
Mean	0.053	0.066	0.077	0.090	**
S.d.	0.079	0.058	0.076	0.099	
N	64	241	304	108	
Vertebrata					
Mean	0.039	0.048	0.057	0.063	*
S.d.	0.080	0.045	0.040	0.037	
N	45	150	152	37	
Invertebrata					
Mean	0.088	0.106	0.101	0.101	ns
S.d.	0.067	0.063	0.100	0.122	
N	19	75	135	62	
Plants					
Mean	--	0.048	0.067	0.132	**
S.d.	--	0.059	0.052	0.077	
N	0	16	17	9	
Mammalia					
Mean	0.026	0.036	0.058	0.054	**
S.d.	0.025	0.028	0.044	0.045	
N	28	52	35	8	
Aves					
Mean	--	0.052	0.045	0.116	ns
S.d.	--	0.035	0.024	--	
N	0	18	18	1	
Reptilia					
Mean	0.097	0.061	0.073	0.049	ns
S.d.	0.166	0.067	0.039	0.005	
N	9	18	24	3	
Amphibia					
Mean	0.016	0.063	0.078	0.038	ns
S.d.	0.010	0.064	0.057	0.008	
N	3	26	21	2	
Pisces					
Mean	0.020	0.044	0.046	0.068	**
S.d.	0.017	0.028	0.027	0.037	
N	5	35	54	23	

Table 11a (contin.)

Taxon	Small	Medium	Large	Very large	Significance
Crustacea					
Mean	0.176	0.109	0.068	0.086	ns
S.d.	0.127	0.090	0.054	0.110	
N	3	10	57	35	
Insecta exc.Dr.					
Mean	0.072	0.091	0.096	0.038	ns
S.d.	0.041	0.055	0.062	0.060	
N	8	32	45	8	
Drosophila					
Mean	0.076	0.142	0.100	0.173	**
S.d.	0.033	0.047	0.047	0.039	
N	6	11	4	7	
Mollusca					
Mean	0.051	0.089	0.209	0.119	ns
S.d.	0.058	0.060	0.214	0.206	
N	2	11	16	9	
Dicotyledoneae					
Mean	--	0.039	0.056	0.054	ns
S.d.	--	0.046	0.048	0.037	
N	0	15	10	2	

Table 11b. Heterozygosity (H) within different life zones among species size categories.

I. Small species size

Taxon	Arctic	Temperate	Tropical	Temperate and Tropical	Cosmopolitan	Significance
All species						
Mean	0.0	0.050	0.063	0.090	--	ns
S.d.	--	0.072	0.113	--	--	
N	1	50	11	1	0	
Vertebrata						
Mean	0.0	0.035	0.059	--	--	ns
S.d.	--	0.066	0.133	--	--	
N	1	36	8	0	0	
Invertebrata						
Mean	--	0.090	0.076	0.090	--	ns
S.d.	--	0.077	0.048	--	--	
N	0	14	3	1	0	
Mammalia						
Mean	0.0	0.030	0.015	--	--	ns
S.d.	--	0.023	0.032	--	--	
N	1	21	6	0	0	

Table 11b (contin.)

Taxon	Arctic	Temperate	Tropical	Temperate and Tropical	Cosmopolitan	Sign.
Reptilia						
Mean	--	0.071	0.190	--	--	ns
S.d.	--	0.146	0.269	--	--	
N	0	7	2	0	0	
Amphibia						
Mean	--	0.016	--	--	--	--
S.d.	--	0.010	--	--	--	
N	0	3	0	0	0	
Pisces						
Mean	--	0.020	--	--	--	--
S.d.	--	0.017	--	--	--	
N	0	5	0	0	0	
Crustacea						
Mean	--	0.176	--	--	--	--
S.d.	--	0.127	--	--	--	
N	0	3	0	0	0	
Insecta exc. *Dr*.						
Mean	--	0.072	--	--	--	--
S.d.	--	0.041	--	--	--	
N	0	8	0	0	0	
Drosophila						
Mean	--	0.070	0.076	0.090	--	ns
S.d.	--	0.023	0.048	--	--	
N	0	2	3	1	0	
Mollusca						
Mean	--	0.010	--	--	--	ns
S.d.	--	--	--	--	--	
N	0	1	0	0	0	
II. Medium species size						
All species						
Mean	0.041	0.068	0.073	0.062	0.074	ns
S.d.	0.040	0.059	0.061	0.053	0.066	
N	14	163	34	18	10	
Vertebrata						
Mean	0.029	0.049	0.058	0.054	0.055	ns
S.d.	0.027	0.044	0.040	0.059	0.074	
N	12	102	18	12	4	
Invertebrata						
Mean	0.108	0.111	0.107	0.076	0.086	ns
S.d.	0.041	0.065	0.070	0.039	0.064	
N	2	50	11	6	6	
Plants						
Mean	--	0.047	0.051	--	--	ns
S.d.	--	0.050	0.081	--	--	
N	0	11	5	0	0	

Table 11b (contin.)

Taxon	Arctic	Temperate	Tropical	Temperate and Tropical	Cosmopolitan	Sign.
Mammalia						
Mean	0.015	0.037	0.091	0.035	0.022	***
S.d.	0.010	0.019	0.042	0.023	0.038	
N	7	33	4	3	3	
Aves						
Mean	0.050	0.043	0.027	0.122	--	**
S.d.	0.031	0.019	0.028	0.040	--	
N	5	9	2	2	0	
Reptilia						
Mean	--	0.063	0.054	0.067	--	ns
S.d.	--	0.074	0.051	0.098	--	
N	0	9	6	3	0	
Amphibia						
Mean	--	0.067	0.046	0.0	--	ns
S.d.	--	0.067	0.007	--	--	
N	0	23	2	1	0	
Pisces						
Mean	--	0.045	0.050	0.034	--	ns
S.d.	--	0.029	0.018	0.029	--	
N	0	28	4	3	0	
Crustacea						
Mean	0.079	0.118	--	0.067	--	ns
S.d.	--	0.100	--	--	--	
N	1	8	0	1	0	
Insecta exc. *Dr*.						
Mean	--	0.102	0.080	0.055	0.031	ns
S.d.	--	0.054	0.061	--	0.019	
N	0	23	5	1	3	
Drosophila						
Mean	--	0.171	0.139	0.105	0.115	ns
S.d.	--	0.038	0.078	0.026	--	
N	0	5	2	3	1	
Mollusca						
Mean	--	0.069	0.124	--	--	ns
S.d.	--	0.034	0.083	--	--	
N	0	7	4	0	0	
Dicotyledoneae						
Mean	--	0.047	0.016	--	--	ns
S.d.	--	0.050	0.022	--	--	
N	0	11	4	0	0	

Table 11b (contin.)

Taxon	Arctic	Temperate	Tropical	Temperate and Tropical	Cosmopolitan	Sign.
III. Large and very large species size						
All species						
Mean	0.050	0.067	0.078	0.141	0.134	***
S.d.	0.033	0.051	0.080	0.161	0.083	
N	18	244	92	47	19	
Vertebrata						
Mean	0.046	0.056	0.060	0.068	0.090	ns
S.d.	0.032	0.040	0.027	0.046	0.057	
N	15	117	35	21	6	
Invertebrata						
Mean	0.069	0.075	0.089	0.201	0.149	***
S.d.	0.038	0.056	0.098	0.194	0.088	
N	3	102	57	26	12	
Plants						
Mean	--	0.084	--	--	0.226	*
S.d.	--	0.063	--	--	--	
N	0	25	0	0	1	
Mammalia						
Mean	0.057	0.057	0.039	0.067	0.074	ns
S.d.	0.036	0.052	0.023	0.035	0.047	
N	7	22	9	7	3	
Aves						
Mean	--	0.044	--	0.061	0.098	ns
S.d.	--	0.020	--	0.077	--	
N	0	15	0	2	1	
Reptilia						
Mean	--	0.056	0.063	0.110	--	*
S.d.	--	0.023	0.025	0.059	--	
N	0	4	18	5	0	
Amphibia						
Mean	--	0.074	0.091	--	--	ns
S.d.	--	0.056	--	--	--	
N	0	22	1	0	0	
Pisces						
Mean	0.037	0.051	0.077	0.040	0.109	**
S.d.	0.026	0.029	0.024	0.017	0.101	
N	8	54	7	7	2	
Crustacea						
Mean	0.084	0.060	0.051	0.148	0.142	***
S.d.	0.039	0.044	0.047	0.143	0.087	
N	2	34	37	15	4	
Insecta exc. *Dr*.						
Mean	--	0.065	0.112	0.152	0.144	**
S.d.	--	0.041	0.069	0.078	0.189	
N	0	38	11	5	2	

Table 11b (contin.)

Taxon	Arctic	Temperate	Tropical	Temperate and Tropical	Cosmopolitan	Sign.
Drosophila						
Mean	--	0.165	0.171	0.065	0.095	ns
S.d.	--	0.069	0.005	--	0.057	
N	0	3	5	1	2	
Mollusca						
Mean	--	0.075	0.285	0.501	--	***
S.d.	--	0.067	0.235	0.254	--	
N	0	17	4	4	0	
Dicotyledoneae						
Mean	--	0.056	--	--	--	--
S.d.	--	0.045	--	--	--	
N	0	12	0	0	0	

Table 11c. Heterozygosity (H) within different habitat types among species size categories.

I. Small species size

Taxon	Under ground	Over ground	Arboreal	Air and Terrestrial	Aquatic	Aquatic littoral $	Aquatic and Terrestrial	Aquatic and Air @	Significance
All species									
Mean	0.028	0.081	--	0.072	0.028	--	0.132	--	*
S.d.	0.035	0.133	--	0.028	0.028	--	0.136	--	
N	25	14	0	10	11	0	4	0	
Vertebrata									
Mean	0.020	0.087	--	--	0.018	--	0.0	--	ns
S.d.	0.021	0.137	--	--	0.017	--	--	--	
N	22	13	0	0	9	0	1	0	
Invertebrata									
Mean	0.087	0.010	--	0.072	0.070	--	0.176	--	ns
S.d.	0.067	--	--	0.028	0.030	--	0.127	--	
N	3	1	0	10	2	0	3	0	
Mammalia									
Mean	0.020	0.043	--	--	0.021	--	0.0	--	ns
S.d.	0.022	0.027	--	--	0.030	--	--	--	
N	18	7	0	0	2	0	1	0	
Reptilia									
Mean	0.020	0.160	--	--	0.016	--	--	--	ns
S.d.	0.019	0.210	--	--	--	--	--	--	
N	3	5	0	0	1	0	0	0	

Table 11c (contin.)

Taxon	Under ground	Over ground	Arboreal	Air and Terrestrial	Aquatic	Aquatic littoral	Aquatic and Terrestrial	Aquatic and Air	Sig.
Amphibia									
Mean	0.018	0.025	--	--	0.006	--	--	--	--
S.d.	--	--	--	--	--	--	--	--	
N	1	1	0	0	1	0	0	0	
Pisces									
Mean	--	--	--	--	0.020	--	--	--	--
S.d.	--	--	--	--	0.017	--	--	--	
N	0	0	0	0	5	0	0	0	
Crustacea									
Mean	--	--	--	--	--	--	0.176	--	--
S.d.	--	--	--	--	--	--	0.127	--	
N	0	0	0	0	0	0	3	0	
Insecta exc. Dr.									
Mean	0.087	--	--	0.066	0.049	--	--	--	ns
S.d.	0.067	--	--	0.023	--	--	--	--	
N	3	0	0	4	1	0	0	0	
Drosophila									
Mean	--	--	--	0.076	--	--	--	--	--
S.d.	--	--	--	0.033	--	--	--	--	
N	0	0	0	6	0	0	0	0	
Mollusca									
Mean	--	0.010	--	--	0.092	--	--	--	--
S.d.	--	--	--	--	--	--	--	--	
N	0	1	0	0	1	0	0	0	
II. Medium species size									
All species									
Mean	0.055	0.055	0.055	0.104	0.055	0.080	0.116	0.093	***
S.d.	0.044	0.052	0.058	0.059	0.048	0.051	0.099	0.074	
N	49	56	22	26	51	18	15	3	
Vertebrata									
Mean	0.039	0.057	0.039	0.064	0.041	0.054	0.091	0.008	*
S.d.	0.023	0.053	0.041	0.042	0.029	0.042	0.088	--	
N	34	35	17	2	40	9	11	1	
Invertebrata									
Mean	0.091	0.060	0.111	0.108	0.106	0.106	0.185	0.136	ns
S.d.	0.058	0.029	0.079	0.060	0.067	0.047	0.106	0.008	
N	15	5	5	24	11	9	4	2	
Plants									
Mean	--	0.048	--	--	--	--	--	--	--
S.d.	--	0.059	--	--	--	--	--	--	
N	0	16	0	0	0	0	0	0	

Table 11c (contin.)

Taxon	Under ground	Over ground	Arboreal	Air and Terrestrial	Aquatic	Aquatic littoral	Aquatic and Terrestrial	Aquatic and Air	Sig.
Mammalia									
Mean	0.037	0.061	0.0	0.034	0.017	--	--	--	***
S.d.	0.020	0.038	0.0	--	0.010	--	--	--	
N	30	10	5	1	5	0	0	0	
Aves									
Mean	--	0.051	0.053	0.094	--	--	--	0.008	ns
S.d.	--	0.019	0.040	--	--	--	--	--	
N	0	5	11	1	0	0	0	1	
Reptilia									
Mean	--	0.063	--	--	--	--	0.043	--	ns
S.d.	--	0.071	--	--	--	--	0.014	--	
N	0	16	0	0	0	0	2	0	
Amphibia									
Mean	0.052	0.031	0.072	--	0.044	0.022	0.101	--	ns
S.d.	0.040	0.013	--	--	0.026	0.006	0.094	--	
N	4	4	1	0	6	2	9	0	
Pisces									
Mean	--	--	--	--	0.044	0.047	--	--	ns
S.d.	--	--	--	--	0.030	0.015	--	--	
N	0	0	0	0	29	6	0	0	
Crustacea									
Mean	--	--	--	--	0.058	0.059	0.185	--	ns
S.d.	--	--	--	--	0.019	0.010	0.106	--	
N	0	0	0	0	3	3	4	0	
Insecta exc.Dr.									
Mean	0.091	--	--	0.078	0.130	--	--	0.136	ns
S.d.	0.058	--	--	0.055	0.047	--	--	0.008	
N	15	0	0	13	2	0	0	2	
Drosophila									
Mean	--	--	--	0.142	--	--	--	--	--
S.d.	--	--	--	0.047	--	--	--	--	
N	0	0	0	11	0	0	0	0	
Mollusca									
Mean	--	0.051	0.111	--	--	0.112	--	--	ns
S.d.	--	0.025	0.079	--	--	0.006	--	--	
N	0	4	5	0	0	2	0	0	
Dicotyledoneae									
Mean	--	0.039	--	--	--	--	--	--	--
S.d.	--	0.046	--	--	--	--	--	--	
N	0	15	0	0	0	0	0	0	

Table 11c (contin.)

Taxon	Under ground	Over ground	Arboreal	Air and Terrestrial	Aquatic	Aquatic littoral	Aquatic and Terrestrial	Aquatic and Air	Sig.
III. Large and very large species size									
All species									
Mean	0.055	0.087	0.066	0.082	0.069	0.109	0.077	--	**
S.d.	0.050	0.060	0.038	0.061	0.074	0.134	0.062	--	
N	42	81	35	52	122	77	12	0	
Vertebrata									
Mean	0.060	0.074	0.057	0.050	0.052	0.061	0.060	--	ns
S.d.	0.052	0.049	0.028	0.028	0.035	0.028	0.019	--	
N	25	38	23	18	76	7	6	0	
Invertebrata									
Mean	0.047	0.112	0.084	0.099	0.098	0.114	0.094	--	ns
S.d.	0.048	0.069	0.047	0.068	0.108	0.140	0.087	--	
N	17	16	12	34	45	70	6	0	
Plants									
Mean	--	0.090	--	--	0.0	--	--	--	ns
S.d.	--	0.065	--	--	--	--	--	--	
N	0	27	0	0	1	0	0	0	
Mammalia									
Mean	0.047	0.068	0.072	0.020	--	--	--	--	ns
S.d.	0.031	0.050	0.057	0.010	--	--	--	--	
N	18	23	2	3	0	0	0	0	
Aves									
Mean	--	--	0.019	0.056	--	--	--	--	*
S.d.	--	--	0.018	0.026	--	--	--	--	
N	0	0	4	15	0	0	0	0	
Reptilia									
Mean	--	0.083	0.064	--	0.078	--	--	--	ns
S.d.	--	0.058	0.019	--	0.063	--	--	--	
N	0	8	17	0	2	0	0	0	
Amphibia									
Mean	0.091	0.084	--	--	0.065	0.006	0.060	--	ns
S.d.	0.082	0.036	--	--	0.093	--	0.019	--	
N	7	7	0	0	2	1	6	0	
Pisces									
Mean	--	--	--	--	0.051	0.070	--	--	ns
S.d.	--	--	--	--	0.033	0.014	--	--	
N	0	0	0	0	72	6	0	0	
Crustacea									
Mean	--	--	--	--	0.088	0.060	0.094	--	ns
S.d.	--	--	--	--	0.107	0.042	0.087	--	
N	0	0	0	0	40	46	6	0	

Table 11c (contin.)

Taxon	Under ground	Over ground	Arboreal	Air and Terrestrial	Aquatic	Aquatic littoral	Aquatic and Terrestrial	Aquatic and Air	Sig.
Insecta exc.Dr.									
Mean	0.063	0.101	0.067	0.091	0.124	--	--	--	ns
S.d.	0.044	0.074	0.042	0.070	0.103	--	--	--	
N	11	8	8	27	2	0	0	0	
Drosophila									
Mean	--	0.173	--	0.131	--	--	--	--	ns
S.d.	--	0.055	--	0.051	--	--	--	--	
N	0	4	0	7	0	0	0	0	
Mollusca									
Mean	0.0	0.075	0.119	--	0.162	0.317	--	--	*
S.d.	0.0	0.025	0.041	--	--	0.253	--	--	
N	5	4	4	0	1	11	0	0	
Dicotyledoneae									
Mean	--	0.057	--	--	0.0	--	--	--	ns
S.d.	--	0.042	--	--	--	--	--	--	
N	0	14	0	0	1	0	0	0	

Abbreviations: N = sample size;
S.d. = standard deviation;
Insecta exc.Dr. = insecta excluding Drosophila.
Significance: * = $p < 0.05$; ** = $p < 0.01$; *** = $p < 0.001$; ns = $p > 0.05$

$ Aquatic, without the littoral species.

@ In all analyses species from the "air + aquatic" habitat type were included in the "aquatic + terrestrial" habitat type.

Species size (Table 11)

a. The different groups of species size differed significantly in _H_ (ANOVA; $p < 0.01$).

b. _H_ increased in the overall species analysis (N = 717) as follows: small < medium < large < very large. This pattern is true for vertebrates as a whole but does not hold in the invertebrates.

c. Several higher taxa displayed a contrasting pattern. For example, in reptiles the highest _H_ occurred in species with small size, whereas the lowest _H_ occurred in the very large one (although the last entry had only 3 species). However, when the 2 highly variable parthenogenetic species were excluded, the lowest H was displayed by the small species size. Even then, the very large group displayed a smaller H than the large and even medium group. In crustaceans, however, both P and H displayed the contrasting pattern, i.e., species with small and medium size displayed the highest H which is significant for P (ANOVA; $p < 0.05$). Obviously, while the level of H was positively correlated with species size in some groups, others contradicted it.

Neutralists assume that heterozygosity is primarily a function of species size (Nei, 1975; Nei and Graur, personal communication, 1982). In order to test this hypothesis, we analyzed separately each category of species size for its potential ecological correlates of two ecological factors: (i) life zone, and (ii) habitat type. The results appear in Table 11 b,c. It is clear cut in most cases that the ecological expectations of H were born out. For example, in the small species group, 25 underground species showed H = 0.028, whereas 14 overground species showed H = 0.081. Similarly, the expected pattern was found in large species size. Similar verification of the ecological hypothesis was seen for the life zone parameters.

Table 12. Heterozygosity (H) within different categories of adult mobility.

Taxon	Almost sedentary	Low	High	Migratory	Significance
All species					
Mean	0.098	0.069	0.063	0.050	***
S.d.	0.102	0.059	0.069	0.040	
N	200	266	177	86	
Vertebrata					
Mean	0.055	0.058	0.043	0.048	ns
S.d.	0.049	0.056	0.039	0.032	
N	56	166	89	74	
Invertebrata					
Mean	0.136	0.088	0.083	0.063	***
S.d.	0.125	0.058	0.085	0.074	
N	92	100	88	12	
Plants					
Mean	0.079	--	--	--	--
S.d.	0.068	--	--	--	
N	52	0	0	0	

Table 12 (contin.)

Taxon	Almost sedentary	Low	High	Migratory	Significance
Mammalia					
Mean	0.037	0.045	0.038	0.013	ns
S.d.	0.026	0.031	0.040	0.005	
N	18	58	47	4	
Aves					
Mean	0.041	0.094	0.085	0.047	ns
S.d.	0.019	--	0.072	0.026	
N	5	1	3	28	
Reptilia					
Mean	0.091	0.068	0.063	0.078	ns
S.d.	0.089	0.086	0.055	0.063	
N	10	34	6	2	
Amphibia					
Mean	0.049	0.082	0.026	0.072	ns
S.d.	0.030	0.070	0.021	0.017	
N	12	30	6	3	
Pisces					
Mean	0.058	0.049	0.045	0.049	ns
S.d.	0.031	0.030	0.026	0.036	
N	10	43	27	37	
Crustacea					
Mean	0.117	0.071	0.073	0.029	*
S.d.	0.086	0.051	0.094	0.018	
N	26	24	49	6	
Insecta exc.Dr.					
Mean	0.087	0.087	0.081	0.097	ns
S.d.	0.035	0.055	0.068	0.094	
N	13	48	29	6	
Drosophila					
Mean	0.113	0.133	0.144	--	ns
S.d.	0.085	0.041	0.063	--	
N	3	14	9	0	
Mollusca					
Mean	0.167	0.027	--	--	ns
S.d.	0.186	0.066	--	--	
N	32	6	0	0	
Dicotyledoneae					
Mean	0.053	--	--	--	--
S.d.	0.047	--	--	--	
N	36	0	0	0	

Abbreviations: N = sample size;
S.d. = standard deviation;
Insecta exc.*Dr*. = insecta excluding *Drosophila*.
Significance: * = $p < 0.05$; ** = $p < 0.01$; *** = $p < 0.001$; ns = $p > 0.05$.

3. Adult mobility (Table 12)

a. Mobility classes differed significantly in their _H_ values (ANOVA; $p < 0.001$).

b. In the overall species analysis ($N = 729$) _H_ decreased as follows: sedentary > low > high > migratory.

c. The above pattern was primarily shown by invertebrates. The most striking examples of this pattern were crustaceans and molluscs, whereas other groups, such as _Drosophila_ and insects in general showed nonsignificantly different patterns.

d. Although vertebrates did not show the clear cut invertebrate pattern, in all 5 vertebrate classes, _H_ decreased from the low to the high category.

Table 13. Heterozygosity (_H_) within different categories of young dispersal.

Taxon	Low	Medium	High	Significance
All species				
Mean	0.082	0.066	0.072	*
S.d.	0.076	0.053	0.094	
N	276	215	207	
Vertebrata				
Mean	0.049	0.057	0.045	ns
S.d.	0.059	0.044	0.033	
N	101	169	95	
Invertebrata				
Mean	0.108	0.093	0.095	ns
S.d.	0.081	0.065	0.123	
N	150	38	98	
Plants				
Mean	0.066	0.121	0.096	ns
S.d.	0.042	0.075	0.092	
N	25	8	14	
Mammalia				
Mean	0.026	0.047	0.040	**
S.d.	0.022	0.034	0.034	
N	39	61	26	

Table 13 (contin.)

Taxon	Low	Medium	High	Significance
Aves				
Mean	0.050	0.060	0.046	ns
S.d.	0.026	0.045	0.024	
N	3	11	23	
Reptilia				
Mean	0.101	0.079	0.078	ns
S.d.	0.145	0.055	0.063	
N	11	25	2	
Amphibia				
Mean	0.057	0.079	--	ns
S.d.	0.042	0.084	--	
N	36	13	0	
Pisces				
Mean	0.049	0.050	0.047	ns
S.d.	0.027	0.028	0.036	
N	12	58	44	
Crustacea				
Mean	0.171	0.084	0.051	***
S.d.	0.127	0.059	0.039	
N	22	14	68	
Insecta exc.*Dr*.				
Mean	0.091	0.054	0.081	ns
S.d.	0.060	0.047	0.069	
N	82	9	2	
Drosophila				
Mean	0.142	0.065	0.109	ns
S.d.	0.055	--	0.023	
N	21	1	4	
Mollusca				
Mean	0.075	0.206	0.295	**
S.d.	0.070	--	0.265	
N	24	1	11	
Dicotyledoneae				
Mean	0.062	0.106	0.049	ns
S.d.	0.042	--	0.056	
N	20	1	10	

Abbreviations: N = sample size;
S.d. = standard deviation;
Insecta exc.*Dr*. = insecta excluding *Drosophila*.
Significance: * = $p < 0.05$; ** = $p < 0.01$;
*** = $p < 0.001$; ns = $p > 0.05$

4. Young dispersal (Table 13)

a. Dispersal classes differed significantly in their H values (ANOVA; $p < 0.05$).

b. In the overall species analysis ($N = 698$) the low dispersal class exhibited the highest H.

c. The above pattern was shown by invertebrates, primarily by crustaceans.

d. A contrasting pattern was indicated by molluscs and plants. Notably, the adults of these taxa are distinctly sedentary, but their young disperse extensively.

Noteworthy, both parameters, adult mobility and young dispersal, which represent a measure of gene flow, displayed the same pattern. This empiric result which derives from the analysis of natural populations conforms with theory (Karlin and McGregor, 1972 a,b; Karlin, 1982 b).

Table 14. Heterozygosity (H) within different categories of sociality.

Taxon	Solitary	Monogamic	Polygamic	Social	Significance
All species					
Mean	0.082	0.040	0.051	0.060	***
S.d.	0.085	0.028	0.047	0.069	
N	384	67	68	121	
Vertebrata					
Mean	0.054	0.042	0.043	0.053	ns
S.d.	0.052	0.028	0.042	0.044	
N	165	58	60	88	
Invertebrata					
Mean	0.104	0.023	0.109	0.080	ns
S.d.	0.098	0.025	0.052	0.109	
N	219	9	8	33	
Mammalia					
Mean	0.034	0.039	0.038	0.065	*
S.d.	0.028	0.025	0.042	0.040	
N	48	24	37	18	

Table 14 (contin.)

Taxon	Solitary	Monogamic	Polygamic	Social	Significance
Aves					
Mean	--	0.046	0.116	0.051	ns
S.d.	--	0.029	--	0.035	
N	0	31	1	4	
Reptilia					
Mean	0.077	0.016	0.046	--	ns
S.d.	0.086	--	0.046	--	
N	32	1	11	0	
Amphibia					
Mean	0.060	--	0.006	0.195	**
S.d.	0.047	--	--	0.180	
N	48	0	1	2	
Pisces					
Mean	0.054	0.040	0.055	0.045	ns
S.d.	0.034	0.057	0.029	0.030	
N	37	2	10	64	
Crustacea					
Mean	0.075	--	--	0.115	ns
S.d.	0.067	--	--	0.145	
N	87	0	0	16	
Insecta exc.*Dr*.					
Mean	0.102	0.023	0.136	0.038	***
S.d.	0.056	0.025	0.008	0.030	
N	68	9	2	15	
Drosophila					
Mean	0.153	--	0.100	--	*
S.d.	0.050	--	0.058	--	
N	16	0	6	0	
Mollusca					
Mean	0.144	--	--	--	--
S.d.	0.182	--	--	--	
N	37	0	0	0	

Abbreviations: N = sample size;
S.d. = standard deviation;
Insecta exc.*Dr*. = insecta excluding *Drosophila*.
Significance: * = $p < 0.05$; ** = $p < 0.01$;
*** = $p < 0.001$; ns = $p > 0.05$.

5. Sociality (Table 14)

a. In general, monogamic species harboured the least genetic variation as compared with either polygamic or solitary species. This pattern was true in all higher taxa analyzed, but only partly true in mammals, where solitary species displayed slightly lower H than monogamic ones.

b. Except for monogamic species which always dispalyed a low H, no consistent pattern was obtained for the other three categories.

Table 15a. Heterozygosity (H) within different body size categories.

Taxon	Small	Medium	Large	Significance
All species				
Mean	0.077	0.070	0.079	ns
S.d.	0.085	0.069	0.070	
N	296	308	90	
Vertebrata				
Mean	0.055	0.053	0.038	ns
S.d.	0.055	0.041	0.037	
N	203	106	49	
Invertebrata				
Mean	0.135	0.081	0.120	***
S.d.	0.118	0.081	0.069	
N	82	180	32	
Plants				
Mean	0.034	0.060	0.155	***
S.d.	0.033	0.054	0.062	
N	11	22	9	
Mammalia				
Mean	0.046	0.036	0.029	ns
S.d.	0.037	0.029	0.035	
N	82	28	18	
Aves				
Mean	0.051	0.064	0.007	ns
S.d.	0.032	0.044	0.001	
N	15	8	2	
Reptilia				
Mean	0.074	0.054	0.058	ns
S.d.	0.085	0.022	0.057	
N	47	4	3	
Amphibia				
Mean	0.065	0.071	0.041	ns
S.d.	0.076	0.056	0.032	
N	14	31	7	
Pisces				
Mean	0.050	0.049	0.040	ns
S.d.	0.032	0.029	0.030	
N	45	35	18	

Table 15a (contin.)

Taxon	Small	Medium	Large	Significance
Crustacea				
Mean	0.134	0.054	0.055	***
S.d.	0.118	0.040	0.037	
N	35	61	9	
Insecta exc.*Dr*.				
Mean	0.097	0.073	0.138	***
S.d.	0.054	0.053	0.064	
N	13	69	14	
Drosophila				
Mean	0.130	--	--	--
S.d.	0.054	--	--	
N	28	0	0	
Mollusca				
Mean	0.314	0.123	0.206	ns
S.d.	0.362	0.144	--	
N	4	33	1	
Dicotyledoneae				
Mean	0.034	0.054	--	ns
S.d.	0.033	0.052	--	
N	11	16	0	

Table 15b. Heterozygosity (*H*) within different life zones among three categories of body size.

I. Small body size

Taxon	Arctic	Temperate	Tropical	Temperate and Tropical	Cosmopolitan	Significance
All species						
Mean	0.074	0.063	0.080	0.128	0.094	**
S.d.	0.026	0.067	0.090	0.136	0.066	
N	7	169	72	35	12	
Vertebrata						
Mean	0.066	0.049	0.062	0.069	0.073	ns
S.d.	0.023	0.054	0.057	0.055	0.062	
N	5	121	52	18	6	
Invertebrata						
Mean	0.095	0.120	0.128	0.190	0.114	ns
S.d.	0.023	0.082	0.135	0.167	0.069	
N	2	37	20	17	6	
Plants						
Mean	--	0.034	--	--	--	--
S.d.	--	0.033	--	--	--	
N	0	11	0	0	0	

Table 15b (contin.)

Taxon	Arctic	Temperate	Tropical	Temperate and Tropical	Cosmopolitan	Significance
Mammalia						
Mean	0.072	0.044	0.044	0.052	0.040	ns
S.d.	0.019	0.037	0.044	0.036	0.031	
N	3	54	15	6	4	
Aves						
Mean	0.080	0.039	--	0.116	0.098	*
S.d.	--	0.022	--	--	--	
N	1	11	0	1	1	
Reptilia						
Mean	--	0.071	0.071	0.099	--	ns
S.d.	--	0.110	0.071	0.080	--	
N	0	15	26	6	0	
Amphibia						
Mean	--	0.066	0.051	--	--	ns
S.d.	--	0.079	--	--	--	
N	0	13	1	0	0	
Pisces						
Mean	0.033	0.041	0.067	0.045	0.181	***
S.d.	--	0.024	0.026	0.011	--	
N	1	28	10	5	1	
Crustacea						
Mean	0.095	0.141	0.042	0.206	0.126	ns
S.d.	0.023	0.099	0.032	0.161	0.099	
N	2	14	7	9	3	
Insecta exc.*Dr*.						
Mean	--	0.088	0.136	0.113	--	ns
S.d.	--	0.059	0.008	--	--	
N	0	10	2	1	0	
Drosophila						
Mean	--	0.149	0.136	0.094	0.102	ns
S.d.	--	0.059	0.056	0.025	0.042	
N	0	10	10	5	3	
Mollusca						
Mean	--	0.0	0.635	0.620	--	--
S.d.	--	0.0	--	--	--	
N	0	2	1	1	0	
Dicotyledcneae						
Mean	--	0.034	--	--	--	--
S.d.	--	0.033	--	--	--	
N	0	11	0	0	0	

Table 15b (contin.)

Taxon	Arctic	Temperate	Tropical	Temperate and Tropical	Cosmopolitan	Significance
II. Medium body size						
All species						
Mean	0.044	0.069	0.058	0.131	0.107	***
S.d.	0.029	0.049	0.057	0.193	0.099	
N	16	212	50	17	10	
Vertebrata						
Mean	0.043	0.057	0.050	0.065	0.0	ns
S.d.	0.031	0.042	0.027	0.058	--	
N	14	76	7	6	1	
Invertebrata						
Mean	0.047	0.077	0.061	0.167	0.119	**
S.d.	0.013	0.052	0.059	0.233	0.097	
N	2	119	38	11	9	
Plants						
Mean	--	0.063	0.051	--	--	ns
S.d.	--	0.046	0.081	--	--	
N	0	17	5	0	0	
Mammalia						
Mean	0.046	0.037	0.043	0.075	0.0	ns
S.d.	0.044	0.024	0.004	--	--	
N	4	18	2	1	1	
Aves						
Mean	0.054	0.056	0.027	0.122	--	ns
S.d.	0.025	--	0.028	0.040	--	
N	3	1	2	2	0	
Reptilia						
Mean	--	0.054	--	--	--	--
S.d.	--	0.022	--	--	--	
N	0	4	0	0	0	
Amphibia						
Mean	--	0.074	0.066	0.0	--	ns
S.d.	--	0.057	0.035	--	--	
N	0	28	2	1	0	
Pisces						
Mean	0.037	0.053	0.074	0.035	--	ns
S.d.	0.028	0.028	--	0.050	--	
N	7	25	1	2	0	
Crustacea						
Mean	0.057	0.050	0.051	0.068	0.190	**
S.d.	--	0.018	0.049	0.040	--	
N	1	27	27	5	1	
Insecta exc. _Dr_.						
Mean	--	0.078	0.027	0.044	0.077	ns
S.d.	--	0.047	0.022	0.015	0.114	
N	0	57	5	2	5	

Table 15b (contin.)

Taxon	Arctic	Temperate	Tropical	Temperate and Tropical	Cosmopolitan	Significance
Mollusca						
Mean	--	0.077	0.133	0.461	--	***
S.d.	--	0.057	0.067	0.296	--	
N	0	23	6	3	0	
Dicotyledoneae						
Mean	--	0.067	0.016	--	--	ns
S.d.	--	0.053	0.022	--	--	
N	0	12	4	0	0	
III. Large body size						
All species						
Mean	0.025	0.059	0.112	0.082	0.157	***
S.d.	0.041	0.049	0.069	0.074	0.074	
N	10	76	15	14	7	
Vertebrata						
Mean	0.013	0.044	0.027	0.048	0.106	***
S.d.	0.010	0.030	0.024	0.040	0.061	
N	9	58	2	9	3	
Invertebrata						
Mean	0.137	0.080	0.125	0.142	0.184	ns
S.d.	--	0.055	0.064	0.087	0.075	
N	1	10	13	5	3	
Plants						
Mean	--	0.146	--	--	0.226	ns
S.d.	--	0.060	--	--	--	
N	0	8	0	0	1	
Mammalia						
Mean	0.013	0.012	0.027	0.061	0.125	***
S.d.	0.011	0.009	0.024	0.043	--	
N	8	4	2	3	1	
Aves						
Mean	0.008	0.048	--	0.007	--	*
S.d.	--	0.017	--	--	--	
N	1	12	0	1	0	
Reptilia						
Mean	--	0.016	--	0.078	--	ns
S.d.	--	--	--	0.063	--	
N	0	1	0	2	0	
Amphibia						
Mean	--	0.041	--	--	--	--
S.d.	--	0.032	--	--	--	
N	0	7	0	0	0	

Table 15b (contin.)

Taxon	Arctic	Temperate	Tropical	Temperate and Tropical	Cosmopolitan	Significance
Pisces						
Mean	--	0.047	--	0.029	0.038	ns
S.d.	--	0.033	--	0.006	--	
N	0	34	0	3	1	
Crustacea						
Mean	--	0.048	0.066	0.049	--	ns
S.d.	--	0.031	0.060	0.017	--	
N	0	4	3	2	0	
Insecta exc.Dr.						
Mean	--	0.047	0.136	0.204	--	*
S.d.	--	0.013	0.054	0.023	--	
N	0	2	9	3	0	
Mollusca						
Mean	--	--	0.206	--	--	--
S.d.	--	--	--	--	--	
N	0	0	1	0	0	
Dicotyledoneae						
Mean	--	--	--	--	--	--
S.d.	--	--	--	--	--	
N	0	0	0	0	0	

Table 15c. Heterozygosity (H) within different categories of habitat type among three categories of body size.

I. Small body size

Taxon	Under ground	Over ground	Arboreal	Air and Terrestrial	Aquatic	Aquatic littoral $	Aquatic and Terrestrial	Aquatic and Air @	Significance
All species									
Mean	0.038	0.074	0.055	0.100	0.080	0.116	0.161	0.136	***
S.d.	0.032	0.077	0.028	0.060	0.094	0.161	0.110	0.008	
N	60	78	30	34	52	25	13	2	
Vertebrata									
Mean	0.032	0.075	0.055	0.052	0.049	0.054	0.104	--	**
S.d.	0.024	0.079	0.028	0.041	0.035	0.016	0.123	--	
N	52	61	30	7	36	10	5	0	
Invertebrata									
Mean	0.077	0.144	--	0.113	0.149	0.157	0.196	0.136	ns
S.d.	0.049	0.065	--	0.058	0.142	0.200	0.093	0.008	
N	8	6	0	27	16	15	8	2	
Plants									
Mean	--	0.034	--	--	--	--	--	--	--
S.d.	--	0.033	--	--	--	--	--	--	
N	0	11	0	0	0	0	0	0	

Table 15c (contin.)

Taxon	Under ground	Over ground	Arboreal	Air and Terrestrial	Aquatic	Aquatic littoral	Aquatic and Terrestrial	Aquatic and Air	Signi-fican-ce
Mammalia									***
Mean	0.033	0.074	0.048	0.023	0.043	--	--	--	
S.d.	0.025	0.045	0.058	0.011	--	--	--	--	
N	47	25	3	4	1	0	0	0	
Aves									*
Mean	--	0.043	0.040	0.091	--	--	--	--	
S.d.	--	0.010	0.027	0.029	--	--	--	--	
N	0	3	9	3	0	0	0	0	
Reptilia									ns
Mean	0.020	0.087	0.064	--	--	--	--	--	
S.d.	0.019	0.109	0.019	--	--	--	--	--	
N	3	27	17	0	0	0	0	0	
Amphibia									ns
Mean	0.038	0.039	0.072	--	--	--	0.104	--	
S.d.	0.008	0.020	--	--	--	--	0.123	--	
N	2	6	1	0	0	0	5	0	
Pisces									ns
Mean	--	--	--	--	0.049	0.054	--	--	
S.d.	--	--	--	--	0.035	0.016	--	--	
N	0	0	0	0	35	10	0	0	
Crustacea									ns
Mean	--	--	--	--	0.146	0.077	0.196	--	
S.d.	--	--	--	--	0.146	0.061	0.093	--	
N	0	0	0	0	15	12	8	0	
Insecta exc.<u>Dr</u>.									*
Mean	0.102	0.085	--	0.037	0.197	--	--	0.136	
S.d.	0.017	0.040	--	0.064	--	--	--	0.008	
N	5	2	0	3	1	0	0	2	
<u>Drosophila</u>									ns
Mean	--	0.173	--	0.122	--	--	--	--	
S.d.	--	0.055	--	0.051	--	--	--	--	
N	0	4	0	24	0	0	0	0	
Mollusca									***
Mean	0.0	--	--	--	--	0.627	--	--	
S.d.	0.0	--	--	--	--	0.011	--	--	
N	2	0	0	0	0	2	0	0	
Dicotyledoneae									--
Mean	--	0.034	--	--	--	--	--	--	
S.d.	--	0.033	--	--	--	--	--	--	
N	0	11	0	0	0	0	0	0	

Table 15c (contin.)

Taxon	Under ground	Over ground	Arboreal	Air and Terrestrial	Aquatic	Aquatic littoral	Aquatic and Terrestrial	Aquatic and Air	Significance
II. Medium body size									
All species									
Mean	0.060	0.067	0.072	0.064	0.059	0.101	0.066	--	*
S.d.	0.055	0.051	0.061	0.042	0.050	0.117	0.043	--	
N	56	51	26	28	77	56	13	0	
Vertebrata									
Mean	0.054	0.065	0.033	0.094	0.048	0.060	0.074	--	ns
S.d.	0.051	0.035	0.049	--	0.031	0.037	0.048	--	
N	29	15	9	1	41	3	8	0	
Invertebrata									
Mean	0.067	0.077	0.092	0.063	0.073	0.104	0.053	--	ns
S.d.	0.059	0.061	0.057	0.042	0.062	0.119	0.035	--	
N	27	15	17	27	36	53	5	0	
Plants									
Mean	--	0.063	--	--	--	--	--	--	--
S.d.	--	0.054	--	--	--	--	--	--	
N	0	21	0	0	0	0	0	0	
Mammalia									
Mean	0.042	0.043	0.0	--	--	--	--	--	*
S.d.	0.028	0.022	0.0	--	--	--	--	--	
N	19	5	4	0	0	0	0	0	
Aves									
Mean	--	0.063	0.059	0.094	--	--	--	--	ns
S.d.	--	0.027	0.054	--	--	--	--	--	
N	0	2	5	1	0	0	0	0	
Reptilia									
Mean	--	0.064	--	--	--	--	0.043	--	ns
S.d.	--	0.029	--	--	--	--	0.014	--	
N	0	2	0	0	0	0	2	0	
Amphibia									
Mean	0.079	0.083	--	--	0.050	0.018	0.084	--	ns
S.d.	0.074	0.042	--	--	0.043	--	0.052	--	
N	10	6	0	0	8	1	6	0	
Pisces									
Mean	--	--	--	--	0.047	0.081	--	--	ns
S.d.	--	--	--	--	0.029	0.011	--	--	
N	0	0	0	0	33	2	0	0	
Crustacea									
Mean	--	--	--	--	0.056	0.053	0.053	--	ns
S.d.	--	--	--	--	0.053	0.029	0.035	--	
N	0	0	0	0	25	31	5	0	

Table 15c (contin.)

Taxon	Under ground	Over ground	Arboreal	Air and Terrestrial	Aquatic	Aquatic littoral	Aquatic and Terrestrial	Aquatic and Air	Significance
Insecta exc. Dr.									
Mean	0.075	0.106	0.067	0.063	0.090	--	--	--	ns
S.d.	0.058	0.085	0.042	0.042	0.054	--	--	--	
N	24	6	8	27	4	0	0	0	
Mollusca									
Mean	0.0	0.057	0.114	--	0.127	0.224	--	--	*
S.d.	0.0	0.030	0.061	--	0.049	0.220	--	--	
N	3	9	9	0	2	10	0	0	
Dicotyledoneae									
Mean	--	0.054	--	--	--	--	--	--	--
S.d.	--	0.052	--	--	--	--	--	--	
N	0	16	0	0	0	0	0	0	

III. Large body size

Taxon	Under ground	Over ground	Arboreal	Air and Terrestrial	Aquatic	Aquatic littoral	Aquatic and Terrestrial	Aquatic and Air	Significance
All species									
Mean	--	0.092	0.007	0.096	0.050	0.090	0.050	0.008	**
S.d.	--	0.073	--	0.066	0.051	0.060	0.034	--	
N	0	22	1	26	55	14	5	1	
Vertebrata									
Mean	--	0.041	0.007	0.048	0.042	0.063	0.050	0.008	ns
S.d.	--	0.043	--	0.017	0.033	0.081	0.034	--	
N	0	10	1	12	48	3	5	1	
Invertebrata									
Mean	--	0.094	--	0.138	0.122	0.098	--	--	ns
S.d.	--	--	--	0.064	0.102	0.056	--	--	
N	0	1	0	14	6	11	0	0	
Plants									
Mean	--	0.138	--	--	0.0	--	--	--	ns
S.d.	--	0.067	--	--	--	--	--	--	
N	0	11	0	0	1	0	0	0	
Mammalia									
Mean	--	0.041	--	--	0.014	--	0.0	--	ns
S.d.	--	0.043	--	--	0.011	--	--	--	
N	0	10	0	0	6	0	1	0	
Aves									
Mean	--	--	0.007	0.048	--	--	--	0.008	*
S.d.	--	--	--	0.017	--	--	--	--	
N	0	0	1	12	0	0	0	1	
Reptilia									
Mean	--	--	--	--	0.058	--	--	--	--
S.d.	--	--	--	--	0.057	--	--	--	
N	0	0	0	0	3	0	0	0	

Table 15c (contin.)

Taxon	Under ground	Over ground	Arboreal	Air and Terrestrial	Aquatic	Aquatic littoral	Aquatic and Terrestrial	Aquatic and Air	Significance
Amphibia									
Mean	--	--	--	--	0.006	0.016	0.063	--	ns
S.d.	--	--	--	--	--	0.014	0.023	--	
N	0	0	0	0	1	2	4	0	
Pisces									
Mean	--	--	--	--	0.046	--	--	--	--
S.d.	--	--	--	--	0.032	--	--	--	
N	0	0	0	0	38	0	0	0	
Crustacea									
Mean	--	--	--	--	0.039	0.062	--	--	ns
S.d.	--	--	--	--	0.028	0.041	--	--	
N	0	0	0	0	3	6	0	0	
Insecta exc.Dr.									
Mean	--	--	--	0.138	--	--	--	--	--
S.d.	--	--	--	0.064	--	--	--	--	
N	0	0	0	14	0	0	0	0	
Mollusca									
Mean	--	--	--	--	--	0.206	--	--	--
S.d.	--	--	--	--	--	--	--	--	
N	0	0	0	0	0	1	0	0	
Dicotyledoneae									
Mean	--	0.062	--	--	0.0	--	--	--	ns
S.d.	--	0.017	--	--	--	--	--	--	
N	0	2	0	0	1	0	0	0	

Abbreviations: N = sample size;
S.d. = standard deviation;
Insecta exc.*Dr*. = insecta excluding *Drosophila*.
Significance: * = $p < 0.05$; ** = $p < 0.01$; *** = $p < 0.001$; ns = $p > 0.05$
$ Aquatic, without the littoral species.
@ In all analyses species from the "air + aquatic" habitat type were included in the "aquatic + terrestrial" habitat type.

Patterns Among Life History Parameters (Fig. 5)

I. Body size (Table 15)

Body size categories were defined separately for each of the higher taxa, and the following within group differences were indicated:

LEVELS OF HETEROZYGOSITY OF BIOTIC FACTORS

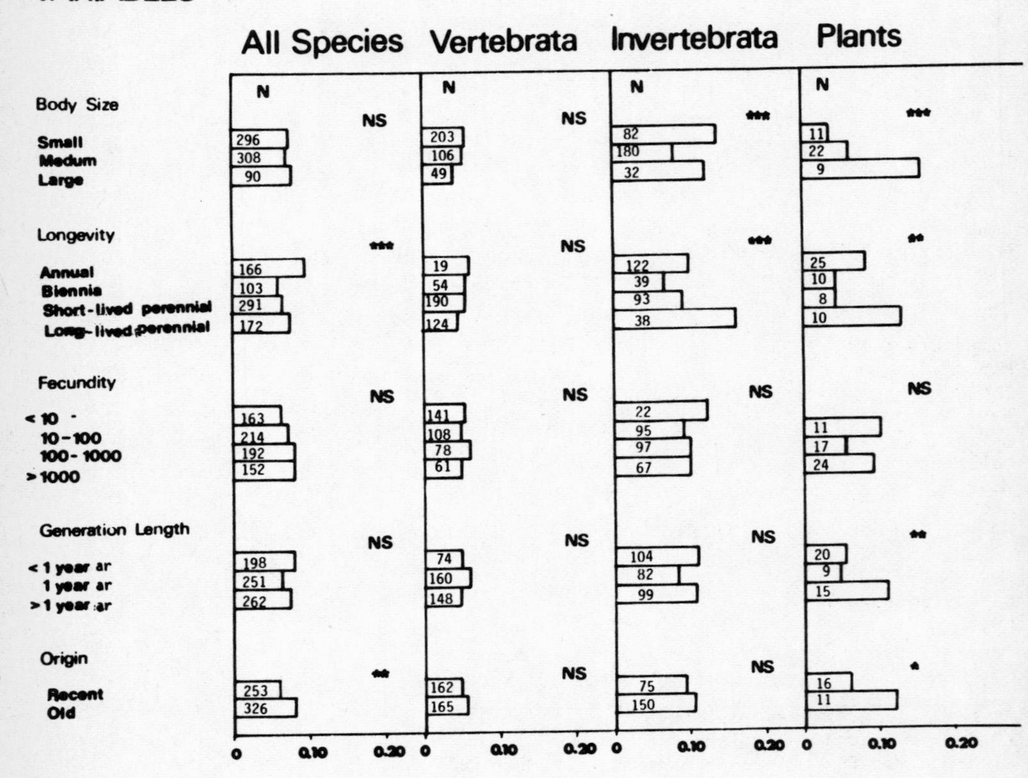

Heterozygosity (H)

N = number of species analyzed
Significance levels as in Tables

Fig. 5a

LEVELS OF HETEROZYGOSITY OF BIOTIC FACTORS

VERTEBRATA

LIFE HISTORY VARIABLES

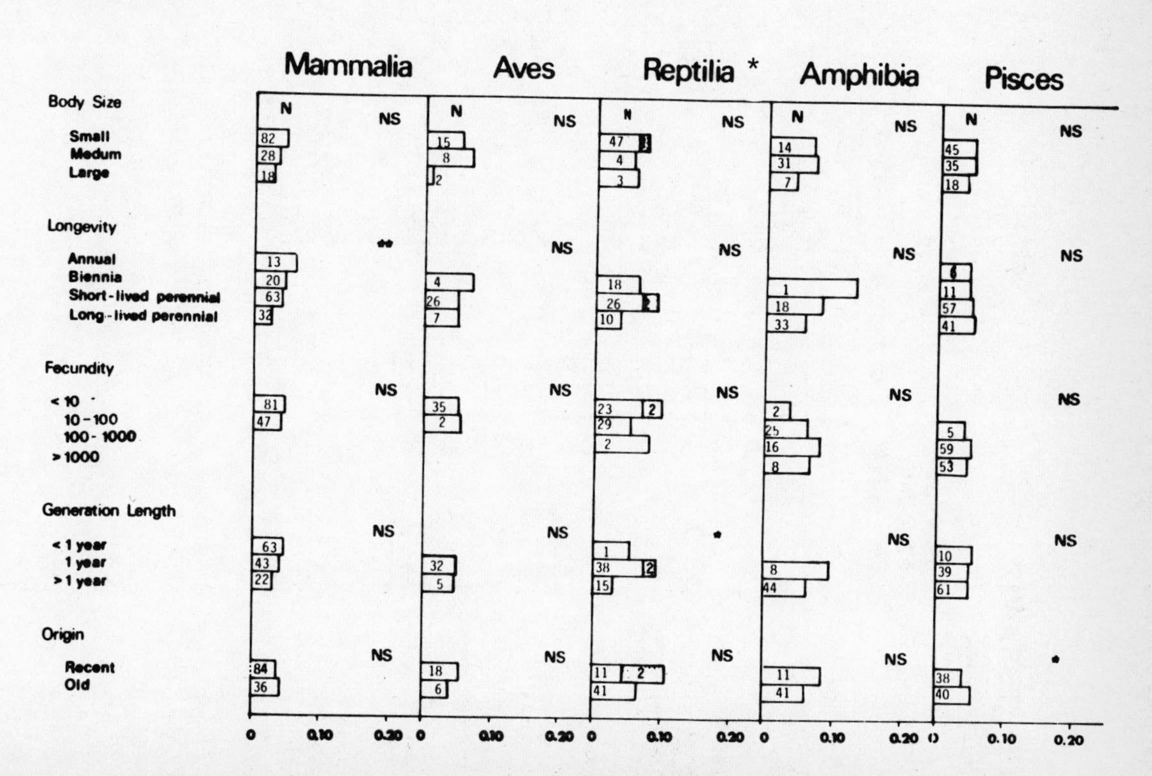

Fig. 5b * Rectangles added to bars are due to parthenogenetic species

Heterozygosity (H̲)

LEVELS OF HETEROZYGOSITY OF BIOTIC FACTORS

INVERTEBRATA

LIFE HISTORY VARIABLES

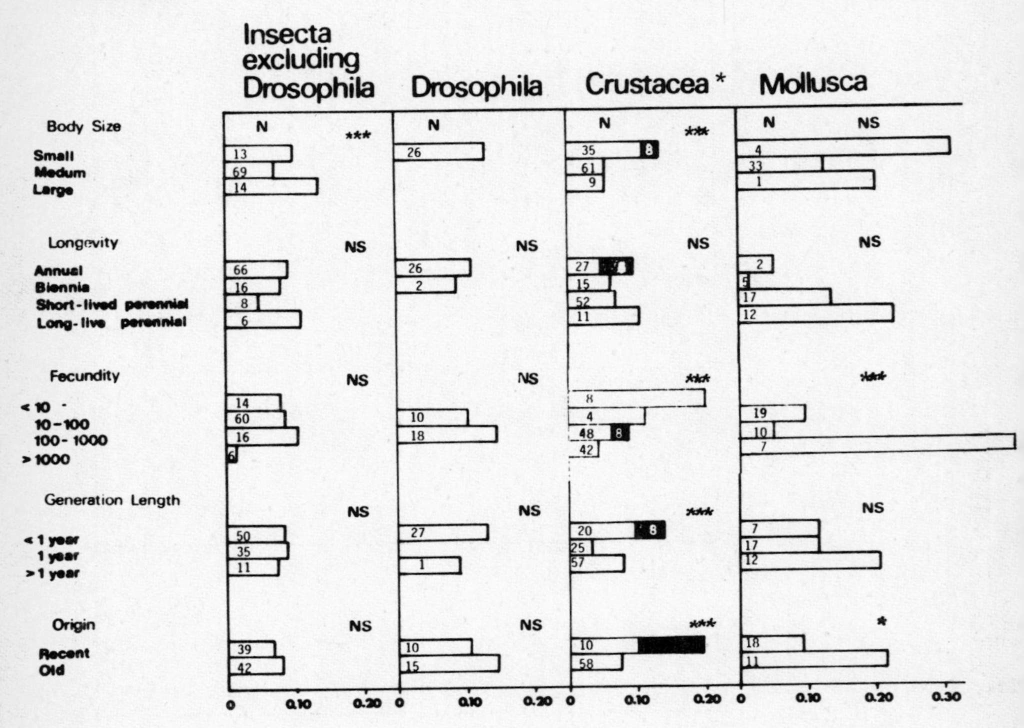

Heterozygosity (H)

* Rectangles added to bars are
due to parthenogenetic species

Fig. 5c

a. Different body size categories did not differ in the levels of *H* in the overall species analysis (N = 694).

b. Multiple comparisons indicated that in vertebrates the larger size category was significantly less variable than the small and medium ones ($p < 0.05$).

c. The invertebrate pattern differed from that of the vertebrate one. The size groups varied in their levels of *H* in different higher taxa. Crustaceans displayed significantly ($p < 0.001$) the highest *H* in the small size group. Multiple comparison analysis indicated that the small body size group was significantly ($p < 0.05$) higher in *H* than either the medium or the large classes, which were similar. A similar pattern was displayed by molluscs. Insects excluding *Drosophila* showed a contrasting pattern ($p < 0.001$) where the highest *H* was exhibited by the largest size category, and the lowest *H* in the medium one.

The suggestion that large mammals harbour less genetic variation than small ones (Simonsen, 1982) is still open to doubt and requires additional testing. Dividing the 18 large mammals into life zone categories (Table 15 b) indicated that the large mammals showed the expected pattern (i.e., temperate and arctic species harbour less genetic variation than species living in other life zones). Since 12 of the 18 species belonged to the arctic and temperate life zones, the interpretation of the phenomenon is questionable. Similarly, large overground mammals harboured more genetic variation than aquatic species (Table 15 c). In any event, in order to either reject or substantiate the claim that large mammals are indeed low in *H* many more large overground and/or tropical mammals should be tested.

Table 16. Heterozygosity (*H*) within different categories of longevity.

Taxon	Annual	Bien-nial	3 to 5 years	> 5 years	Significance
All species					
Mean	0.093	0.059	0.066	0.076	***
S.d.	0.076	0.041	0.074	0.088	
N	166	103	291	172	
Vertebrata					
Mean	0.060	0.056	0.055	0.046	ns
S.d.	0.047	0.032	0.055	0.042	
N	19	54	190	124	
Invertebrata					
Mean	0.100	0.066	0.090	0.160	***
S.d.	0.079	0.048	0.100	0.134	
N	122	39	93	38	
Plants					
Mean	0.084	0.043	0.043	0.130	**
S.d.	0.070	0.048	0.052	0.063	
N	25	10	8	10	
Mammalia					
Mean	0.067	0.048	0.042	0.025	**
S.d.	0.052	0.039	0.029	0.030	
N	13	20	63	32	
Aves					
Mean	--	0.067	0.048	0.050	ns
S.d.	--	0.033	0.031	0.036	
N	0	4	26	7	
Reptilia					
Mean	--	0.064	0.092	0.035	ns
S.d.	--	0.021	0.109	0.034	
N	0	18	26	10	
Amphibia					
Mean	--	0.131	0.079	0.056	ns
S.d.	--	--	0.070	0.051	
N	0	1	18	33	
Pisces					
Mean	0.045	0.045	0.048	0.053	ns
S.d.	0.034	0.024	0.028	0.036	
N	6	11	57	41	
Crustacea					
Mean	0.099	0.064	0.071	0.107	ns
S.d.	0.124	0.034	0.063	0.091	
N	27	15	52	11	
Insecta exc.*Dr*.					
Mean	0.089	0.079	0.048	0.112	ns
S.d.	0.061	0.057	0.039	0.050	
N	66	16	8	6	

Table 16 (contin.)

Taxon	Annual	Biennial	3 to 5 years	> 5 years	Significance
Drosophila					
Mean	0.133	0.087	--	--	ns
S.d.	0.055	0.004	--	--	
N	26	2	0	0	
Mollusca					
Mean	0.052	0.017	0.136	0.229	ns
S.d.	0.074	0.037	0.185	0.202	
N	2	5	17	12	
Dicotyledoneae					
Mean	0.067	0.043	0.028	0.076	ns
S.d.	0.049	0.048	0.031	--	
N	18	10	7	1	

Abbreviations: N = sample size;
S.d. = standard deviation;
Insecta exc.Dr. = insecta excluding Drosophila.
Significance: * = $p < 0.05$; ** = $p < 0.01$; *** = $p < 0.001$;
ns = $p > 0.05$.

2. Longevity (Table 16)

a. Longevity categories differed significantly in the overall species analysis (N = 732) in their H values (ANOVA; $p < 0.001$).

b. In general, high H values characterized both the annual and long-lived species. This pattern was prominent particularly in crustaceans and insects. The vertebrates, however, generally displayed a nonsignificant decrease in H from annuals to long-lived perennials. Only mammals showed a significant pattern ($p < 0.01$).

c. Multiple comparison testing indicated that in the overall species analysis annuals differed significantly from biennials or short-lived perennials ($p < 0.01$). In addition, long-lived invertebrates differed significantly from all other groups ($p < 0.01$).

Noteworthy, the highest level of H appears to characterize the shortest and the longest life cycles, in plants and in the invertebrates. This pattern may relate to the swift turnover in both space and time of the annuals in contrast to the permanence and persistence over changing time in the longest life cycle.

Table 17, Heterozygosity (H) within different categories of generation length.

Taxon	< 1 year	1 year	> 1 year	Significance
All species				
Mean	0.082	0.065	0.074	ns
S.d.	0.081	0.067	0.078	
N	198	251	262	
Vertebrata				
Mean	0.048	0.058	0.047	ns
S.d.	0.037	0.058	0.040	
N	74	160	148	
Invertebrata				
Mean	0.111	0.080	0.108	ns
S.d.	0.097	0.082	0.102	
N	104	82	99	
Plants				
Mean	0.056	0.044	0.113	**
S.d.	0.052	0.056	0.071	
N	20	9	15	
Mammalia				
Mean	0.047	0.038	0.031	ns
S.d.	0.038	0.033	0.031	
N	63	43	22	
Aves				
Mean	--	0.051	0.045	ns
S.d.	--	0.031	0.040	
N	0	32	5	
Reptilia				
Mean	0.052	0.089	0.028	*
S.d.	--	0.088	0.031	
N	1	38	15	
Amphibia				
Mean	--	0.097	0.060	ns
S.d.	--	0.097	0.049	
N	0	8	44	
Pisces				
Mean	0.053	0.048	0.048	ns
S.d.	0.028	0.029	0.033	
N	10	39	61	

Table 17 (contin.)

Taxon	< 1 year	1 year	> 1 year	Significance
Crustacea				
Mean	0.142	0.033	0.082	***
S.d.	0.123	0.025	0.072	
N	20	25	57	
Insecta exc.*Dr*.				
Mean	0.086	0.088	0.076	ns
S.d.	0.068	0.046	0.055	
N	50	35	11	
Drosophila				
Mean	0.131	--	0.090	ns
S.d.	0.054	--	--	
N	27	0	1	
Mollusca				
Mean	0.115	0.115	0.207	ns
S.d.	0.227	0.149	0.204	
N	7	17	12	
Dicotyledoneae				
Mean	0.040	0.044	0.069	ns
S.d.	0.030	0.056	0.051	
N	14	9	6	

Abbreviations: N = sample size;
S.d. = standard deviation;
Insecta exc.*Dr*. = insecta excluding *Drosophila*.
Significance: * = $p < 0.05$; ** = $p < 0.01$; *** = $p < 0.001$;
ns = $p > 0.05$.

Generation length (Table 17)

a. Although the overall species analysis (N = 711) displayed a concave pattern of *H*, it essentially hid two contrasting patterns, those of vertebrates and invertebrates.

b. In general, vertebrates displayed a decrease in *H* with generation length, or a convex pattern, which is significant only in reptiles ($p < 0.05$). The other four vertebrate classes showed a nonsignificant decrease in *H*, with increasing generation length.

c. Invertebrates displayed a largely concave pattern which exists and is highly significant only in crustaceans. In molluscs, H increased towards the > 1 year generation length category, whereas in insects excluding Drosophila, H decreased with increasing generation length.

d. Plants displayed a concave pattern, whereas in dicotyledons H increased with generation length.

Table 18. Heterozygosity (H) within different categories of fecundity.

Taxon	< 10	10-100	100-1000	> 1000	Significance
All species					
Mean	0.063	0.070	0.080	0.079	ns
S.d.	0.067	0.057	0.068	0.107	
N	163	214	192	152	
Vertebrata					
Mean	0.053	0.048	0.060	0.048	ns
S.d.	0.059	0.038	0.044	0.037	
N	141	108	78	61	
Invertebrata					
Mean	0.124	0.092	0.101	0.102	ns
S.d.	0.081	0.065	0.079	0.147	
N	22	95	97	67	
Plants					
Mean	--	0.101	0.051	0.091	ns
S.d.	--	0.067	0.054	0.073	
N	0	11	17	24	
Mammalia					
Mean	0.042	0.039	--	--	ns
S.d.	0.038	0.031	--	--	
N	81	47	0	0	
Aves					
Mean	0.050	0.053	--	--	ns
S.d.	0.031	0.064	--	--	
N	35	2	0	0	
Reptilia					
Mean	0.096	0.052	0.078	--	ns
S.d.	0.115	0.027	0.063	--	
N	23	29	2	0	
Amphibia					
Mean	0.038	0.060	0.080	0.065	ns
S.d.	0.008	0.056	0.076	0.042	
N	2	25	16	8	

Table 18 (contin.)

Taxon	< 10	10-100	100-1000	> 1000	Significance
Pisces					
Mean	--	0.040	0.052	0.045	ns
S.d.	--	0.031	0.026	0.035	
N	0	5	59	53	
Crustacea					
Mean	0.196	0.113	0.088	0.043	***
S.d.	0.092	0.098	0.093	0.031	
N	8	4	48	42	
Insecta exc.Dr.					
Mean	0.082	0.087	0.106	0.029	ns
S.d.	0.029	0.067	0.045	0.018	
N	14	60	16	6	
Drosophila					
Mean	--	0.104	0.144	--	ns
S.d.	--	0.045	0.054	--	
N	0	10	18	0	
Mollusca					
Mean	--	0.094	0.053	0.412	***
S.d.	--	0.066	0.057	0.273	
N	0	19	10	7	
Dicotyledoneae					
Mean	--	0.107	0.041	0.047	**
S.d.	--	0.030	0.042	0.043	
N	0	6	15	14	

Abbreviations: N = sample size;
S.d. = standard deviation;
Insecta exc.Dr. = insecta excluding *Drosophila*.
Significance: * = $p < 0.05$; ** = $p < 0.01$; *** = $p < 0.001$;
ns = $p > 0.05$.

Fecundity (Table 18)

a. Although the overall species analysis displayed a general increase of *H* with fecundity, no consistent pattern was found by analyzing individual higher taxa.

b. A general increase in *H* with fecundity was revealed in birds and *Drosophila*.

c. A nonsignificant convex pattern was displayed by insects excluding *Drosophila*, fishes and amphibians.

d. A significant concave pattern was displayed by molluscs, and a nonsignificant one by reptiles.

e. The most significant pattern was revealed by crustaceans. *H* decreased significantly ($p < 0.01$) with increasing fecundity. A similar weak pattern was also revealed by mammals.

The above diverse patterns did not reveal any straightforward general explainable structure. It would be worthwhile to test critically whether the generally positive correlation revealed by the overall species analysis is empirically genuine and whether it supports the theoretical expectation.

Table 19. Heterozygosity (*H*) of old and recent species.

Taxon	Recent	Old	Significance
All species			
Mean	0.063	0.081	**
S.d.	0.067	0.076	
N	253	326	
Vertebrata			
Mean	0.048	0.056	ns
S.d.	0.053	0.046	
N	162	165	
Invertebrata			
Mean	0.095	0.105	ns
S.d.	0.085	0.092	
N	75	150	
Plants			
Mean	0.062	0.121	*
S.d.	0.045	0.075	
N	16	11	
Mammalia			
Mean	0.039	0.042	ns
S.d.	0.030	0.038	
N	84	36	

Table 19 (contin.)

Taxon	Recent	Old	Significance
Aves			
Mean	0.054	0.038	ns
S.d.	0.035	0.042	
N	18	6	
Reptilia			
Mean	0.105	0.063	ns
S.d.	0.143	0.055	
N	11	41	
Amphibia			
Mean	0.085	0.060	ns
S.d.	0.085	0.051	
N	11	41	
Pisces			
Mean	0.039	0.057	*
S.d.	0.023	0.035	
N	38	40	
Crustacea			
Mean	0.195	0.076	***
S.d.	0.154	0.075	
N	10	58	
Insecta exc.*Dr*.			
Mean	0.069	0.079	ns
S.d.	0.056	0.048	
N	39	42	
Drosophila			
Mean	0.107	0.144	ns
S.d.	0.060	0.054	
N	8	15	
Mollusca			
Mean	0.092	0.213	*
S.d.	0.053	0.216	
N	18	11	
Dicotyledoneae			
Mean	0.061	0.064	ns
S.d.	0.049	0.050	
N	13	3	

Abbreviations: N = sample size;
S.d. = standard deviation;
Insecta exc.*Dr*. = insecta excluding *Drosophila*.
Significance: * = $p < 0.05$; ** = $p < 0.01$;
*** = $p < 0.001$; ns = $p > 0.05$.

5. Origin (Table 19)

a. In the overall species analysis (N = 579), older species were significantly more variable than recent ones (ANOVA; $p < 0.01$).

b. The above mentioned pattern was revealed as a trend in mammals, fishes, molluscs, insects and plants.

c. The reverse pattern, i.e., that recent species harbour more genetic variation, was highly significant in crustaceans ($p < 0.001$) as well as in amphibians, reptiles and birds.

Table 20. Heterozygosity (H) within different categories of chromosome number.

Taxon	chromosome number < 18	20 - 38	40 - 58	60 - 78	> 80	Significar
All species						
Mean	0.101	0.061	0.062	0.066	0.065	***
S.d.	0.067	0.052	0.052	0.048	0.102	
N	58	152	127	35	20	
Vertebrata						
Mean	--	0.051	0.047	0.051	0.029	ns
S.d.	--	0.041	0.033	0.036	0.024	
N	0	115	96	23	14	
Invertebrata						
Mean	0.134	0.092	0.109	0.094	0.149	ns
S.d.	0.058	0.060	0.070	0.056	0.162	
N	29	17	30	12	6	
Plants						
Mean	0.069	0.092	0.040	--	--	ns
S.d.	0.060	0.075	--	--	--	
N	29	20	1	0	0	
Mammalia						
Mean	--	0.037	0.043	0.059	--	ns
S.d.	--	0.039	0.034	0.038	--	
N	0	33	57	16	0	
Aves						
Mean	--	--	--	0.060	--	--
S.d.	--	--	--	--	--	
N	0	0	0	1	0	

Table 20 (contin.)

Taxon	< 18	20 - 38	40 - 58	60 - 78	> 80	Signif.
Reptilia						
Mean	--	0.050	0.044	--	--	ns
S.d.	--	0.034	0.041	--	--	
N	0	39	6	0	0	
Amphibia						
Mean	--	0.059	--	--	--	--
S.d.	--	0.042	--	--	--	
N	0	42	0	0	0	
Pisces						
Mean	--	0.181	0.054	0.030	0.029	***
S.d.	--	--	0.030	0.021	0.024	
N	0	1	33	6	14	
Crustacea						
Mean	--	0.107	0.074	--	0.160	ns
S.d.	--	0.069	0.076	--	0.207	
N	0	7	9	0	4	
Insecta exc.*Dr*.						
Mean	0.102	0.082	0.103	0.094	0.125	ns
S.d.	0.059	0.055	0.053	0.056	0.032	
N	6	10	12	12	2	
Drosophila						
Mean	0.137	--	--	--	--	--
S.d.	0.052	--	--	--	--	
N	22	0	0	0	0	
Mollusca						
Mean	0.250	--	0.133	--	--	ns
S.d.	--	--	0.067	--	--	
N	1	0	6	0	0	
Dicotyledoneae						
Mean	0.063	0.031	--	--	--	ns
S.d.	0.048	0.033	--	--	--	
N	26	9	0	0	0	

Abbreviations: N = sample size;
S.d. = standard deviation;
Insecta exc.*Dr*. = insecta excluding *Drosophila*.
Significance: * = $p < 0.05$; ** = $p < 0.01$;
*** = $p < 0.001$; ns = $p > 0.05$.

6. Chromosome number (Table 20)

Chromosome numbers were categorized as follows: (i) 6-18; (ii) 20-38; (iii) 40- 58; (iv) 60-78; and (v) > 80.

a. The analysis involves only 392 species for which chromosome numbers were recorded. The resulting pattern was, therefore, sketchy particularly when trends were sought in each higher taxon separately, and because there was relatively little consistency among taxa.

b. H increased highly significantly ($p < 0.001$) with a decrease in chromosome numbers in fishes. When the relationship between the level of H and chromosome number within the first three categories was analyzed the same trends were indicated also in reptiles, molluscs and crustaceans.

Table 21. Heterozygosity (H) within different mating type categories.

Taxon	Selfed	Mixed	Out crossed	Parthenogenetic	Significance
All species					
Mean	0.031	0.073	0.079	0.213	***
S.d.	0.045	0.062	0.080	0.168	
N	23	13	183	12	
Vertebrata					
Mean	--	--	0.065	0.390	***
S.d.	--	--	0.049	0.014	
N	0	0	90	2	
Invertebrata					
Mean	0.0	0.098	0.086	0.178	*
S.d.	0.0	0.139	0.107	0.162	
N	5	2	73	10	
Plants					
Mean	0.040	0.069	0.114	--	**
S.d.	0.047	0.050	0.071	--	
N	18	11	20	0	
Mammalia					
Mean	--	--	0.035	--	--
S.d.	--	--	0.028	--	
N	0	0	22	0	
Aves					
Mean	--	--	0.120	--	--
S.d.	--	--	0.028	--	
N	0	0	3	0	
Reptilia					
Mean	--	--	0.097	0.390	***
S.d.	--	--	0.072	0.014	
N	0	0	13	2	

Table 21 (contin.)

Taxon	Selfed	Mixed	Out crossed	Parthenogenetic	Significance
Amphibia					
Mean	--	--	0.087	--	--
S.d.	--	--	0.069	--	
N	0	0	11	0	
Pisces					
Mean	--	--	0.062	--	--
S.d.	--	--	0.030	--	
N	0	0	41	0	
Crustacea					
Mean	--	--	0.035	0.210	***
S.d.	--	--	0.025	0.166	
N	0	0	20	8	
Insecta exc. *Dr*.					
Mean	--	0.197	0.056	0.048	**
S.d.	--	--	0.039	0.023	
N	0	1	31	2	
Drosophila					
Mean	--	--	0.117	--	--
S.d.	--	--	0.060	--	
N	0	0	7	0	
Mollusca					
Mean	0.0	0.0	0.232	--	ns
S.d.	0.0	--	0.205	--	
N	5	1	11	0	
Dicotyledoneae					
Mean	0.037	0.056	0.076	--	ns
S.d.	0.051	0.032	0.047	--	
N	14	9	11	0	

Abbreviations: N = sample size;
S.d. = standard deviation;
Insecta exc. *Dr*. = insecta excluding *Drosophila*.
Significance: * = $p < 0.05$; ** = $p < 0.01$; *** = $p < 0.001$;
ns = $p > 0.05$.

7. Mating system (Table 21)

a. Parthenogenetic species (N = 12) exhibited significantly higher ($p < 0.001$) genetic variation than primarily outcrossing, i.e., sexual species.

b. In molluscs, primarily selfed species displayed zero genetic variation (H and P), whereas primarily outcrossing species exhibited remarkably high H.

c. In plants, genetic polymorphism (P) increased significantly in the following order: selfers (0.16) < mixed (0.32) < outcrossers (0.55). The same trend was displayed significantly both in dicotyledons and monocotyledons.

Obviously, selfing leads, as is well known, to individual homozygosity, whereas obligatory parthenogenesis may lead to high heterozygosity due to suppression of recombination.

Table 22. Heterozygosity (H) within different modes of reproduction.

Taxon	Asexual	Sexual	Mixed	Significance
All species				
Mean	0.301	0.069	0.065	***
S.d.	0.132	0.071	0.064	
N	8	622	20	
Vertebrata				
Mean	0.390	0.048	--	***
S.d.	0.014	0.037	--	
N	2	327	0	
Invertebrata				
Mean	0.272	0.096	0.064	***
S.d.	0.142	0.093	0.066	
N	6	249	16	
Plants				
Mean	--	0.076	0.067	ns
S.d.	--	0.067	0.063	
N	0	46	4	
Mammalia				
Mean	--	0.034	--	--
S.d.	--	0.028	--	
N	0	100	0	
Aves				
Mean	--	0.051	--	--
S.d.	--	0.031	--	
N	0	36	0	

Table 22 (contin.)

Taxon	Asexual	Sexual	Mixed	Significance
Reptilia				
Mean	0.390	0.062	--	***
S.d.	0.014	0.052	--	
N	2	42	0	
Amphibia				
Mean	--	0.059	--	--
S.d.	--	0.046	--	
N	0	38	0	
Pisces				
Mean	--	0.049	--	--
S.d.	--	0.031	--	
N	0	111	0	
Crustacea				
Mean	0.272	0.070	0.024	***
S.d.	0.142	0.064	0.015	
N	6	97	2	
Insecta exc.*Dr*.				
Mean	--	0.090	0.034	**
S.d.	--	0.060	0.028	
N	0	80	10	
Drosophila				
Mean	--	0.131	--	--
S.d.	--	0.049	--	
N	0	21	0	
Mollusca				
Mean	--	0.143	0.250	ns
S.d.	--	0.184	--	
N	0	36	1	
Dicotyledoneae				
Mean	--	0.054	0.028	ns
S.d.	--	0.048	0.040	
N	0	33	2	

Abbreviations: N = sample size;
S.d. = standard deviation;
Insecta exc.*Dr*. = insecta excluding *Drosophila*.
Significance: * = $p < 0.05$; ** = $p < 0.01$; *** = $p < 0.001$;
ns = $p > 0.05$.

8. Mode of reproduction (Table 22)

a. Asexual species (N = 8) which are essentially parthenogenetic, displayed very high genetic variation.

b. Species combining both sexual and asexual mode of reproduction (N = 20) displayed the lowest levels of H.

Table 23. Heterozygosity (H) within different modes of pollination.

Taxon	Selfed	Animals	Wind	Significance
Plants				
Mean	0.041	0.064	0.154	***
S.d.	0.047	0.042	0.070	
N	18	19	11	
Monocotyledoneae				
Mean	0.050	--	0.203	**
S.d.	0.032	--	0.057	
N	4	0	3	
Dicotyledoneae				
Mean	0.038	0.064	--	ns
S.d.	0.051	0.042	--	
N	14	19	0	
Gymnospermeae				
Mean	--	--	0.146	--
S.d.	--	--	0.065	
N	0	0	7	
Pteridophyta				
Mean	--	--	0.060	--
S.d.	--	--	--	
N	0	0	1	

Abbreviations: N = sample size;
S.d. = standard deviation;
Significance: * = $p < 0.05$; ** = $p < 0.01$; *** = $p < 0.001$;
ns = $p > 0.05$.

9. Pollination mechanism (Table 23)

The level of genetic diversity increased with the following means of pollination: selfed < animals < wind ($p < 0.001$). Similar results were obtained by Hamrick et al., (1979).

Table 24. Pearsonian correlation matrix (r) among fifteen biotic and two genetic variables for (a) all animal species (b) vertebrates (c) invertebrates. The three correlation matrices were the data plotted in the SSA diagrams (fig.7 a - c).

a. All animal species (n = 458)

	Lz	Gr	Ht	Hr	Ar	Tr	Ps	Ss	Mo	Gf	So	Bs	Lo	Gl	Fe	H
Life zone (Lz)	1.000															
p	--															
Geographical range (Gr)	0.150	1.000														
p	***	--														
Habitat type (Ht)	0.104	0.142	1.000													
p	*	**	--													
Habitat range (Hr)	0.132	0.413	0.128	1.000												
p	**	***	**	--												
Aridity (Ar)	0.146	0.228	0.185	0.261	1.000											
p	**	***	***	***	--											
Territoriality (Tr)	0.016	0.179	0.187	0.113	0.125	1.000										
p	ns	***	***	*	**	--										
Population structure (Ps)	0.082	0.245	-0.102	0.372	0.192	-0.066	1.000									
p	ns	***	*	***	***	ns	--									
Species size (Ss)	0.207	0.482	0.109	0.502	0.247	0.189	0.336	1.000								
p	***	***	*	***	***	***	***	--								
Adult mobility (Mo)	-0.041	0.267	-0.033	0.356	0.165	0.068	0.344	0.166	1.000							
p	ns	***	ns	***	***	ns	***	***	--							
Young dispersal (Gf)	0.129	0.117	0.097	0.343	0.026	0.015	0.278	0.283	0.279	1.000						
p	**	*	*	***	ns	ns	***	***	***	--						
Sociality (So)	0.038	0.217	-0.128	0.057	0.016	-0.167	0.120	0.099	0.277	0.139	1.000					
p	ns	***	**	ns	ns	***	**	*	***	**	--					
Body size (Bs)	-0.000	0.233	0.066	0.097	0.109	0.273	0.088	0.093	0.277	0.039	0.071	1.000				
p	ns	***	ns	*	*	***	ns	*	***	ns	ns	--				
Longevity (Lo)	-0.155	-0.071	-0.116	0.040	-0.072	-0.117	-0.011	-0.216	0.091	0.130	0.046	0.251	1.000			
p	***	ns	*	ns	ns	*	ns	***	ns	**	ns	***	--			
Generation length (Gl)	-0.145	0.022	0.118	0.167	0.120	0.027	0.091	-0.048	0.213	0.245	0.077	0.348	0.692	1.000		
p	**	ns	*	***	**	ns	ns	ns	***	***	ns	***	***	--		
Fecundity (Fe)	0.138	0.248	0.053	0.424	0.155	0.395	0.165	0.393	0.205	0.284	0.042	0.262	0.032	0.256	1.000	
p	**	***	ns	***	***	***	***	***	***	***	ns	***	ns	***	--	
Heterozygosity (H)	0.232	0.111	0.306	0.009	0.178	0.256	-0.081	0.121	-0.206	-0.057	-0.197	-0.018	-0.095	-0.072	0.076	1.000
p	***	*	***	ns	***	***	ns	**	***	ns	***	ns	*	ns	ns	--
Polymorphism (P)	0.228	0.067	0.272	0.010	0.171	0.273	-0.019	0.116	-0.224	-0.088	-0.253	-0.066	-0.147	-0.079	0.035	0.795
p	***	ns	***	ns	***	***	ns	*	***	ns	***	ns	**	ns	ns	***

Table 24 (contin.)

b. Vertebrate species (n = 242)

	Lz	Gr	Ht	Hr	Ar	Tr	Ps	Ss	Mo	Gf	So	Bs	Lo	Gl	Fe	H
Life zone (lz)	1.000															
p	--															
Geographical range (Gr)	0.079	1.000														
p	ns	--														
Habitat type (Ht)	0.123	0.195	1.000													
p	ns	**	--													
Habitat range (Hr)	0.152	0.440	0.142	1.000												
p	*	***	*	--												
Aridity (Ar)	0.110	0.240	0.281	0.355	1.000											
p	ns	***	***	***	--											
Territoriality (Tr)	-0.065	0.137	0.042	0.193	0.088	1.000										
p	ns	*	ns	**	ns	--										
Population structure (Ps)	0.065	0.170	0.016	0.367	0.216	-0.044	1.000									
p	ns	**	ns	***	**	ns	--									
Species size (Ss)	0.154	0.460	0.134	0.492	0.319	0.063	0.308	1.000								
p	*	***	*	***	***	ns	***	--								
Adult mobility (Mo)	-0.066	0.320	0.083	0.339	0.224	0.322	0.211	0.179	1.000							
p	ns	***	ns	***	***	***	***	**	--							
Young dispersal (Gf)	0.074	0.161	-0.114	0.148	0.120	0.096	0.299	0.273	0.357	1.000						
p	ns	*	ns	*	ns	ns	***	***	***	--						
Sociality (So)	0.072	0.395	0.030	0.104	0.062	0.032	0.102	0.269	0.237	0.281	1.000					
p	ns	***	ns	ns	ns	ns	ns	***	***	***	--					
Body size (Bs)	-0.013	0.286	0.080	0.197	0.170	0.360	0.072	0.070	0.378	0.007	0.172	1.000				
p	ns	***	ns	**	**	***	ns	ns	***	ns	**	--				
Longevity (Lo)	-0.149	0.181	0.060	0.070	0.038	0.316	-0.101	-0.177	0.328	-0.168	-0.088	0.552	1.000			
p	*	**	ns	ns	ns	***	ns	**	***	**	ns	***	--			
Generation length (Gl)	-0.128	0.206	0.253	0.204	0.289	0.396	0.085	0.055	0.372	-0.069	0.006	0.554	0.692	1.000		
p	*	***	***	***	***	***	ns	ns	***	ns	ns	***	***	--		
Fecundity (Fe)	-0.022	0.274	-0.008	0.449	0.178	0.514	0.078	0.286	0.285	0.061	0.120	0.322	0.191	0.366	1.000	
p	ns	***	ns	***	**	***	ns	***	***	ns	ns	***	**	***	--	
Heterozygosity (H)	0.139	0.227	0.198	0.310	0.221	0.080	0.077	0.346	0.003	0.071	-0.063	-0.027	-0.056	0.083	0.169	1.000
p	*	***	**	***	***	ns	ns	***	ns	ns	ns	ns	ns	ns	**	--
Polymorphism (P)	0.128	0.120	0.185	0.230	0.211	0.090	0.172	0.294	0.007	0.097	-0.143	-0.107	-0.083	0.068	0.116	0.817
p	*	ns	**	***	***	ns	**	***	ns	ns	*	ns	ns	ns	ns	***

Table 24 (contin.)

c. Invertebrate species (n = 216)

	Lz	Gr	Ht	Hr	Ar	Tr	Ps	Ss	Mo	Gf	So	Bs	Lo	Gl	Fe	H
Life zone (Lz)	1.000															
p	--															
Geographical range (Gr)	0.190	1.000														
p	**	--														
Habitat type (Ht)	-0.064	-0.004	1.000													
p	ns	ns	--													
Habitat range (Hr)	0.103	0.381	0.114	1.000												
p	ns	***	ns	--												
Aridity (Ar)	0.159	0.181	-0.002	0.119	1.000											
p	*	**	ns	ns	--											
Territoriality (Tr)	-0.115	0.134	-0.000	0.001	0.089	1.000										
p	ns	*	ns	ns	ns	--										
Population structure (Ps)	0.164	0.381	-0.111	0.394	0.215	0.043	1.000									
p	*	***	ns	***	***	ns	--									
Species size (Ss)	0.168	0.482	-0.116	0.541	0.103	0.101	0.471	1.000								
p	*	***	ns	***	ns	ns	***	--								
Adult mobility (Mo)	0.081	0.269	-0.005	0.406	0.127	-0.082	0.464	0.297	1.000							
p	ns	***	ns	***	ns	ns	***	***	--							
Young dispersal (Gf)	0.192	0.091	0.283	0.514	-0.070	-0.038	0.264	0.329	0.221	1.000						
p	**	ns	***	***	ns	ns	***	***	***	--						
Sociality (So)	0.103	0.031	-0.157	0.008	-0.005	-0.287	0.086	0.038	0.265	0.003	1.000					
p	ns	ns	*	ns	ns	***	ns	ns	***	ns	--					
Body size (Bs)	-0.059	0.106	-0.076	-0.055	-0.054	0.003	0.166	0.037	0.204	0.086	-0.013	1.000				
p	ns	ns	ns	ns	ns	ns	*	ns	**	ns	ns	--				
Longevity (Lo)	0.009	-0.209	0.097	0.048	-0.107	-0.105	-0.081	-0.035	-0.349	0.337	-0.038	0.173	1.000			
p	ns	**	ns	ns	ns	ns	ns	ns	***	***	ns	*	--			
Generation length (Gl)	-0.083	-0.135	0.189	0.148	-0.056	-0.208	0.043	-0.033	-0.055	0.493	0.072	0.181	0.688	1.000		
p	ns	*	**	*	ns	**	ns	ns	ns	***	ns	**	***	--		
Fecundity (Fe)	0.252	0.136	-0.125	0.422	0.039	-0.059	0.398	0.435	0.250	0.591	0.080	0.063	0.209	0.303	1.000	
p	***	*	ns	***	ns	ns	***	***	***	***	ns	ns	**	***	--	
Heterozygosity (H)	0.226	0.018	0.243	-0.127	0.167	0.210	-0.085	-0.077	-0.273	-0.091	-0.217	-0.107	0.108	-0.043	-0.110	1.000
p	***	ns	***	ns	*	**	ns	ns	***	ns	***	ns	ns	ns	ns	--
Polymorphism (P)	0.212	-0.045	0.161	-0.150	0.112	0.216	-0.049	-0.140	-0.333	-0.173	-0.267	-0.155	0.068	-0.057	-0.229	0.778
p	**	ns	*	*	ns	***	ns	*	***	*	***	*	ns	ns	***	***

Abbreviations: N = sample size;
Significace: * = $p < 0.05$; ** = $p < 0.01$; *** = $p < 0.001$; ns = $p > 0.05$.

Correlation Matrix (Table 24)

The Pearsonian correlation matrix of the 15 biotic factors including ecological, demographic, and life history characteristics is given in Table 24. Approximately 75% of the 105 correlations among variables were statistically significant. The highest correlations were among the two genetic indices, H and P (r = 0.80) indicating that both estimators similarly describe genetic diversity within populations. The highest correlations between H and representatives of the 3 biotic subdivisions were with Habitat type (Ht), Adult mobility (Mo) and Longevity (Lo) (r = 0.31, 0.21, 0.10, respectively). The average correlation levels within the ecological, demographic, and life history biotic subdivisions were r = 0.17, 0.23, and 0.31, respectively. Finally, the average correlation between the ecological and demographic factors was r = 0.17, range: 0.02 - 0.50; ecological and life history factors, r = 0.14, range: 0.0 - 0.42; and demographic and life history, r = 0.14, range: 0.01 - 0.39.

Table 25. Coefficients of multiple regression (R^2) with dependent variable heterozygosity (H), and as independent variables (a) 15 biotic variables, (b) a subset of 6 ecological, (c) 5 demographic and (d) 4 life history variables. The analysis has been conducted for all species, for two major and for the nine higher taxa separately.

	Stepwise					Model		
	X1	X2	X3	X4	R^2 1	R^2 12	R^2 123	R^2 1234
(1) All species								
All variables N = 493	Tr	Lz	Ht	Mo	0.072***	0.132***	0.170***	0.199***
Ecological variables N = 576	Ht	Lz	Tr	Hr	0.076***	0.131***	0.174***	0.180***
Demographic variables N = 593	Mo	Ss	So	Ps	0.037***	0.060***	0.073***	0.078***
Life history variables N = 666	Fe	Lo			0.007*	0.014		
(2) vertebrata								
All variables N = 267	Ht	Hr	Bs	Tr	0.040**	0.061***	0.072***	0.087***
Ecological variables N = 317	Ht	Hr	Lz	Tr	0.032**	0.047***	0.057***	0.069***
Demographic variables N = 330	Ss	Gf			0.020*	0.025*		
Life history variables N = 348	Bs				0.009			

Table 25 (contin.)

	X1 X2 X3 X4	R^2 1	R^2 12	R^2 123	R^2 1234
(3) Mammalia					
All variables N = 102	Ss Mo Ht Gf	0.148***	0.168***	0.200***	0.232***
Ecological variables N = 106	Gr	0.121***			
Demographic variables N = 116	Ss Mo So	0.100***	0.134***	0.191***	
Life history variables N = 128	Lo	0.108***			
(4) Aves					
All variables N = 20	Lz Gf Bs Ar	0.153	0.272	0.416*	0.519*
Ecological variables N = 20	Lz Tr Gr Ht	0.153	0.247	0.375	0.463*
Demographic variables N = 24	no explanation				
Life history variables N = 25	no explanation				
(5) Reptilia					
All variables N = 34	Tr Bs Lz Gf	0.134*	0.252*	0.311**	0.411**
Ecological variables N = 54	Tr Lz	0.087*	0.116*		
Demographic variables N = 34	Ss	0.058			
Life history variables N = 54	Gl Lo	0.094*	0.125*		
(6) Amphibia					
All variables N = 42	Hr	0.171**			
Ecological variables N = 47	Hr	0.058			
Demographic variables N = 47	So	0.153**			
Life history variables N = 51	Lo	0.054			
(7) Pisces					
All variables N = 69	Ss So Ps Gr	0.040	0.098*	0.155*	0.196**
Ecological variables N = 90	Lz Gr	0.033	0.058		
Demographic variables N = 109	Ss So Ps Gf	0.091**	0.128**	0.159***	0.177***
Life history variables N = 89	no explanation				

Table 25 (contin.)

	X1	X2	X3	X4	R^2 1	R^2 12	R^2 123	R^2 1234
(8) Invertebrata								
All variables N = 226	Lz	Mo	Tr	Ht	0.071***	0.124***	0.169***	0.213***
Ecological variables N = 259	Lz	Tr	Ht	Hr	0.080***	0.128***	0.168***	0.200***
Demographic variables N = 263	Mo				0.049			
Life history variables N = 276	Bs	Lo	Gl		0.029**	0.053***	0.075***	
(9) Crustacea								
All variables N = 92	Hr	Lz	Fe	Ss	0.390***	0.479***	0.535***	0.576***
Ecological variables N = 98	Hr	Lz			0.327***	0.417***		
Demographic variables N = 101	Gf	So			0.322***	0.360***		
Life history variables N = 99	Fe	Gl			0.281***	0.362***		
(10) Insecta excluding *Drosophila*								
All variables N = 76	So	Lz	Ar	Ht	0.185***	0.300***	0.388***	0.445***
Ecological variables N = 77	Tr	Lz	Ar	Ht	0.127**	0.238***	0.283***	0.308***
Demographic variables N = 91	So	Mo	Gf		0.180***	0.198***	0.213***	
Life history variables N = 96	Bs				0.038			
(11) *Drosophila*								
All variables N = 22	Tr	Lz	Gf	Gr	0.336**	0.554***	0.626***	0.660***
Ecological variables N = 26	Lz	Tr	Ar	Gr	0.263**	0.497***	0.592***	0.668***
Demographic variables N = 22	Ps				0.226*			
Life history variables N = 28	Fe	Lo			0.134	0.228*		
(12) Mollusca								
All variables N = 34	Lz	Ht	Lo	Gf	0.588***	0.692***	0.722***	0.737***
Ecological variables N = 53	Lz	Ht	Hr		0.589***	0.686***	0.700***	
Demographic variables N = 36	Gf				0.312***			
Life history variables N = 35	Fe	Lo			0.328***	0.538***		

Abbreviations: N = sample size;
Variables names as in Table 24;
Significance: * = $p < 0.05$; ** = $p < 0.01$; *** = $p < 0.001$;
All figures without asterics are nonsignificant ($p > 0.05$).

EXPLAINED PORTION OF GENETIC DIVERSITY BY BIOTIC VARIABLES

Coefficient of multiple regression (R^2)

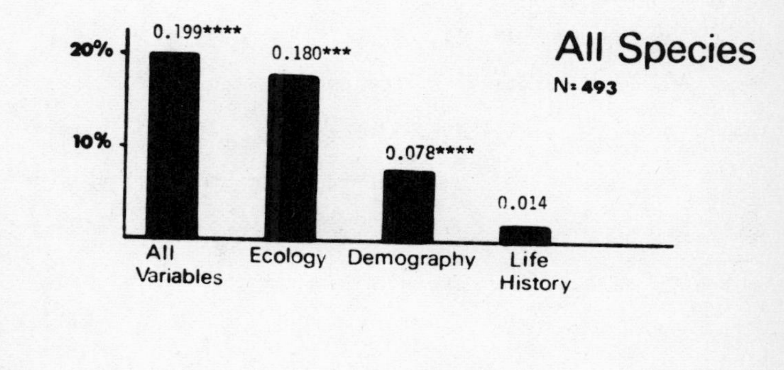

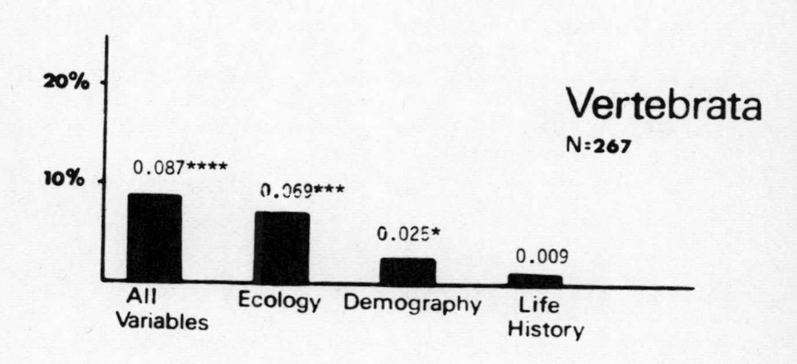

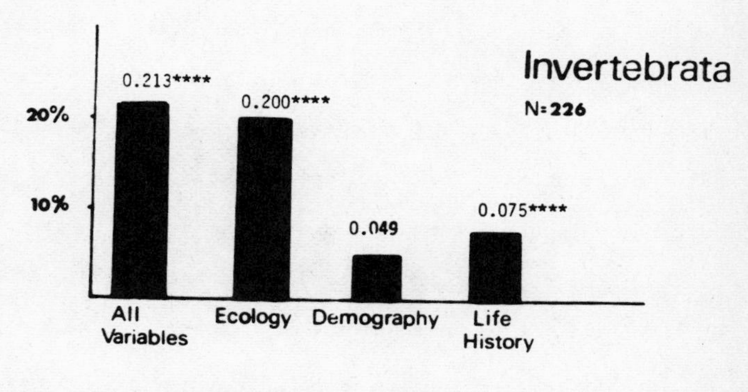

Levels of Significance: * $p < 0.05$
** $p < 0.01$
*** $p < 0.001$
**** $p < 0.0001$

Fig. 6

Multiple Regression Analysis (Table 25; Fig. 6)

A test for the best predictors of heterozygosity (H) was conducted by stepwise multiple regression analysis, MR (Hull and Nie, 1981), employing 15 of the biotic variables, consisting of 6 ecological, 5 demographic and 4 life history factors as independent variables. In general, biotic parameters explain significantly a substantial amount of genetic variance in H (Table 25; Fig. 6). For example, 0.199 of the variance in H was explained significantly ($p < 0.001$) by a 4-variable combination when all 15 biotic variables entered into the analysis. Remarkably, out of the 4-variable combination, the first 3 were ecological (Territoriality, Life zone, and Habitat type) and only the fourth was demographic (Adult mobility). Similarly, the ecological, demographic, and life history variables separately explained significantly 0.180 (90%), 0.078 (39%) and 0.007 (3.5%) of the variance in H, respectively. Here again, the ecological factors explained by far a higher proportion of the overall explained genetic variance of H (90%) ! Similarly, the ecological factors were predominant in the separate analysis of invertebrates and vertebrates, as well as in the 4 invertebrate higher taxa, birds and reptiles.

Averaging over the 9 taxa, the percent of genetic variance in H explained by a maximum of 4-variable combination for the overall, ecological, demographic, and life history subdivisions was 0.439, 0.323, 0.188, 0.161, or in percent (when the overall is considered 100%), 100%, 74%, 43%, and 37%, respectively. Evidently, the subdivision into 9 taxa increases substantially the level of the explained portion of genetic variance in H. Furthermore, both analyses described above emphasize the predominant contribution of the ecological factors in explaining genetic diversity. The explanatory scores of the 6 ecological parameters in the MR over the taxa were: Life zone (7), Geographical range and Territoriality (4), Habitat range and Habitat type (3) and Aridity (2). This ranking may express the relative importance of these ecological factors in genetic differentiation. The ranking of the 5 demographic variables was: Sociality (5), Young dispersal (4), Species size (3), and Adult mobility and Population structure (2). Finally, the ranking of the life history characters was: Longevity (5), Fecundity (3), Generation length (2), and Body size (1).

Smallest Space Analysis (SSA): Genetic and Biotic Relationships (Fig. 7)

The multivariate pattern of genetic-biotic relationships of 458 animal species having complete biotic profiles is represented graphically by 3 SSA-I diagrams (Fig. 7 A-C) based on 15 biotic factors (6 ecological, 5 demographic, and 4 life history characteristics). Proximity of points in the diagrams reflects higher correlations, as is seen by the contiguity of P and H in all 3 diagrams. The following pattern was indicated in the overall diagram (Fig. 7 A). Variables of each of the three biotic subdivisions can be organized in three polygons. The ecological polygon is the smallest and the closest of the three polygons to the genetic indices. The other two polygons, the demographic and life history are much larger and further apart from the genetic indices. Within the ecological polygon three variables, Habitat type (represented by species living in broader habitat ranges, i.e., in aquatic-littoral or aquatic + terrestrial habitats), Territoriality (represented by nonterritorial species), and Life zone (represented by species living in either temperate + tropical, or are cosmopolitan) appear in close proximity to high P and H in the diagram reflecting their relatively strong correlations (r = 0.31, 0.26, 0.23, respectively) to genetic indices. In other words, species characterized by the above mentioned features harbour higher levels of genetic diversity. By contrast, Habitat range appears as the farthest ecological factor in relation to the genetic diversity (r = 0.01). This is due to the antagonistic patterns of vertebrates and particularly crustaceans, which is exemplified by the location of generalist species in the vertebrates (Fig. 7 B) and invertebrates (Fig. 7 C) SSA diagrams.

Within the demographic polygon, large species size is located equidistantly from high P and H as are the sedentary species, although their correlation coefficients were different (r = 0.12 and 0.22, respectively). This is one example of the distortion in the SSA diagram of the true correlation matrix. The amount of overall

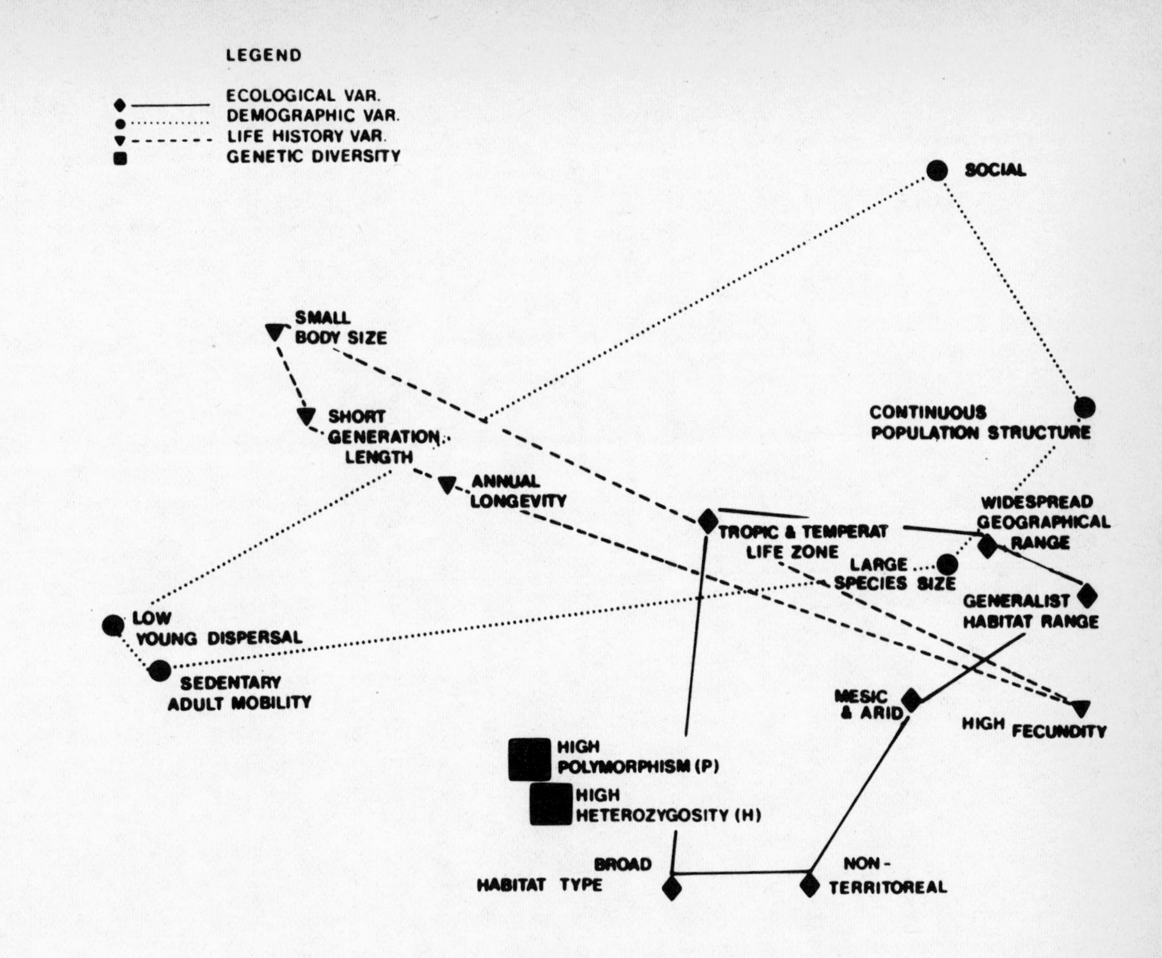
LEGEND
ECOLOGICAL VAR.
DEMOGRAPHIC VAR.
LIFE HISTORY VAR.
GENETIC DIVERSITY
SOCIAL
SMALL BODY SIZE
SHORT GENERATION LENGTH
ANNUAL LONGEVITY
CONTINUOUS POPULATION STRUCTURE
WIDESPREAD GEOGRAPHICAL RANGE
TROPIC & TEMPERAT LIFE ZONE
LARGE SPECIES SIZE
GENERALIST HABITAT RANGE
LOW YOUNG DISPERSAL
SEDENTARY ADULT MOBILITY
MESIC & ARID
HIGH FECUNDITY
HIGH POLYMORPHISM (P)
HIGH HETEROZYGOSITY (H)
BROAD HABITAT TYPE
NON-TERRITOREAL
SSA-I All Species

Fig. 7a

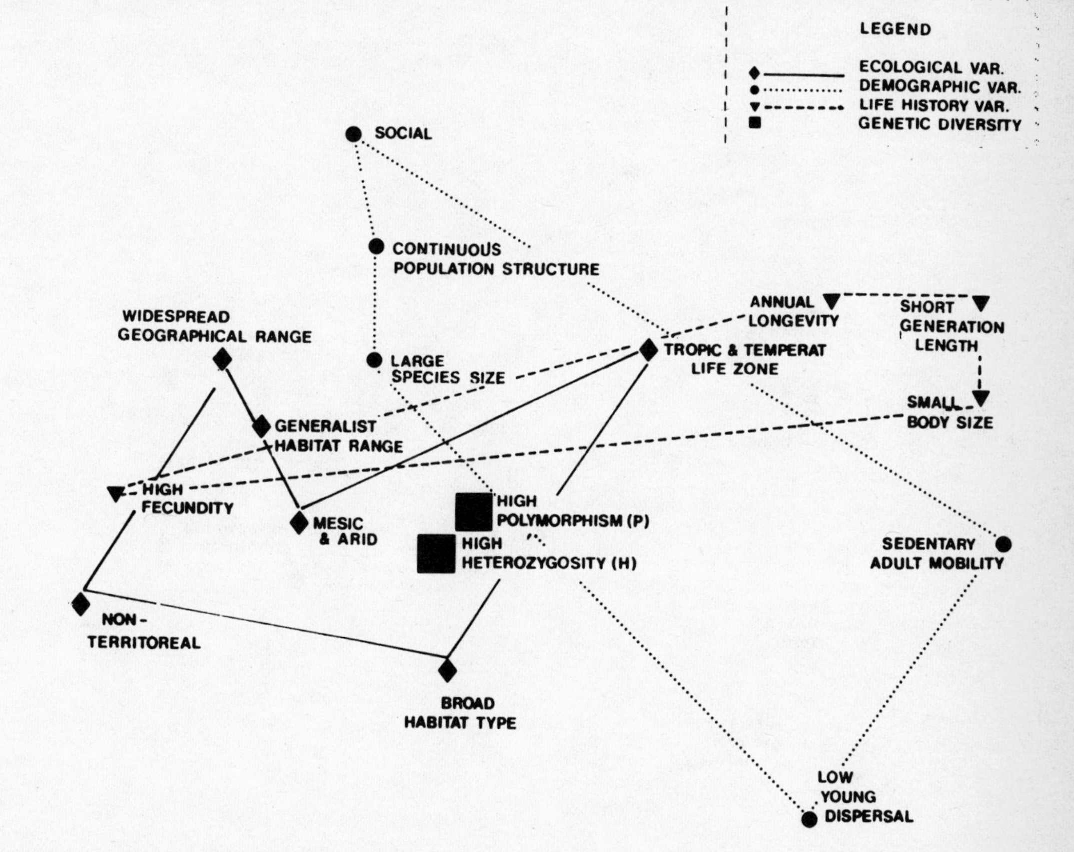

SSA-I Vertebrata

Fig. 7b

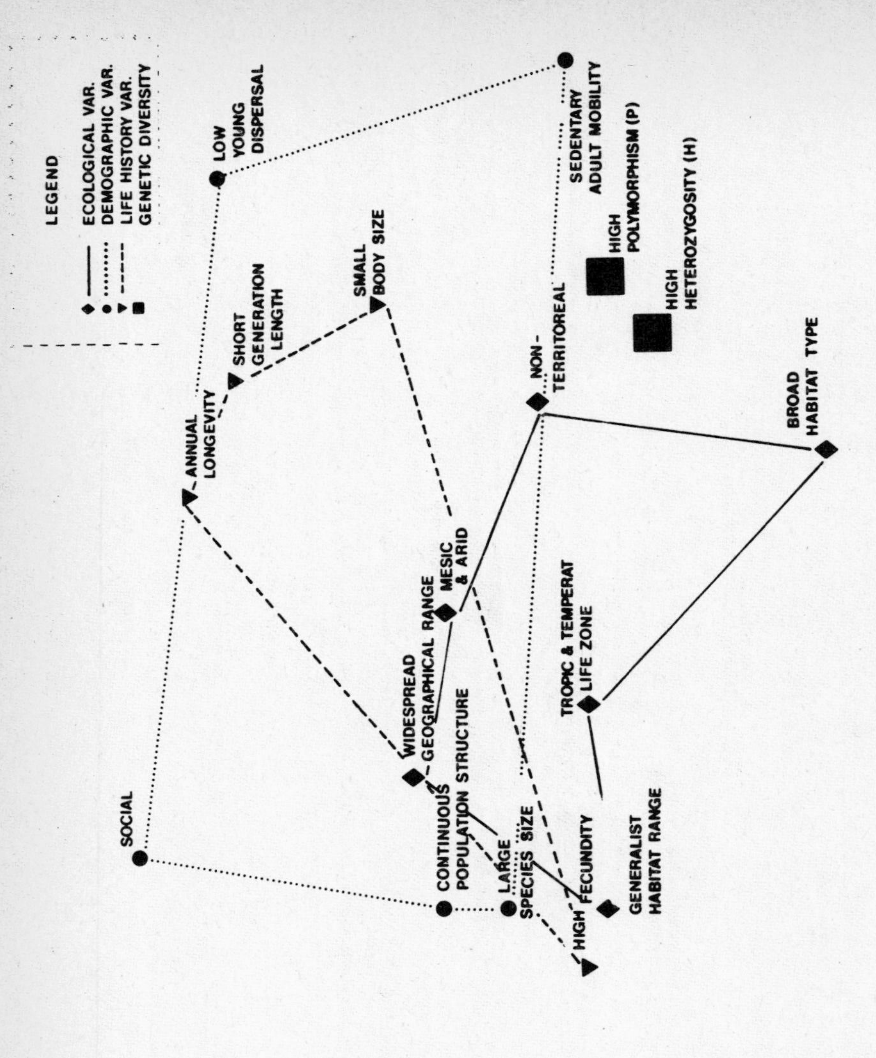

LEGEND
ECOLOGICAL VAR.
DEMOGRAPHIC VAR.
LIFE HISTORY VAR.
GENETIC DIVERSITY
LOW YOUNG DISPERSAL
SEDENTARY ADULT MOBILITY
HIGH POLYMORPHISM (P)
HIGH HETEROZYGOSITY (H)
SMALL BODY SIZE
SHORT GENERATION LENGTH
ANNUAL LONGEVITY
NON-TERRITOREAL
BROAD HABITAT TYPE
MESIC & ARID
WIDESPREAD GEOGRAPHICAL RANGE
TROPIC & TEMPERAT LIFE ZONE
CONTINUOUS POPULATION STRUCTURE
LARGE SPECIES SIZE
HIGH FECUNDITY
GENERALIST HABITAT RANGE
SOCIAL
SSA-I
Invertebrata

Fig. 7c

distortion is given by the Guttman-Lingoes' coefficient of alienation which indicates the deviation from the goodness of fit.

Within the life history polygon, longevity (represented by the category of annual species) appears closest ($r = 0.12$) to the genetic indices. The other two SSA diagrams representing the genetic-biotic relationships of vertebrates and invertebrates (Fig. 7 b,c), should be interpreted along the same lines as described above. The general pattern in these diagrams reflects the close relationships between ecological parameters and genetic diversity.

Summary of Evidence

Our results indicate the following general patterns and associations across 400 - 1100 species, representing diverse taxa from several phyla:

1. The levels of genetic diversity of enzymes and proteins vary nonrandomly among populations, species, higher taxa, ecological parameters (life zone, geographical range, habitat type and range, and climatic region); demographic parameters (species size and population structure, gene flow and sociality); and a series of life history characteristics (longevity, generation length, fecundity, origin, and parameters related to the mating system and mode of reproduction).

2. Protein diversity of species analyzed here is generally positively correlated with broader geographic, climatic, and habitat spectra. Thus generally, species are more polymorphic if they live in the following ecological conditions: several life zones, nonendemics, nonterritorial, overground versus underground habitats (or generally compound or complex habitats versus "simple" or relatively more constant and buffered habitats) and in broader climatic ranges (e.g., mesic + xeric). Likewise, genetic diversity is higher in large-sized species as well as in those having patchy population structure, low adult mobility, and low young dispersal. Finally, genetic variation is higher in species characterized by the

following life history and biological characteristics: small body size, annuals or long-lived perennials, older in time, with smaller diploid chromosome numbers, primarily outcrossed or ameiotic parthenogens, and plant species reproducing sexually and pollinated by wind. Species with the above characteristics harbour generally more genetic diversity than their opposite counterparts.

3. Genetic diversity is partly correlated and predictable by 3 - 4 variable combinations of ecological, demographic and life history variables, largely in this order over most higher taxa reviewed here. Ecological factors explain by far the highest proportion of the genetic variance as compared with demographic or life history factors (90%, 39%, 3.5%, respectively). However, significant low intercorrelations mostly (r = 0.1 - 0.3) occur both within and between the subdivided biotic factors, as reflected by the above percentages. Therefore, only critical experimentation (e.g., Nevo et al., 1983) may disentangle some of the biotic intercorrelations and thereby their relative individual contribution to fitness.

THEORETICAL CONSIDERATIONS

Selective and Nonselective Processes

What is the role and relative importance of deterministic and stochastic processes in the genetic structure and differentiation of natural populations? To what extent are the levels of genetic diversity determined by natural selection, migration, and other biological parameters as against random factors involving sampling error, initial conditions, founder and small size population effects, including historical bottlenecking, and general random drift fluctuations? This question haunted evolutionary biology for a long time and was even augmented by the new data of allozymic variation derived from electrophoretic studies. The discovery of vast protein polymorphisms in nature (reviewed by Powell, 1975; Selander, 1976; Nevo,

1978; Hamrick et al., 1979), primarily due to achievements of molecular biology and the application of protein electrophoresis in population genetics (and now the ever-growing field of DNA polymorphisms), did not resolve the problem. The debate was only transferred from the phenotypic to the molecular level, where evolution was explained contrastingly in Darwinian and non-Darwinian terms. Two major dichotomous alternatives emerged to explain allozymic variation: the neutral (Kimura, 1968; Kimura and Ohta, 1971; Nei, 1975, and modifications in Kimura, 1979 a,b) and selectionist (Ayala, 1977; Milkman, 1978; Clarke, 1979; Wills, 1981) theories, each claiming priorities in explaining polymorphism and molecular evolution in nature.

Recently, even some Darwinian evolutionary biologists fell into the human tendency to explain nature in dichotomous alternatives, now at the molecular level. Thus, Carson (1982: 429) believes that electrophoretically detected variability is irrelevant to adaptive evolution, and that "indeed much of it may be neutral to selection". Others (e.g., Schnell and Selander, 1981: 60) claim that research on protein polymorphism "enhanced our understanding of the genetic structure of populations of mammals and other organisms, but has yielded little understanding of the mechanisms by which the polymorphisms are maintained in natural populations". Furthermore, they go on to claim that in mammals "it is extremely difficult to define relationships between ecological factors and genetic, karyotypic, and morphological variables other than those that might be reasonably expected on the basis of historical demographic principles" (Schnell and Selander, 1981: 98). They found no convincing evidence of adaptive relationships between genetic variability and temporal or spatial environmental heterogeneity, and conclude that "electrophoretic studies of proteins may still be useful in providing evidence relating to population structure. But they will not tell us much that is new about the genetics of adaptive evolution in anatomy, physiology, behavior, and ecology". These conclusions are opposed to those drawn by Nevo (1978) for animals and plants, and by Hamrick et al. (1979) for plants, as well as other reviews of this highly debated field of evolutionary biology (e.g., Nevo, 1983 a,b), including the present review.

Theoretical and Empirical Methodologies

Various theoretical as well as empirical attempts were made to resolve the neutralist-selectionist controversy. Theoretical testing of neutral models was done by Nei (1975) and Nei et al. (1976) and critically reviewed by Ewens (1979). Nei and collaborators have persistently argued that testing supports the neutral model. Recently, Nei and Graur (personal communication, 1982) slightly modified the theory concluding "that available data on protein polymorphism are most easily explained by a modified form of neutral mutation hypothesis in which the effects of bottlenecks and fluctuations selection are taken into account". However, their analysis completely ignores genetical demographic-environmental interactions (Ayala, 1968), and concentrates primarily on effective species size, as if the latter was an isolated independent variable. In contrast to Nei's analysis, the critical review of Ewens (1979) concludes that so far in all tests neutrality is still rejected. A major problem in discriminating between selective and neutral models of electrophoretic profiles is their ability to explain the data equally well, often leading to ambiguous and nondiscriminating conclusions (Moran, 1976; Ewens and Feldman, 1976; Hartl, 1980).

Since current theoretical tests appear to have insufficient power to resolve the neutrality question, a possible alternative approach is the development of a theory of genotype-phenotype interaction with selection on the phenotype (Lewontin, 1974). This approach as rightly pointed out by Ewens and Feldman (1976) "clearly involves more ecology than population geneticists have been willing to use". The genetic-environmental approach has been employed successfully by several authors (e.g., Hedrick et al., 1976; Nevo, 1978, 1983 a; Hamrick et al., 1979). It has been utilized here on a much larger scale and incorporates in the data set ecological, demographic, and life history profiles kindly provided by 135 individual researchers. Without the active response to our questionaire this study could not have avoided a substantive amount of our subjective biases. The present analysis is, therefore, an attempt to overcome the individual biases of the present authors in order to arrive at generalized patterns in organic nature, as objectively as possible, despite the justified criticism detailed below.

Our analysis includes crude and quasi-relative estimates. Besides some objective parameters, it also involves undefined and vague concepts, as well as ambiguous and subjective interpretations. As critically concluded by Karlin (1982 b), real valued index for measuring heterogeneity essentially compels a single scaling over all environments. Since intrinsically an environment is complex, it can not be summarized in a single value and not all environments are comparable. The dichotomous descriptions of the environment, e.g., fine versus course grained, variable versus constant, unstable versus stable, are certainly simplistic and open to subjective interpretations. There appears to be no universal vector of indices for some of the biotic parameters used here. For example, what is meant by characterizing low versus medium migration patterns? Is there a consistent scale by which to assess rate of mobility or degree of isolation within and between taxa? How can spatial and temporal variances be meaningfully compared? How is habitat range defined and what is the biological meaning of generalists and specialists? Is it a universal concept or rather taxon dependent? Is it defined on the trophic axis alone or on several biological axes? If the latter is true, are the axes comparable among taxa? How is species size estimated? How should one define the extent of geographic range? Is there a universal scaling among taxa, or are they rather taxon-dependent? How can one objectively categorize population structure and degree of isolation? Can all biotic parameters be compared not only within but also between higher taxa? Finally, average heterozygosity estimates may improperly scale the effects of deme sizes, differences in local migration rates, and the spectrum of selection influences (Karlin, 1982 b).

We recognize the basic validity of the above criticisms. Nevertheless, our approach attempts to overcome these clear difficulties empirically by basing our analysis on the intuitive field knnowledge of the naturalist which often contains important biological insights which can hardly be overestimated in defining organic nature. This knowledge can never substitute for objective quantification, if the latter is available. It should only serve as a preliminary qualitative analysis, and ideally should be used in concert with optimal quantification. A better classification and standardization of biotic variables should involve more alleles, loci, species, higher taxa, habitats, specific simple and compound environmental and migration patterns, and more precise specifications of the biotic parameters. In particular, it should be rewarding to design critical tests approximating as far as possible controlled experimentation. Such an analysis may achieve a better elucidation of the genetic-biotic functional relations, and reduce the background noise which necessarily enters into the present analysis. We, therefore, consider our analysis as a first step towards this goal.

Critical Analysis of the Methodology of Genetic-Environmental Correlations

Genetic-environmental correlations demonstrate <u>inferentially</u> the adaptive significance of enzyme polymorphisms. Yet, if conducted on many loci and over many populations and unrelated species involving varied ecologies, demographies, and life histories, any emerging pattern, if repetitive and consistent over many taxa, space and time and involving diverse biological attributes, must be robust (Nevo, 1983 a). Furthermore, they may roughly indicate the relative contributions of each of the biotic subdivisions, i.e., ecological, demographic, and life histiory to adaptive evolution, at least for structural genes coding for soluble enzymes. Recall that the neutral theory never questioned specific examples of adaptive evolution at the molecular level. Neutralists believe that adaptive evolution occurs by positive Darwinian selection only in a small proportin of genes, while most molecular evolutionary change is propelled by random drift (Nei, 1975; Kimura, 1979 b). Therefore, only the demonstration of genetic diversity-environmental correlations over <u>many</u> loci and species (since individual counter-examples can always be found) may enable us to assess the relative importance of the various evolutionary forces of natural selection, migration, mutation, and genetic drift in adaptive evolution.

The assumption of the neutral theory that only a small proportion of genes are involved in adaptive molecular evolution is inconsistent with our data. Clearly, the correlations found are primarily descriptive rather than deterministic. They do not establish directly causal relations between biotic and genetic diversity. Yet, their consistency in many unrelated taxa ranging globally strongly suggests that some at least score genuine associations between biotic factors and protein diversity. These are presumably mediated at least partly by ecological selection apparently affecting the overall level of genetic diversity, apart from their operation at allele frequencies of the specific loci studied (Nevo, 1983 a). Furthermore, since many of the correlations found are independent evolutionary trials due to the reproductive isolation between species, the ecological-biotic correlates of genetic diversity may suggest, either directly or indirectly, causative mechanisms of genetic differentiation. Obviously, the final verification or falsification of these presumed causes depends on future critical fitness testing (e.g., Nevo et al., 1983; see Hedrick et al., 1976 for critical review of genetic-environmental correlations). Other correlations of genetic diversity (i.e., quarternary structure, subunit molecular weight, and others) do not exclude, but rather add another dimension to the biotic correlates at the molecular level analyzed here.

The unequivocal demonstration of the adaptive significance of an enzyme polymorphism depends on interrelated multidisciplinary evidence. The latter must show that the enzyme phenotypic diversity varies with the environment and involves in vivo biochemical and physiological differences among allozyme variants that contribute to fitness. This has been demonstrated for leucine amino peptidase (LAP) polymorphism in the marine bivalve Mytilus edulis by Koehn and colleagues (Koehn, 1978; Koehn and Immerman, 1981) or for glutamate pyruvate transaminase (GPT) (Burton and Feldman, 1981). Yet, such a comprehensive approach depends on extensive efforts in several experimental fields, finally demonstrating a high probable adaptiveness in a specific case, rather than substantiating a general pattern. Furthermore, it will usually prove very difficult to obtain all the experimental information an ideal analysis might require (Johnson, 1979). Thus, while this approach is imperative for revealing the molecular structures and processes contributing to fitness as indicated earlier, it does not answer the basic selectionist-neutralist dilemma: how much of each? This dilemma must be analyzed first statistically preferably at the multilocus level, and then, complementarily, at the biochemical and physiological levels of individual genes.

The methodology of genetic-environmental correlations was extensively criticized due to its seemingly multiple inherent faults including special choice of environments, spuriousness, noncausality, nongenetic mechanisms, etc.(Schnell and Selander, 1981; Nei and Graur, personal communication, 1982; among others). Certainly, precaution is advisable (Levins and Lewontin, 1980), but not a total dismissal of this important methodology. Taken individually, any of the cases analyzed here could be dismissed as a spurious genetic-biotic correlation. However, these massive observations collectively constitute coherent corroborating evidence of parallel genetic-biotic associations in many unrelated species and higher taxa, differing in their biological and phylogenetic records but sharing similar varying biotic backgrounds. The emerging shared patterns of genetic diversity in reproductively isolated species, which are otherwise historically different in their place and time of divergence and their subsequent evolution, strongly suggest that levels of enzyme diversity, far from being mostly neutral, contain a substantial amount of adaptive differentiation by natural selection.

Stochasticity, including founder effects and initial conditions and constraints, genetic drift, sampling, bottlenecking, and neutrality, certainly occur in nature and probably on a substantial scale. Nevertheless, all these factors, singly or in combination, are unlikely to primarily explain parallel genetic patterns of the levels of heterozygosity and polymorphism on a massive and global scale. Stochastic factors certainly interact with natural selection and their importance in both adaptation and speciation, particularly in small peripheral populations and during historical bottlenecks, is undeniable. They seem, however, to be largely secondary in importance to natural selection in genetic differentiation of most populations and species. This conclusion is true on a global as well as on a regional scale. In Israel, genetic parallelism was demonstrated in 38 unrelated plant and animal species involving 162 populations distributed along a stress of increasing aridity. The level and pattern of genic diversity, either of 15 individual loci shared by all species, r that of the overall genetic indices of polymorphism and heterozygosity, vary onrandomly between loci, populations, species, and habitats, and generally increase owards the desert over geographically short distances (Nevo, 1983 a).

The neutralist-selectionist controversy is yet unresolved. It will certainly not e settled by any single study, no matter how exhaustive, whether correlative or iochemical-physiological. In fact, it may never be resolved completely since the harp dichotomy of either school may be unrealistic. The problem is how much daptiveness is implicated at the molecular level. Even this target may be nrealistic since, as Wills (1981) noted, "it seems unlikely that we will ever be ble to determine what proportion of all alleles in a population is neutral and what roportion is subject to selection". Moreover, the issue was oversimplified and the harp dichotomy between neutral and selected alleles is artificial, primarily since lleles may change direction of selection not only with environment, but also with he genetic background (Hartl and Dykhuizen, 1981) and hierarchy of selective units Bargiellc and Grossfield, 1979), analogous to Wright's shifting balance theory of volution (Wright, 1970, 1982). In any event, our understanding of the relative roles f stochastic and deterministic factors at the molecular level depends on extensive nd intensive field and critical laboratory studies at single and multilocus systems n many species with full consideration of the ecological, biochemical, physiological nd historical factors.

The genetic-biotic correlation methodology employed here is neither exhaustive nor ufficient to resolve the adaptive nature of protein polymorphism. Nevertheless, it oes deal with the statistical nature of the selectionist-neutralist controversy, and t sheds some light on the relative roles of selective and nonselective factors at ne molecular level. Likewise, it ranks the ecological, demographic, and life history iotic subdivisions, presumably in their right order and appropriate perspective of nteraction. Notably, the proportion of genetic variance out of the total explained y each of the 3 biotic subdivisions, ecological, demographic, and life history is)%, 39%, 3.5%, respectively. Ecological factors explain a much higher proportion of ne genetic variance than demographic ones. Moreover, ecological and life history factors combined explain more than 90% of the explained portion of genetic variance. Of course, in the overall analysis only 20% of the genetic variance was explained. The rest is unexplained under the constraints of the present crude analysis which avoids such important factors as competition, parasites, other ecological factors that did not enter as separate categories into the analysis, as well as historical and general drift forces. Likewise, in the overall analysis opposite trends exist between taxa which are resolved, hence increase the explained portion of genetic diversity, when higher taxa are analyzed separately. Nevertheless, within the explained portion of 20% genetic variance, our analysis rejects the conclusion that population differentiation is better explained by demographic rather than by ecological factors. It is indeed remarkable how relatively little of the genetic variance is accounted for in our stepwise multiple regression analysis by the factor of species size, the demographic factor supposed to best explain the level of heterozygosity by neutral theory (Nei and Graur, personal communication, 1982). In fact, we do subscribe to Selander's (1975: 328) final conclusion, but in a reversed order, i.e., that the structure of populations ultimately will be explained in terms of interactions of deterministic and stochastic factors.

The genetic-biotic correlation methodology paves the way for a better critical testing of the proteins analyzed here and to direct evaluation of their hypothesized biotic relations. This unquestionably imperative complementary analysis must be conducted through biochemical kinetics, both *in vitro* and *in vivo*, of allozymes in question, (reviewed in Nevo, 1983 a,b) as well as through their physiological function (Johnson, 1979). This biochemical-physiological testing coupled with biological testing of differential fitness of allozymes (Nevo et al., 1983) may reveal the molecular basis for maintenance of enzyme polymorphisms in natural populations.

Maintenance of Polymorphism

The neutral theory

Neutrality theory allows for negative or purifying selection of deleterious mutations and also for the positive selection of the small proportion of advantageous mutations which cause slow adaptive evolution, hence are primarily noticeable over geologic time (Nei, 1975). However, it asserts that most molecular diversity in nature is nonselective and is maintained in populations through mutational input and random fixation. Hence, for neutrality theory polymorphism is simply an incidental and unimportant phase of molecular evolution. the level of observed genic diversity in nature is believed to be in rough agreement with the expected value of neutral mutations, when mutation rate is inferred from the rate of amino acid subtitutions (Kimura, 1968, 1969, 1979 a,b; Kimura and Crow, 1964; Ohta and Kimura, 1973; King and Ohta, 1975; Nei, 1975, 1980; Yamazaki and Maruyama, 1975).

The selection theories

The selection balance theory. Theoretically, many types of natural selection can maintain polymorphisms, in addition to the classical model of heterosis. Most of these types involve some sort of balance selection in which one homozygote is favoured under some conditions (of niche, habitat, season, life cycle, density, frequency, gametes, sexes, zygotes, groups, etc.), but disfavoured under others (see Hartl, 1980: 225, Table X, and references therein, for classification of some mechanisms that can maintain genetic polymorphism). Thus, polymorphism may be maintained not only through the simple model of overdominance (e.g. Wilton and Sved, 1979), but also even when the heterozygote is always intermediate in fitness between the homozygote of varying fitnesses. The most likely general mechanisms for maintaining genetic polymorphisms are spatially heterogeneous environments (Levene, 1953; Hedrick et al., 1976; Bryant, 1976; Gillespie, 1978). However, many additional specific mechanisms may result in genetic diversity including temporally varying environments (Dempster, 1956; Haldane and Jayakar, 1963; Gillespie, 1972; Hartl and Cook, 1973; Bryant, 1974; Karlin and Liberman, 1974; Karlin and Levikson, 1974; Turelli, 1977), epistasis (Karlin and Feldman, 1970; Feldman et al., 1974), nonepistasis (Karlin and Liberman, 1979 a,b) and symmetric heterozygosity selection regime (Karlin and Avni, 1981). Compound models such as multiple niche and habitat selection (Powell and Taylor, 1979) prove very promising in explaining the maintenance of genetic polymorphism. Finally, the suggestion that overall fitness improves with increasing total heterozygosity was substantiated experimentally (Wallace, 1958; Zouros et al., 1981), as well as theoretically, for multilocus theory (Karlin and Avni, 1981; Turelli and Ginzburg, 1983) though theoretical objections were raised on genetic and load arguments (Mukai et al., 1974; Simmons et al., 1978). The data analyzed here are in line with some of the theoretical models mentioned above, as will be discussed later.

The maintenance of genetic polymorphism under various natural structured viability regimes versus random fitness assignments was theoretically compared by Karlin (1981). An increased likelihood for a globally stable equilibrium is predicted for the more structured viability models. Accordingly, "if observed allele frequency data exhibit a reasonably consistent common set over different populations and epochs, the contingencies for a structured selection mechanism may be relevant. On the other hand, where the allele frequency observations vary significantly in space or time with few segregating alleles in any particualr sample, an explanation of the observed variability based on fitness interactions is unlikely. Other forces, such as migration, population structure, mating pattern, genetic frequency and/or ecological density factors and strong randomizing recombination interactions, may be important". In general, the more structure in the viability matrix, the greater is the level of polymorphism (Karlin and Feldman, 1981). Furthermore, a mixture of underdominance, directional, and overdominant spatially varying selection can produce a wide variety of stable polymorphic, and/or fixation states. Spatial is more effective than temporal variation in protecting polymorphisms (Karlin, 1982 b).

Gillespie and colleagues (Gillespie, 1974 a,b, 1977, 1978; Gillespie and Langley, 1974) concluded, on theoretical grounds, that allozymic polymorphism is primarily due to selection acting on environmental varition in gene function. Genetic variation will be more likely in spatiotemporally more variable environments than in constant ones. Furthemore, heterozygote intermediacy plus random environmental fluctuations are sufficient elements to explain genetic variability, and the conditions for polymorphism in heterogeneous environments are less stringent than those of overdominant selection in multiple allelic systems, as also concluded by Lewontin et al. (1978). Theoretically, stable equilibrium may be generated even in a single niche if selection coefficients vary from generation to generation in a specific nonoverdominant manner in infinitely large populations (Dempster, 1955; Haldane and Jayakar, 1963; Karlin and Levikson, 1974).

The idea of positive correlation between genetic and environmental variation, the niche width variation hypothesis (Van Valen, 1965), is widespread in evolutionary biology (i.e.,Levene, 1953; Levins, 1968). It was critically tested and confirmed for karyotypic, and suggested for electrophoretic (Powell, 1971; McDonald and Ayala, 1974; Powell and Wistrand, 1978), as well as for quantitative traits (Mackay, 1981, and references therein). It was also supported by allozyme studies comparing habitat specialist and generalist species (Nevo, 1978, 1983 c, and this study) and by life history analysis (Hamrick et al., 1979). The present extensive analysis largely supports the niche width variation hypothesis, and corroborates much of the theory discussed above.

The selection-migration theory. Theory predicts that there are marked differences in the way selection, migration, and population structure influence the nature and levels of genetic polymorphism. The effects of different types of spatially and temporally varying selection regimes, based on different population structures coupled with migration patterns on the existence and nature of polymorphism, have been summarized by Karlin (1982 b) following extensive studies (Karlin, 1976, 1977 a,b; Karlin and Richter-Dyn, 1976; Karlin and Campbell, 1978, 1979, 1980; see also an earlier review by Hedrick et al., 1976). These studies establish the conditions for the existence of a protected polymorphism for a hierarchy of migration patterns, thereby permitting qualitative comparisons of the influence of different structures of migration exchange in contributing to the maintenance of polymorphism. Elucidation of the genetic- biotic relationships will be advanced with a more complete classification of selection-migration structures and the conditions for a protected polymorphism (Karlin, 1982 b). Some of the theoretical expectations will be compared later with the empirical results reviewed here.

The multilocus theory. The two niche Levene model (1953) has been extended recently to multilocus and multiniche structures (reviewed in Hedrick et al., 1976, 1978; Felsenstein, 1976; Karlin, 1979, 1982 a,b; Karlin and Avni, 1981; Turelli and Ginzburg, 1983).

The multilocus generalized symmetric selection regime (Karlin and Avni, 1981) is characterized by the property that fitness depends on the specific loci that are homozygous or heterozygous, but is not influenced by the allelic composition at the loci. The multilocus regime extends earlier conclusions based on two locus (Lewontin and Kojima, 1969; Kojima and Lewontin, 1970; Bodmer and Felsenstein, 1967) and three locus (Feldman et al., 1974). The following robust conclusions were derived by Karlin and Avni (1981) for the important case where the fitness assignments depends only on the number of heterozygous loci (complete symmetric aggregate heterozygosity selection, CSHS-model). (a) A stable central polymorphism is most easily expressed under free recombination when enhanced fitness is concordant with increasing heterozygosity; (b) The stability of the central type polymorphism is ameliorated with "more recombination"; (c) It suffices to be able to achieve relatively high fitness with only a few heterozygous loci in order to achieve stability for the central equilibrium; (d) Many or few loci may be decisive to fitness depending on whether the realized number of gamete types is high or low, respectively; (e) An increased number of loci facilitates the stability of the central type polymorphism (see also Karlin, 1979); (f) "More recombination" enhances the existence of polymorphism. Finally, at stable, viability maintained, multilocus polymorphisms, mean fitness typically increases with heterozygosity (Turelli and Ginzburg, 1983). Some of these conclusions are supported by the present review as discussed later.

THE MATCHING OF EVIDENCE AND THEORETICAL MODELS

To what extent does the evidence presented in this review match the theoretical models mentioned above and/or the neutral mutation or mutation-drift hypothesis? Furthermore, is the correlative evidence presented here sufficient to discriminate between these and other hypotheses? Obviously, this review can not replace critical testing (in the sense of Platt, 1964). While its major strength lies in its extensive coverage of species, i.e., the robustness over and above individual counterexamples, its weaknesses are easily specified. It was not predesigned or planned as a *prehoc* analysis along ideal lines involving proper standardization and a controlled experimental design. Essentially, it represents a *posthoc* "holistic" approach which tests the outcome of multiple factors in space and time, some supporting each other, others antagonistic, and still others intermediate in their interaction patterns. Such *posthoc* analysis is bound to involve much unexplained evolutionary noise, partly due to our ignorance in specifying the ensemble of interacting forces and variables, partly due to intractable and unique historical processes, and partly due to unexplained stochastic noise. It appears remarkable that explicable and expected patterns should emerge at all, despite these noise-generating multiple unknowns, which should militate against the discovery of patterns. Our attempt to match the significantly explained portion of genetic variance with theoretical models, must therefore, necessarily, be crude and suggestive for further discriminatory analysis.

Predictions of the neutral theory

The neutral theory predicts that heterozygosity is a function of effective population size (Ne) and mutation rate (Nei, 1980; Nei and Graur, personal communication, 1982). The expectation of neutrality theory, that polymorphism should be higher in large and continuous populations and smallest in isolates, is negated by the invertebrate species analyzed here, and is only born out weakly by vertebrates and strongly by the small sample of plants. Even in vertebrates, the increase in H with species size is much less than expected by neutrality theory (Nei, 1980). The attempt made by neutralists (Nei and Graur, personal communication, 1982) to explain this discrepancy by Pleistocene bottlenecking due to climatic shifts, is not applicable to the data analyzed here. If such an extensive bottlenecking effect indeed occurred, it should have had a much stronger impact on vertebrates (which are characterized by a lower species size and longer generation length) as compared to invertebrates, in contrast to our results. Therefore, in contrast to the expectation of neutrality theory (Nei, 1980), species size does not enter into the overall stepwise multiple regression analysis as a major component. Even within the set of 5 demographic variables alone, it enters as a first variable only in 2 out of 9 higher taxa. In all higher invertebrate taxa, species size hardly plays any role in explaining genetic variance. When tested alone (Table 24) the correlation between species size and H is significant as also indicated by others (e.g., Soule', 1976; Nei, 1980), but when entering into the analysis with other ecological, demographic, and life history variables, it is largely rejected as an explanatory variable. Notably, as we have shown earlier (Table 11 B,C) when species size was standardized, heterozygosity was shown to be correlated distinctly with ecological factors.

The data analyzed in this review is inconsistent with the expectation of random unstructured fitness assignments specified by Karlin (1981). The patterns revealed are structured taxonomically, ecologically, demographically, and life historically, and display nonrandomness on a massive scale. Such global, regional and local structures, intimately associated with the ecological and biological background are hardly explicable by random forces alone. In contrast, the data exhibit a substantial patterning over different populations and species, hence to a large extent the contingencies for structured selection mechanisms may be relevant (Karlin, 1981).

Predictions of selection theories

The selection balance theory. The niche width variation hypothesis (Van Valen, 1965) is generally supported by our genetic-ecological relationships. The hypothesis suggests that the amount of genetic variation may be regarded as an adaptive strategy for increasing population fitness in a spatiotemporally heterogeneous environments. Allozymic diversity of species analyzed here is, in general, positively correlated with broader spectra of life zones, geographic range, habitat type, habitat range, as well as climates. Genetic polymorphism is higher in general with increasing ecological heterogeneity, instability, and uncertainty in space or time. This empirical results conform with theoretical expectation that the existence of a protected polymorphism is more likely in a more heterogeneous environment (Hedrick et al., 1976; Karlin, 1982 a,b). However, it should be recalled that the maintenance of a polymorphism is proabably due to a complex interaction of environmental heterogeneity and genetic factors, the relative importance of which varies over the species range (Karlin, 1982 b).

The niche width problem can also be resolved by "general purpose" genotypes (Baker, 1965) that are broad niched ecological generalists and physiologically and developmentally flexible (Karlin and McGregor, 1972 a; Hochachka and Somero, 1973), as conceived theoretically and reported empirically (e.g., in parthenogenetic earthworms, Jaenike and Selander, 1979). Obviously, such generalized genotypes will display lower polymorphism in broader environmental amplitudes thereby opposing the trend hypothesized by the niche width variation hypothesis. Such generalized genotypes, where single alleles have selective advantage throughout the species range, are likely to evolve in migratory species, for example in birds. Indeed, the migratory category in our analysis displays a lower level of heterozygosity in the variable of Adult mobility. Evidently, there is a full scale genetic- ecological continuum between the sedentary and migratory species, as is presumably true for most biological parametrization, such as coarse versus fine grain, constant versus fluctuating etc. Dichotomy is facile to analyze but for most biotic parameters it is artificial and unrealistic.

The selection migration theory. In general, theory predicts that limited migration ("less" but nonzero migration) and increased population clustering are very effective in maintaining global polymorphism in a finite population (Karlin, 1982 b, and references therein). Theory generally predicts considerable polymorphism with low migration rates and more monomorphism for moderate and higher migration. For strongly oscillating migration patterns protection of polymorphism is more likely than with uniform mixing. There appears to be a threshold level of medium migration flow such that the maintenance of a stable polymorphism is minimal at that rate of migration. Spatial, rather than temporal, heterogeneity of the environment is a more powerful force for producing and maintaining polymorphism. Our empirical data is rather crude for any attempt of matching with the specific mathematical models analyzed by Karlin. However, the more general theoretical results are supported by our data analysis. Our results generally indicate a negative correlation between genetic polymorphism and increasing gene flow, as well as a decrease in heterozygosity in continuous or uniform population structures in animals. Plants, however, indicate a reversed trend where genetic diversity increases with gene flow (pollination mechanisms and seed dispersal) and population continuity (compare also with Hamrick et al., 1979).

The multilocus theory. Some of the predictions of multilocus theory at least in the symmetric models (Karlin, 1979; Karlin and Avni, 1981) are validated by the present analysis. Heterozygosity increases with more recombination. Likewise, heterozygosity per se, seems to be of adaptive significance in addition to the adaptive contribution of individual genes (see also Turelli and Ginzburg, 1983).

The sexuality theory. The relationship between breeding systems and adaptive patterns was reviewed by Maynard Smith (1978). Theory predicts that more recombination, bisexuality, and multideme interactions facilitate the establishment of central type polymorphisms (Karlin, 1979; Karlin and Avni, 1981). These theoretical expectations are validated by our analysis. Fully sexual species display higher level of H as compared with species combining both sexual and asexual mode of reproduction. Only the fixed heterozygosity displayed by some of the parthenogenetic species derived from hybrid origin such as lizards (e.g., Parker and Selander, 1976) and fish (Vrijenhoek et al., 1976) may display H values excessively higher than those characterizing sexual species (for the ecological correlates of parthenogenesis in earthworms see Jaenike and Selander, 1979). Total selfing occurring in some molluscs results, as expected, in no genetic diversity, whereas primarily outcrossing species exhibit remarkably high levels of heterozygosity.

Biotic selection through predators and pathogens is believed to favour sexual reproduction and thereby genetic diversity (Clarke, 1979; Hamilton et al., 1979). The evolution and genetics of sexuality was critically reviewed by Bell (1982). He argues that spatial variation of habitats in terms of parasites and competitors varying geographically has a prime role in the evolution of sexuality, in contrast to the temporal variation model as contended by Hamilton (1982). Noteworthy, a similar spatial-temporal dichotomy relates to the evolution of genetic polymorphism where theoretically either space (Karlin, 1982 b) or time (Bryant, 1974) variation is considered of prime importance in safeguarding polymorphism. Our data underscore the importance of spatial heterogeneity in genetic diversification with sexual reproduction and genetic recombination, leading to higher levels of genetic polymorphism. This syndrome appears to be adaptive to life in complex environments both in plants (Levin, 1975) and in animals (Glesener and Tilman, 1978; Maynard Smith, 1978).

CONCLUSIONS AND PROSPECTS

Organic nature is extremely complex and is in a dynamic flux. The understanding of the origin and evolution of this complexity is the central issue of evolutionary biology. It must explore interactions between parts and wholes at all levels analyzing the particular in order to understand the general. Diversity and unity, chance and determinism, constancy and change, continuity and discontinuity, are all ingredients of organic nature. Evolutionary biology must cope with all. As pointed out by Whitehead, we must seek simplicity but never believe in it. Evolutionary biology, as science in general, must keep up its never ending search for unravelling the patterns and processes of organic nature. However, cognizant of the extreme complexity of reality it must also practice humility and modesty in its claims which must always remain relative rather than absolute truths of the structure of nature.

Populations and species are finite, hence are subject to continuous interaction of random and nonrandom evolutionary forces. Elucidation of organic nature will not advance by holding tight to either of the dichotomous alternatives of random versus nonrandom evolutionary world views. This appears to be true for any of the hierarchical organizational levels of complexity in nature, and is particularly true in biology. The idea of neutrality theory that organic nature is sharply dichotomized in terms of the operating forces into the phenotypic and molecular levels is unrealistic. It asserts that at the phenotypic level, largely deterministic positive Darwinian selection acts on phenotypes primarily through environmental effects on polygenes. In contrast, the neutral theory believes that at the genotypic level, molecular polymorphisms are almost invisible to natural selection and are governed primarily by randonm genetic drift (Kimura, 1979 b). In other words, for neutrality theory protein variation in nature essentially reflects evolutionary noise and is maintained in populations through mutational input and random fixation.

The evidence of genetic structure of populations and species summarized in this review indicates that genetic polymorphism and heterozygosity in nature are structured on a massive scale. This finding suggests that various forms of selection, primarily through the mechanisms of spatiotemporally varying environments and epistasis, as well as balancing, directional, diversifying, frequency dependent, and purifying selection regimes, are massively involved in genetic structure and differentiation of populations. The selectionist-neutralist controversy is essentially a quantitative problem: "How much of each?" rather than a qualitative either/or problem. About 80% of the overall analysis of genetic variance of all species analyzed in this review is yet open to further analysis and interpretation, and certainly involves both stochastic and deterministic interactions, yet to be unravelled. However, the 20% of the genetic variance explored here is primarily explicable by ecological heterogeneity. Notably, within individual higher taxa, the explained portion of genetic diversity increases (see Table 25; mean 44%, and maximum of 74% in molluscs).

Different levels of polymorphism and heterozygosity cut across taxonomic and biotic borders, indicating that ecological parameters within each taxon and biotic entity may play predominant roles in the genetic structure and differentiation of populations. The amounts of genetic diversity are often nonrandom among populations, species, life zones, habitats and other biological characteristics. The correlative method employed in this study provides inferential evidence supporting the hypothesis that environmental heterogeneity is a major factor in maintaining and structuring genetic diversity in natural populations. However, direct experimental evidence establishing causal relationships between ecological and genetic structures should be urgently sought by critical tests at the populational and microgeographical level, complemented by studies of biochemical kinetics and physiological function of protein variation.

The problems raised in interpreting protein polymorphisms are bound to increase in our attempts to interpret the manyfold higher diversity found at the DNA level. The dichotomous stochastic and deterministic world views, which flourished earlier in the interpretation of the phenotypic and protein levels, will prosper even more dramatically at the DNA level, as witnessed by current studies and interpretations (e.g., Orgel and Crick, 1980; Kimura, 1981). We suggest that in order to avoid much fruitless polemics, now at the DNA level, many more critical experiments involving natural populations should be designed including ecology, in an attempt to discriminate between the alternative hypotheses. Understanding of organic nature at the DNA level may be much more complex to unravel than that of the protein level. However, even if stochasticity is greater at the DNA level than that observed at the protein one, interaction with deterministic factors are bound to be of paramount importance. Our future task is to attempt to quantify this complex interaction not by avoiding but rather by including the biotic forces operating in the real world in our analysis of nature.

SUMMARY

The roles and relative importance of the major evolutionary forces causing evolutionary change in the protein level, i.e., mutation, natural selection, migration and random genetic drift, are still unclear and controversial. In an attempt to highlight this problem we analyzed the correlates of biotic factors involving ecological, demographic, and life history variables with genetic diversity in natural populations of animals and plants. This review involves 1111 species studied electrophoretically for protein, primarily allozymic variation, for an average of 23 gene loci each. For 815 of the 1111 species we obtained from individual researchers a biotic profile characterized by 21 variables (7 ecological, 5 demographic, and 9 life history and other biological characteristics). We then estimated the levels of genetic diversity (indexed by heterozygosity and polymorphism for (a) all species, three major taxa (vertebrates, invertebrates and plants) and 10 different higher taxa; (b) the categorized 21 biotic factors; then (c) correlated the levels of genetic diversity with the biotic factors, and, lastly (d) matched some of the evidence obtained with theoretical predictions.

The following results are indicated:

1. The levels of genetic diversity vary nonrandomly among populations, species, higher taxa; ecological parameters (life zone, geographic range, habitat type and range, climatic region); demographic parameters (species size and population structure, gene flow, and sociality); and a series of life history characteristics (longevity, generation length, fecundity, origin, and parameters related to the mating system and mode of reproduction).

2. Genetic diversity is higher (i) in species living in broader environmental spectra, (ii) in large species with patchy population structure and limited migration, as well as in solitary or social species, and (iii) in species with small body size, annuals or long-lived perennials, older in time, with smaller diploid chromosome numbers, primarily outcrossed; and plant species reproducing sexually and pollinated by wind. Species with the above characteristics harbour generally more genetic diversity than their opposite counterparts.

3. Genetic diversity is partly correlated and predictable by 3-4 variable combination of ecological, demographic, and life history variables, largely in this order. Ecological factors account for the highest proportion of the 20% explained genetic variance of all species as compared with demographic and life history factors (90%, 39%, and 3.5%, respectively). Within individual higher taxa the explained portion of genetic diversity increases considerably (mean 44%, and maximum of 74% in molluscs). However, significant small intercorrelations (r = mostly 0.1 - 0.3) occur both within and between the subdivided biotic variables. Therefore, additional critical tests at the population microgeographical levels, complemented by biological, biochemical and physiological experimentation, may verify the inferences of causal relationships between biotic factors and genetic diversity.

The patterns and correlates of genetic diversity revealed here over many unrelated species subdivided into different biotic regimes, strongly implicate selection in the genetic differentiation of species. Natural selection in several forms, but most likely through the mechanisms of spatiotemporally varying environments and epistasis at the various life cycle stages of organisms, appears to be an important evolutionary force causing change at the molecular level in many species. Other evolutionary forces, including mutation, migration, and genetic drift, certainly interact with natural selection, either directly or indirectly, and thereby contribute differentially, according to circumstance, to population genetic differentiation at the molecular level.

The role and relative importance of each evolutionary force and its interactive patterns, and the establishment of direct cause- effect relationships between biotic and genetic factors, is a future challenge at both the protein and DNA levels. The ability to approach generalization depends on many more carefully designed field observations on many populations and species, coupled with critical field and laboratory testing of many loci, to assess the contribution of protein and DNA polymorphisms to fitness.

ACKNOWLEDGEMENTS

Our deep gratitude is extended to the numerous scientists who kindly cooperated with us and provided the indispensable biotic profiles for the species they tested genetically. They amount to 135 individual scientists without whose generous help this work could not have achieved its relative unbiased structure. They certainly contributed the objective spirit so important for our analysis. Too numerous to be listed here, their individual names can be found in the reference list next to the species they studied. We also thank N. Storch and D. Kaplan for technical assistance and D. Kaplan, K.E. Holsinger, M.W. Feldman, E. Golenberg and T. Stockheim for critical comments on the manuscript. This study was supported by grants from the United States-Israel Binational Science Foundation, BSF, Jerusalem, Israel.

REFERENCES OF THE DATA

1. Abreu-Grobois,F.A., Beardmore,J.A.: International study on Artemia II. Genetic characterization of Artemia populations - an electrophoretic approach. In: The Brine Shrimp Artemia. Personne,G., Sorgedoos,P., Rods,O., Jaspers,E., (eds.), University Press, Wetteren, Belgium (1980).

2. Adams,S.E., Smith,M.H., Baccus,R.: Biochemical variation in the American Alligator. Herpetologica 36 (1980) 289-296.

3. Adest,G.A.: Genetic relationships in the genus Uma (Iguanidae). Copeia (1977) 47-52.

4. Ahmad,M., Skibinski,D.O.F., Beardmore,J.A.: An estimate of the amount of genetic variation in the common mussel Mytilus edulis. Biochem.Genet. 15 (1977) 833-846.

5. Allendorf,F.W., Christiansen,F.B., Dobson,T., Eanes,W.F., Frydenberg,O.: Electrophoretic variation in large mammals. I. The polar bear, Thalarctos maritimus. Hereditas 91 (1979) 19-22.

6. Allendorf,F.W., Knudsen,K.L., Blake,G.M.: Frequencies of null alleles at enzyme loci in natural populations of ponderosa and red pine. Genetics 100 (1982) In press.

7. Allendorf,F.W., Utter,F.M.: Population genetics. In: Fish Physiology, Vol VIII, Academic Press, (1978) pp. 407-454.

8. Altukhov,Y.P., Salmenkova,E.A., Omelchenko,,V.T., Sachko,G.D., Slynko, V.I.: The number of monomorphic and polymorphic loci in the population of the tetraploid salmon species Oncorynchus keta (in Russian). Genetika 8 (1972) 67-75.

9. Anderson,J.E., Giblett,E.R.: Intraspecific red cell enzyme variation in the pigtailed macaque (Macaca nemestrina). Biochem.Genet. 13 (1975) 189-211.

10. Andersson,L., Ryman,N., Rosenberg,R., Stahl,G.: Genetic variability in Atlantic herring (Clupea harengus harengus): Description of protein loci and population data. Hereditas 95 (1981) 69-78.

11. Angelici,M.C., Matthaeis,De E., Cobolli Sbordoni,M., Sbordoni,V.: Biochemical systematics of genus Zyganea. (1982) Personal communication.

12. Aquadro,C.F., Kilpatrick,C.W.: Morphological and biochemical variation and differentiation in insular and mainland deer mice (Peromyscus maniculatus) In: Mammalian Population Genetics, Smith,M.N., Joule, J., (eds.) Univ. of Georgia Press (Athens) (1981) pp. 214-230.

13. Arntzen,J.W.: Electrophoretic analysis of French crested newts (Triturus cristatus) and marbled newts (Triturus marmoratus). (1982), Personal communication.

14. Ashton,R.E.Jr., Braswell,A.L., Guttman,S.I.: Electrophoretic analysis of three species of Necturus (Amphibia: Proteidae), and the taxonomic status of Necturus lewisi (Brimley). Brimleyana 4 (1980) 43-46

15.Aspinwall,N.: Genetic analysis of North American populations of the pink salomon, *Oncorhynchus gorbuscha*. Possible evidence for the neutral mutation-random drift hypothesis. Evolution 28 (1974) 295-305.

16.Autem,M., Bonhomme,F.: Elements de systematique biochemique chez les muglides de Mediterranee. Biochem.Syst.Ecol. 8 (1980) 305-308.

17.Avise,J.C., Ayala,F.J.: Genetic differentiation in speciose versus depauperate phylads: Evidence from the California minnows. Evolution 30 (1976) 46-58.

18.Avise,J.C., Giblin-Davidson,C., Laerm,J., Patton,J.C., Lansman,R.A.: Mitochondrial DNA clones and matriarchal phylogeny within and among geographic populations of the pocket gopher, *Geomys pinetis*. Proc.Natl.Acad.Sci.USA 76 (1979) 6694-6698.

19.Avise,J.C., Patton,J.C., Aquadro,C.F.: Evolutionary genetics of birds. I. Relationships among North American thrushes and allies. The Auk 97 (1980) 135-147.

20.Avise,J.C., Patton,J.C., Aquadro,C.F.: Evolutionary genetics of birds. II. Conservative protein evolution in North American sparrows and relatives. Syst.Zool. 29 (1980) 323-334.

21.Avise,J.C., Patton,J.C., Aquardo,C.F.: Evolutionary biology of the birds: Comparative molecular evolution in New World warblers and rodents. J.Hered. 71 (1980) 303-310.

22.Avise,J.C., Selander,R.K.: Evolutionary genetics of cave-dwelling fishes of the genus *Astyanax*. Evolution 26 (1972) 1-19.

23.Avise,J.C., Smith,J.J., Ayala,F.J.: Adaptive differentiation with little genic change between two native California minnows. Evolution 29 (1975) 411-426.

24.Avise,J.C., Smith,M.H.: Biochemical genetics of sunfish. II. Genic similarity between hybridizing species. Amer.Natur. 108 (1974) 458-472.

25.Avise,J.C., Smith,M.H.: Biochemical genetics of sunfish. I. Geographic variation and subspecific intergradation in the bluegill, *Lepomis macrochirus*. Evolution 28 (1974) 42-56.

26.Avise,J.C., Smith,M.H., Selander,R.K.: Biochemical polymorphism and systematics in the genus *Peromyscus* VII. Geographic differentiation in members of the *truei* and *maniculatus* species group. J.Mamm. 60 (1979) 177-192.

27.Avise,J.C., Smith,M.H., Selander,R.K.: Biochemical polymorphism and systematics in the genus *Peromyscus*. VI. The *boylii* species group. J.Mamm. 55 (1974) 751-763.

28 Avise,J.C., Smith,M.H., Selander,R.K., Lawlor,T.E., Ramsey,P.R.: Biochemical polymorphism and systematics in the genus *Peromyscus*. V. Insular and mainland species of the subgenus *Haplomylomys*. Syst. Zool. 23 (1974) 226-238

29.Avise,J.C., Straney,D.O., Smith,M.H.: Biochemical genetics of sunfish IV. Relationships of centrarchid genera. Copeia (1977) 250-258.

30.Ayala,F.J., Hedgecock,D., Zumwalt,G.S., Valentine,J.W.: Genetic variation in Tridacna maxima, an ecological analog of some unsuccessful evolutionary lineages. Evolution 27 (1973) 177-191.

31.Ayala,F.J., Powell,J.R.: Enzyme variability in the Drosophila willistoni group. VI. Levels of polymorphism and the physiological function of enzymes. Biochem.Genet. 7 (1972) 331-345.

32.Ayala,F.J., Powell,J.R., Dobzhnasky,Th.: Polymorphisms in continental and island populations of Drosophila willistoni. Proc.Nat.Acad.Sci.USA 68 (1971) 2480-2483.

33.Ayala,F.J., Tracey,M.L.: Genetic differentiation within and between species of the Drosophila willistoni group. Proc.nat.Acad.Sci.USA. 71 (1974) 999-1003.

34.Ayala,F.J., Tracey,M.L., Barr,L.G., Ehrenfeld,J.G.: Genetic and reproductive differentiation of the subspecies, Drosophila equinoxialis carribenesis. Evolution 28 (1974) 24-41.

35.Ayala,F.J., Tracey,M.L., Barr,L.G., McDonald,J.F., Perez-Salas,S.: Genetic variation in natural populations of five Drosophila species and the hypothesis of the selective neutrality of protein polymorphisms. Genetics 77 (1974) 343-384.

36.Ayala,F.J., Valentine,J.W.: Genetic variability in the cosmopolitan deep-water ophiuran Ophiomusium lymani. Marine Biology 27 (1974) 51-57.

37.Ayala,F.J., Valentine,J.W.: Genetic variability in the pelagic environment: A paradox? Ecology 60 (1979) 24-29.

38.Ayala,F.J., Valentine,J.W., Barr,L.G., Zumwalt,G.S.: Genetic variability in a temperate intertidal phoronid, Phoronopsis viridis. Biochem.Genet. 11 (1974) 413-427.

39.Ayala,F.J., Valentine,J.W., DeLaca,T.E., Zumwalt,G.S.: Genetic variability of the Antarctic brachiopod Liothyrella notorcadensis and its bearing on mass extinction hypotheses. Journal of Paleontology 49 (1975) 1-9.

40.Ayala,F.J., Valentine,J.W., Hedgecock, D., Barr,L.G.: Deep-sea asteroids: High genetic variability in a stable environment. Evolution 29 (1975) 203-212.

41.Ayala,F.J., Valentine,J.W., Zumwalt,G.S.: An electrophoretic study of the antarctic zooplankter Euphausia superba. Limnology and Oceanography 20 (1975) 635-640.

42.Baccus,R., Joule,J., Kimberling,W.J.: Linkage and selection analysis of biochemical variants in Peromyscus maniculatus. J.Mamm. 61 (1980) 423-435

43.Baker,C.M.A., Manwell,C.: Molecular genetics of avian proteins. VIII. Egg white proteins of the migratory quail, Coturnix coturnix - new concepts of "Hybrid Vigour". Comp.Biochem.Physiol. 23 (1967) 21-42.

44.Baker,C.M.A., Manwell,C., Labisky,R.F., Harper,J.A.: Molecular genetics of avian protein. V. Egg blood and tissue protein of the ring-necked pheasant, Phasinus colchicus L. Comp.Biochem.Physiol. 17 (1966) 467-499.

45.Baker,J., Maynard Smith,J., Strobeck,C.: Genetic polymorphism in the bladder campion, Silene maritima. Biochem.Genet. 13 (1975) 393-410.

46.Baker,M.C.: Song dialects and genetic differences in white-crowned sparrows (Zonotrichia leucophrys). Evolution 29 (1975) 226-241.

47.Baker,M.C., Fox,S.F.: Dominance, survival, and enzyme polymorphism in dark-eyed juncos, Junco hyemalis. Evolution 32 (1978) 697-711.

48.Baker,M.C., Fox,S.F.: Differential survival in common grackles sprayed with Turgitol. Amer.Natur. 112 (1978) 675-682.

49.Baker,R.J., Honeycutt,R.L., Arnold,M.L., Sarich,V.M., Genoways,H.H.: Electrophoretic and immunological studies on the relationship of the Brachyphyllinae and the Glossophaginae. J.Mamm. 62 (1981) 665-672.

50.Barker,J.S.F., Mulley,J.C.: Isozyme variation in natural populations of Drosophila buzatii. Evolution 30 (1976) 213-233.

51.Barret,V.A., Vyse,E.R.: Comparative genetics of three trumpeter swan populations. The Auk 99 (1982) 103-108.

52.Barrowclough,G.F., (1982) Personal communication.

53.Barrowclough,G.F., Corbin,K.W.: Genetic variation and differentiation in the Parulidae. The Auk 95 (1978) 691-702.

54.Battaglia,B., Bisol,P.M., Fava,G.: Genetic variability in relation to the environment in some marine invertebrates. In: Marine Organisms: Genetic, Ecology and Evolution. Battaglia,B., Beardmore,J., (eds.) Plenum Press New York (1978) pp. 53-70.

55 Battaglia,B., Bisol,P.M., Varotto,V.: Variabilite genetique dans des populations de Tisbe holothuriae (Copepoda, Harpacticoida) marines et d'eau saumatre. Arch.Zool.Exp.gen. 119 (1978) 251-264.

56 Bellemin,J., Adest,G., Gorman,G.C.: Genetic uniformity in northern populations of Thamnophis sirtalis (Serpentes: Colubridae). Copeia (1978) 150-151.

57 Benado,M., Aguilera,M., Reig,O.A., Ayala,F.J.: Biochemical genetics of chromosome forms of Venezuelan spiny rats of the Proechimys guairae and Proechimys trinitatis superspecies. Genetica 50 (1979) 89-97.

58 Benharrat,K., Quignard,J-.P., Pasteur,N.: Les gobies noirs (Gobius niger Linne, 1758) de la cote Mediterraneenne Francaise: Variation du polymorphisme enzymatique des populations lagunaires et marines. Cybium 3 (1981) 29-33.

59.Berrebi,P., Britton-Davidian,J.: Enzymatic survey of four populations of Atherina boyeri based on electrophoresis and the occurence of a microsporidiosis. J.Fish.Biol. 16 (1980) 149-157.

60.Berry,R.J.: Population dynamics of the house mouse. Symp.Zoo.Soc.Lond. 47 (1981) 395-425.

61.Berry,R.J., Bonner,W.N., Peters,J.: Natural selection in house mice (Mus musculus) from South Georgia (south Atlantic Ocean). J.Zool.,Lond. 189 (1979) 385-398.

62. Berry, R.J., Peters, J.: Macquarie Island house mice: A genetical isolate on a sub-antarctic island. J.Zool., Lond. 176 (1975) 375-389.

63. Berry, R.J., Peters, J.: Heterogeneous heterozygosities in Mus musculus populations. Proc.R.Soc.Lond.B. 197 (1977) 485-503.

64. Berry, R.J., Peters, J., Van Aarde, R.J.: Sub-antarctic house mice: colonization, survival and selection. J.Zool., Lond. 184 (1978) 127-141.

65. Berry, R.J., Sage, R.D., Lidicker, W.Z., Jackson, W.B.: Genetical variation in three Pacific house mouse (Mus musculus) populations. J.Zool., Lond. 193 (1981) 391-404

66. Bezy, R.L., Gorman, G.C., Adest, G.A., Kim, Y.J.: Divergence in the island night lizard Xantusia riversiana (Sauria: Xantusiidae). In: The California Islands: Proceedings of a Multidisciplinary Symposium. Power, D.M. (ed.), Santa Barbara Nat.Hist.Mus., Santa Barbara (1980) pp. 565-583.

67. Bezy, R.L., Gorman, G.C., Kim, Y.J., Wright, J.W.: Chromosomal and genetic divergence in the fossorial lizards of the family Anniellidae. Syst. Zool. 26 (1977) 57-71.

68. Bisol, P.M., Battaglia, B., Varotto, V.: Variabilita genetica in animali marini: Grammarus aequicauda stock (Amphipoda). Bollettino di Zoologia 45 (1978).

69. Bisol, P.M., Marigo, N.: Variabilita' genetica in Ophiotrix fragilis (Ophiuroidea, Echinodermata). Atti Associazione Genetica Italiana. XXV (1980) 46-48.

70. Blanc, F., Cariou, M.-L.: High genetic variability of lizards of the sand-dwelling lacertid genus Acanthodactylus. Genetica 54 (1980) 141-147.

71. Bohlin, R.G., Zimmerman, E.G.: Genic differentiation of two chromosome races of the Geomys bursarius complex. J.Mamm. 63 (1982) 218-228.

72. Bonhomme, F., Selander, R.K.: Estimating total genic diversity in the house mouse. Biochem.Genet. 16 (1978) 287-297.

73. Bonnel, M.L., Selander, R.K.: Elephant seals: Genetic variation and near extinction. Science 184 (1974) 908-909.

74. Bowen, B.S., Yang, S.Y.: Genetic control of enzyme polymorphisms in the California vole, Microtus californicus. Biochem.Genet. 16 (1978) 455-467.

75. Brittnacher, J.G., Sims, S.R., Ayala, F.J.: Genetic differentiation between species of the genus Speyeria (Lepidoptera: Nymphalidae). Evolution 32 (1978) 199-210.

76. Britton, J., Thaler, L.: Evidence for the presence of two sympatric species of mice (Genus Mus L.) in southern France based on biochemical genetics. Biochem.Genet. 16 (1978) 213-225.

77. Britton-Davidian, J., Bonhomme, F., Croset, H., Capanna, E., Thaler, L.: Variabilite genetique chez les populations de souris (Genre Mus L.) a nombre chromosique reduit. C.R.Acad.Sc.Paris 290 (1980) 195-198.

78. Brown,K.: Low genetic variability and high similarities in the crayfish genera Cambarus and Procambarus. Am.Midl.Nat. 105 (1981) 225-232.

79. Browne,R.A.: Genetic variation in island and mainland populations of Peromyscus leucopus. Am.Midl.Nat. 97 (1977) 1-9.

80. Bruce,E.J., Ayala,F.J.: Phylogenetic relationships between man and the apes: Electrophoretic evidence. Evolution 33 (1979) 1040-1056.

81. Bryant,E.H., Dijk,Van H., Delden,Van W.: Genetic variability of the face fly, Musca autumnalis de Deer, in relation to a population bottleneck. Evolution 35 (1981) 872-881.

82. Bucklin,A., Hedgecock,D.: Biochemical genetic evidence for a third species of Metridium (Coelenterata: Actiniaria). Marine Biology 66 (1982) 1-7.

83. Bullini,L., Nascetti,G., Ciafre,S., Rumore,F., Biocca,E., Montalenti,S.G., Rita,G.: Ricerche cariologiche ed elettroforetiche su Parascaris univalens e Parascaris equrum. Academia Nazionale dei Lincei. Serie VIII, Vol. LXV, (1978) 151-159.

84. Busack,C.A., Thorgaard,G.H., Bannon,M.P., Gall,G.A.E.: An electrophoretic, karyotypic and meristic characterization of the eagle lake trout, Salmo gairdneri aquilarum. Copeia (1980) 418-424.

85. Buth,D.G.: Genetic relationships among the torrent suckers, genus Thoburnia. Biochem.Syst.Ecol. 7 (1979) 311-316.

86. Buth,D.G.: Biochemical systematics of the cyprinid genus Notropis - I. The subgenus Luxilus. Biochem.Syst.Ecol. 7 (1979) 69-79.

87. Buth,D.G.: Evolutionary genetics and systematic relationships in the catostomid genus Hypentelium. Copeia (1980) 280-290.

88. Buth,D.G., Burr,B.M.: Isozyme variability in the cyprinid genus Campostoma. Copeia (1978) 298-311.

89. Buth,D.G., Burr,B.M., Schenck,J.R.: Electrophoretic evidence for relationships and differentiation among members of the percid subgenus Microperca. Biochem.Syst.Ecol. 8 (1980) 297-304.

90. Buth,D.G., Crabtree,C.B.: Genetic variability and population structure of Catostomus santaanae in the Santa Clara drainage. Copeia (1982) 439-444.

91. Cabrera,V.M., Gonzalez, A.M., Gullon,A.: Enzymatic polymorphism in Drosophila subobscura populations from the Canary Islands. Evolution 34 (1980) 875-887.

92. Cabrera,V.M., Gonzalez, A.M., Larruga,J.M., Gullon,A.: Electrophoretic variability in natural populations of Drosophila melanogaster and Drosophila simulans. Genetica (1982) In press.

93. Cameron,D.G., Vyse,E.R.: Heterozygosity in Yellowstone Park elk, Cervus canadensis. Biochem.Genet. 16 (1978) 651-657.

94. Campbell,C.A., Valentine,J.W., Ayala,F.J.: High genetic variability in a population of Tridacna maxima from the Great Barrier Reef. Marine Biology 33 (1975) 341-345.

95. Carlson,D.M., Kettler,M.K., Fisher,S.E., Whitt,G.S.: Low genetic variability in paddlefish populations. Isozyme Bulletin 14 (1981) 79.

96. Carlson,D.M., Kettler,M.K., Fisher,S.E., Whitt,G.S.: Low genetic variability in paddlefish populations. Copeia (1982) 721-725.

97. Carson,H.L., Johnson,W.E., Nair,P.S., Sene,F.M.: Allozymic and chromosomal similarity in two Drosophila species. Proc.Nat.Acad.Sci.USA 72 (1975) 4521-4525.

98. Case,S.M.: Biochemical systematics of members of the genus Rana native to western North America. Syst.Zool. 27 (1978) 299-311.

99. Case,S.M.: Electrophoretic variation in two species of ranid frogs, Rana boylei and R. muscosa. Copeia (1978) 311-320.

100. Case,S.M., Haneline,P.G., Smith,M.F.: Protein variation in several species of Hyla. Syst.Zool. 24 (1975) 281-295.

101. Catzeflis,F., Graf,J.-D., Hausser,J., Vogel,P.: Comparaison biochimique des musaraignes du genre Sorex en Europe occidentale (Soricidae, Mammalia). Z.f.zool.Systematik u.Evolutionsforchung 20 (1982) 223-233.

102. Cervelli,M., Fava,G.: Variabilita' genetica in Idotea baltica (Isopoda, valvifera) della laguna di Venezia. Mem.Biol.Marina e Oceanogr.,Suppl. (1980) 381-382.

103. Cesaroni,D., Allegrucci,G., Caccone,A., Cobolli Sbordoni,M., Matthaeis, De E., DiRao,M., Sbordoni,V.: Genetic variability and divergence between populations and species of Nesticus cave spiders. Genetica 56 (1981) 81-92.

104. Chakraborty,R., Haag,M., Ryman,N., Stahl,G.: Hierarchical gene diversity analysis and its application to brown trout population data. Hereditas 97 (1982) 17-21.

105. Cianchi,R., Maini,S., Bullini,L.: Genetic distance between pheromone strains of the European corn borer, Ostrinia nubilalis: Different contribution of variable substrate, regulatory and non regulatory enzymes. Heredity 45 (1980) 383-388.

106. Cianchi,R., Urbanelli,S., Coluzzi,M., Bullini,L.: Genetic distance between two sibling species of the Aedes mariae complex (Diptera, Cullicidae). Parassitologia 20 (1978) 39-46.

107. Clark,R.L., Templeton,A.R., Sing,C.F.: Studies of enzyme polymorphisms in the Kamuela population of D. mercatorum. I. Estimation of the level of polymorphism. Genetics 98 (1981) 597-611.

108. Corbin,K.W., Sibley,C.G., Ferguson,A.: Genic changes associated with the establishment of sympatry in orioles of the genus Icterus. Evolution 33 (1979) 624-633.

109. Corbin,K.W., Sibley,C.G., Ferguson,A., Wilson,A.C., Brush,A.H., Ahlquist, J.E.: Genetic polyphormism in New Guinea starlings of the genus Aplonis. The Condor 76 (1974) 307-318.

110. Costa,R., (1982) Personal communication.

111. Costa, R., Bisol, P.M.: Genetic variability in deep-sea organisms. Biol. Bull. 155 (1978) 125-133.

112. Costa, R., Bisol, P.M., Sibuet, M.: Genetic variability in deep-sea holothurians. International Echinoderms Conference, Tampa Bay, (1981).

113. Cothran, E.G.: Personal communication.

114. Cothran, E.G., Zimmerman, E.G., Nadler, C.F.: Genic differentiation and evolution in the ground squirrel subgenus _Ictidomys_ (genus _Spermophilus_) J.Mamm. 58 (1977) 610-622.

115. Crabtree, C.B.: Subspecific genetic differentiation in the topsmelt (_Atherinops affinis_). Isozyme Bulletin 14 (1981) 86-87.

116. Crawford, D.J., Bayer, R.J.: Allozyme divergence in _Coreopsis cyclocarpa_ (Compositae). Systematic Botany 6 (1981) 373-379.

117. Crawford, D.J., Smith, E.B.: Allozyme variation in _Coreopsis nuecensoides_ and _C. nuecensis_ (Compositae), a progenitor-derivative species pair. Evolution 36 (1982) 379-386.

118. Crease, T.J., Herbert, P.D.N.: Genetic divergence between metapopulations of _Daphnia magna_. (1982) Personal communication.

119. Cross, T.F., Ward, R.D.: Protein variation and duplicate loci in the Atlantic salmon, _Salmo salar_ L. Genet.Res.Cambr. 36 (1980) 147-165.

120. Czikeli, H., Miller, I., Gemeiner, M.: Liver-isozyme-patterns in three subspecies of the yellow wagtail (_Motacilla flava_ ssp.; Passeriformes, Aves). 4th Int. Congress on Isozymes. The University of Texas, Austin, 1982.

121. Daly, J.C., Richardson, B.J.: Allozyme variation between populations of baitfish species _Stolephorus heterolobus_ and _St. devisi_ (Pisces: Engraulidae) and _Spratelloides gracilis_ (Pisces: _Dussumieriidae_) from Papua New Guinea waters. Aust.J.Mar.Freshwater Res. 31 (1980) 701-711.

122. Dando, P.R., Southward, A.J.: Enzyme variation in _Chthamalus stellatus_ and _Chthamalus montagui_ (Crustacea: Cirripedia): Evidence for the presence of _C. montagui_ in the Adriatic. J.Mar.Biol.Ass.U.K. 59 (1979) 307-320.

123. Dando, P.R., Southward, A.J.: A new species of _Chthamalus_ (Crustacea: Cirripedia) characterized by enzyme electrophoresis and shell morphology: With a revision of other species of _Chthamalus_ from the western shores of the Atlantic Ocean. J.Mar.Biol.Ass.U.K. 60 (1980) 187-831.

124. Daugherty, C.H., Bell, B.D., Adams, M., Maxson, L.R.: An electrophoretic study of genetic variation in the New Zealand frog genus _Leiopelma_. New Zealand Journal of Zoology 8 (1981) 543-550.

125. Dehring, T.R., Brown, A.F., Daugherty, C.H., Phelps, S.R.: Survey of the genetic variation among Eastern Lake Superior lake trout (_Salvelinus namaycush_). Can.J.Fish.Aquat.Sci. 38 (1981) 1738-1746.

126. Delay, B., Sbordoni, V., Cobolli-Sbordoni, M., Matthaeis, De E.: Divergences genetiques entre les populations de _Speonomus delarouzeei_ du massif du canigou (Coleoptera, Bathysciinae). Mem. Biospeol. 7 (1980) 235-247.

127. Densmore,L.D., Lester,L.J.: Protein heterogeneity in an isopod. Isozyme Bulletin 13 (1980) 97.

128. Dessauer,H.C., Braun,M., (1982) Personal communication.

129. Dessauer,H.C., Cole,J.M., (1982), Personal communication.

130. Dessauer,H.C., Gartside,D.F., Gans,C.: Protein evidence on the genetic diversity and affinites of Uropeltid snakes. American Zoologist 16 (1976) 268.

131. Dessauer,H.C., Gartside,D.F., Gans,C., (1982), Personal communication.

132. Dessauer,H.C., Gartside,D.F., Zweifel,R.G.: Protein electrophoresis and the systematics of some New Guinea hylid frogs (genus Litoria). Syst. Zool. 26 (1977) 426-436.

133. Dessauer,H.C., Kaltenbach,R., (1982), Personal communication.

134. Dessauer,H.C., Nevo,E.: Geographic variation of blood and liver proteins in cricket frogs. Biochem.Genet. 3 (1969) 171-188.

135. Dessauer,H.C., Nevo,E., Chuang,K.C.: High genetic variability in an ecological variable vertebrate, Bufo viridis. Biochem.Genet. 13 (1975) 651-661.

136. Duncan,R., Highton,R.: Genetic relationships of the eastern large Plethodon of the Ouachita Mountains. Copeia (1979) 95-110.

137. Dunlap,D.G.: Geographic variation of proteins and call in Rana pipiens from the Northcentral United States. Copeia (1981) 876-879.

138. Ellstrand,N.C., Levin,D.A.: Recombination system and population structure in Oenothera. Evolution 34 (1980) 923-933.

139. Ellstrand,N.C., Levin,D.A.: Evolution of Oenothera laciniata (Onagraceae), a permanent translocation heterozygote. Systematic Biology 5 (1980) 6-16.

140. Eriksson,K., Halkka,O., Lokki,J., Saura,A.: Enzyme polymorphism in feral, outbred and inbred rats (Rattus norvegicus). Heredity 37 (1967) 341-349.

141. Feder,J.H.: Natural hybridization and genetic divergence between the toads Bufo boreas and Bufo punctatus. Evolution 33 (1979) 1089-1097.

142. Feder,J.H., Wurst,G.Z., Wake,D.B.: Genetic variation in western salamanders of the genus Plethodon, and the status of Plethodon gordoni. Herpetologica 34 (1978) 64-69.

143. Felley,J.D., Avise,J.C.: Genetic and morphological variation of bluegill populations in Florida lakes. Trans.Am.Fish.Soc. 109 (1980) 108-115.

144. Ferris,S.D., Buth,D.G., Whitt,G.S.: Substantial genetic differentiation among populations of Catostomus plebeius. Copeia (1982) 444-449.

145. Ferris,S.D., Whitt,G.S.: Genetic variability in species with extensive gene duplication: The tetraploid catostomid fishes. Amer.Natur. 115 (1980) 650-666.

146. Fisher, P.W., Browne, D., Cameron, D.G., Vyse, E.R.: Genetics of rainbow trout in a geothermally heated stream. Trans.Am.Fish.Soc. (1982) In press.

147. Fisher, R.A., Putt, W., Hackel, E.: An investigation of the products of 53 gene loci in three species of wild Canidae: Canis lupus, Canis latrans, and Canis familiaris. Biochem.Genet. 14 (1976) 963-974.

148. Frydenberg, O., Simonsen, V.: Genetics of zoarces populations. V. Amount of protein polymorphism and degree of genetic heterozygosity. Hereditas 75 (1973) 221-232.

149. Fujio, Y., Kato, Y.: Genetic variation in fish populations. Bulletin of the Japanese Society of scientific Fisheries 45 (1979) 1169-1178.

150. Fuller, B., Lester, L.J.: Correlations of allozymic variation with habitat parameters using the grass shrimp, Palaemonetes pugio. Evolution 34 (1980) 1099-1104.

151. Galleguillos, R.A., Ward, R.D.: Genetic and morphological divergence between populations of the flatfish Platichthys flesus (L.) (Pleuronectidae). Biol. J. Linn. Soc. 17 (1982) In press.

152. Gardenal, C.N., (1982), Personal communication.

153. Gardenal, C.N., Sabattini, M.S., Blanco, A.: Enzyme polymorphism in a population of Calomys musculinus (Rodentia, Cricetidae). Biochem.Genet. 18 (1980) 563-575.

154. Garten, T.Jr.,: Relationships between agressive behavior and genic heterozygosity in the oldfield mouse, Peromyscus polionotus. Evolution 30 (1976) 59-72.

155. Gartside, D.F., Dessauer, H.C., Joanen, T.: Genic homozygosity in an ancient reptile (Alligator mississippiensis). Biochem.Genet. 15 (1977) 655-663.

156. Gartside, D.F., Rogers, J.S., Dessauer, H.C.: Speciation with little genic and morphological differentiation in the ribbon snakes Thamnophis proximus and T. sauritus (Colubridae). Copeia (1977) 697-705.

157. Gemmeke Von, H.: Proteinvariation und taxonomie in der gattung Apodemus (Mammalia, Rodentia). Z.Saugetiere 45 (1980) 348-365.

158. Gill, A.E.: Genetic divergence of insular populations of deer mice. Biochem.Genet. 14 (1976) 835-848.

159. Glover, D.G., Smith, M.H., Ames, L., Joule, J., Dubach, J.M.: Genetic variation in pika populations. Can. J. Zool. 55 (1977) 1841-1845.

160. Gonzalez, A.M., Cabrera, V.M., Larruga, J.M., Gullon, A.: Genetic distance in the sibling species Drosophila melanogaster, Drosophila simulans and Drosophila mauritiana. Evolution, 36 (1982) 517-522.

161. Gonzalez, A.M., Cabrera, V.M., Larruga, J.M., Gullon, A.: Molecular variation in insular endemic Drosophila species of the Macronesian Archipelagos. (1982) Personal communication.

162. Gooch, J.L.: Allozyme genetics of life cycle stages of brachyurans. Chesapeake Science 18 (1977) 284-289.

163.Gooch,J.L., Schopf,T.J.M.: Genetic variability in the deep sea: Relation to environmental variability. Evolution 26 (1972) 545-552.

164.Gorman,G.C., Buth,D.G., Soulè,M., Yang,S.Y.: The relationships of the Anolis cristatellus species group: Electrophoretic analysis. J. Herpetol. 14 (1980) 269-278.

165.Gorman,G.C., Kim,Y.J.: Genetic variation and genetic distance among populations of Anolis lizards on two lesser Antillean island banks. Syst.Zool. 24 (1975) 369-373.

166.Gorman,G.C., Kim,Y.J.: Anolis lizards of the eastern Caribbean: A case study in evolution. II. Genetic relationships and genetic variation of the bimaculatus group. Syst. Zool. 25 (1976) 62-77.

167.Gorman,G.C., Kim,,Y.J.: Genotypic evolution in the face of phenotypic conservativeness: Abudefduf (Pomacentridae) from the Atlantic and Pacific sides of Panama. Copeia (1977) 694-697.

168.Gorman,G.C., Kim,Y.J., Rubinoff,R.: Genetic relationship of three species of Bathygobius from the Atlantic and Pacific sides of Panama. Copeia (1976) 361-364.

169.Gorman,G.C., Kim,Y.J., Yang,S.Y.: The genetics of colonization: Loss of variability among introduced populations of Anolis lizards (Reptilia, Lacertilia, Iguanidae) J.Herpetol. 12 (1978) 47-51.

170.Gorman,G.C., Soulè,M., Yang,S.Y., Nevo,E.: Evolutionary genetics of insular Adriatic lizards. Evolution 29 (1975) 52-71.

171.Gottlieb,L.D.: Allelic diversity in the outcrossing annual plant Stephanomeria exigua ssp. carotifera (Compositae). Evolution 29 (1975) 213-225.

172.Gould,J.S., Woodruff,D.S., Martin,J.P.: Genetics and morphometrics of Cerion at Pongo Carpet: A new systematic approach to this enigmatic land snail. Syst.Zool. 23 (1974) 518-535.

173.Graf,J.-D.: Genetique biochimique,zoogeographie et taxonomie des Arvicolidae (Mammalia, Rodentia). Revue Suisse Zool. 89 (1982) 749-787.

174.Graf,J.-D., Meylan,A.: Polymorphisme chromosomique et biochimique chez Pitymys multiplex (Mammalia, Rodentia). Z.Saugetierkunde 45 (1980) 133-148.

175.Graves,J.E., Somero,G.N.: Electrophoretic and functional enzymic evolution in four species of Eastern Pacific barracudas from different thermal environments. Evolution 36 (1982) 97-106.

176.Greenbaum,I.F.: Genetic interactions between hybridizing cytotypes of the tent-making bat (Uroderma bilobatum). Evolution 35 (1981) 306-321.

177.Greenbaum,I.F., Baker,R.J.: Evolutionary relationships in Macrotus (Mammalia: Chiroptera): Biochemical variation and karyology. Syst. Zool. 25 (1976) 15-25.

178.Guries,R.P., Ledig,F.T.: Genetic diversity and population structure in pitch pine (Pinus rigida Mill.). Evolution 36 (1982) 387-402.

179. Guttman,S.I.: Genetic variation in the genus Bufo II. Isozymes in northern allopatric populations of the American toad, Bufo americanus. Isozymes, IV Genetics and Evolution, Academic press,Inc. (1975) 679-697.

180. Guttman,S.I.: Electrophoretically derived estimates of genetic variation in natural populations of anurans. (1981), Personal communication.

181. Guttman,S.I., Grau,G.A., Karlin,A.A.: Genetic variation in Lake Erie great blue herons (Ardea herodias). Comp.Biochem.Physiol. 66B (1980) 167-169.

182. Guttman,S.I., Karlin,A.A., Labanick,G.M.: A biochemicl and morphological analysis of the relationship between Plethodon longicrus and Plethodon yonahlossee (Amphibia, Urodela, Plethodontidae). Journal of Herpetology 12 (1978) 445-454.

183. Guttman,S.I., Wood,T.K., Karlin,A.A.: Genetic differentiation along host plant lines in the sympatric Enchenopa binotata Say complex (Homoptera: Membracidae). Evolution 35 (1981) 205-217.

184. Gyllensten,U., Reuterwall,C., Ryman,N.: Genetic variability in Scandinavian populations of willow grouse (Lagopus lagopus L.) and rock ptarmigan (Lagopus mutus L.). Hereditas 91 (1979) 301.

185. Gyllensten,U., Ryman,N., Reuterwall,C.: Allozyme differentiation within and between European red deer subspecies and the management of red deer genetic resources in Sweden. SNV PM (1982) In Press.

186. Hafner,D.J., Hafner,J.C., Hafner,M.S.: Systematic status of kangaroo mice, genus Microdipodops: morphometric, chromosomal, and protein analysis. J.Mamm. 60 (1979) 1-10.

187. Hafner,D.J., Petersen,K.E., Yates,T.L.: Evolutionary relationships of jumping mice (genus Zapus) of the southwestern United States. J.Mamm. 62 (1981) 501-512.

188. Hall,W.P., Selander,R.K.: Hybridization of karyotypically differentiated populations in the Sceloporus grammicus complex (Iguanidae). Evolution 27 (1973) 226-242.

189. Halliday,R.B.: Heterozygosity and genetic distance in sibling species of meat ants (Iridomyrmex purpureus group). Evolution 35 (1981) 234-242.

190. Hamrick,J.L., Linhart,Y.B., Mitton,J.B.: Relationships between life history characteristics and electrophoretically detectable genetic variation in plants. Ann.Rev.Ecol.Syst. 10 (1979) 173-200.

191. Harris,H., Hopkinson,D.A.: Average heterozygosity per locus in man: An estimate based on the incidence of enzyme polymorphisms. Ann.Hum.Genet. Lond. 36 (1972) 9-19.

192. Hebert,P.D.N., Moran,C.: Enzyme variability in natural populations of Daphnia carinata king. Heredity 45 (1980) 313-321.

193. Hedgecock,D.: Genetic variation in two widespread species of salamanders, Taricha granulosa and Taricha torosa. Biochem.Genet. 14 (1976) 561-576.

194. Hedgecock,D.: Population subdivision and genetic divergence in the red-bellied newt, Taricha rivularis. Evolution 32 (1978) 271-286.

195. Hedgecock,D.: Biochemical genetic variation and evidence of speciation in Chthamalus barnacles of the tropical eastern Pacific Ocean. Marine Biology 54 (1979) 207-214.

196. Hedgecock,D., Ayala,F.J.: Evolutionary divergence in the genus Taricha (Salamandridae). Copeia (1974) 738-747.

197. Hedgecock,D., Stelmach,D.J., Nelson,K., Lindenfelser,M.E., Malecha,S.R.: Genetic divergence and biogeography of natural populations of Macrobrachium rosenbergii. Proc.World Maricul.Soc. 10 (1979) 873-879.

198. Herrebout,W.M., Menken,S.B.J., Povel,G.D.E., Walter,Van De T.P.N.: The position of Yponomeuta yanagawanus Matsumura (Lepidoptera, Yponomeutidae). Netherlands Journal of Zoology 32 (1982) 313-324.

199. Highton,R., Webster,T.P.: Geographic protein variation and divergence in populations of the salamander Plethodon cinereus. Evolution 30 (1976) 33-45.

200. Honeycutt,R.L., Williams,S.L.: Genic differentiation in pocket gophers of the genus Pappogeomys, with comments on intergeneric relationships in the subfamily Geomyinae. J.Mamm. 63 (1982) 208-217.

201. Johns,P.E., Baccus,R., Manlove,M.N., Pinder,J.E., Smith,M.H.: Reproductive patterns, productivity and genetic variability in adjacent white-tailed deer populations. Proc.Annual Conf.S.E.Assoc.Fish ` Wildlife Agencies 31 (1977) 167-172.

202. Johnson,A.G., Utter,F.M., Hodgins,H.O.: Electrophoretic investigation of the family Scorpaenidae. Fishery Bulletin 70 (1972) 403-413.

203. Johnson,A.G., Utter,F.M., Hodgins,H.O.: Estimate of genetic polymorphism and heterozygosity in three species of rockfish (genus Sebastes). Comp. Biochem.Physiol. 44B (1973) 397-406.

204. Johnson,G.L., Packard,R.L.: Electrophoretic analysis of Peromyscus comanche Blair, with comments on its systematic status. Occasional Papers, The Museum, Texas Tech University 24 (1974) 1-16.

205. Johnson,M.S.: Comparative geographic variation in Menidia. Evolution 28 (1974) 607-618.

206. Johnson,M.S.: Biochemical systematics of the atherinid genus Menidia. Copeia (1975) 662-691

207. Johnson,M.S., (1982), Personal communication.

208. Johnson,M.S., Brown,J.L.: Genetic variation among trait groups and apparent absence of close inbreeding in grey-crowned babblers. Behav. Ecol.Sociobiol. 7 (1980) 93-98.

209. Johnson,M.S., Clarke,B., Murray,J.: Genetic variation and reproductive isolation in Partula. Evolution 31 (1977) 116-126.

210. Johnson,W.E., Selander,R.K.: Protein variation and systematics in kangaroo rats (genus Dipodomys). Syst.Zool. 20 (1971) 377-405.

211. Johnson,W.E., Selander,R.K., Smith,M.H., Kim,Y.J.: XIV. Biochemical genetics of sibling species of the cotton rat (Sigmodon). The University

of Texas Publication - Studies in Genetics VII. Publ.No.7213 (1972) 297-305.

212. Joseph, S., Singh, R.S.: Evolutionary genetics of aphids. I. Genetic variation in natural populations of Rose aphids (Macrosiphum rosae). (1982) Personal communication.

213. Kahler, A.L., Allard, R.W., Krzakowa, M., Wehrhahn, C.F., Nevo, E.: Associations between isozyme phenotypes and environment in the slender wild oat (Avena barbata) in Israel. Theor.Appl.Genet. 56 (1980) 31-47.

214. Kalezic, M.L., Hedgecock, D.: Genetic variation and differentiation of three common European newts (Triturus) in Yugoslavia. British Journal of Herpetology 6 (1979) 49-57.

215. Karlin, A.A., Guttman, S.I.: Hybridization between Desmognathus fuscus and Desmognathus ochrophaeus (Amphibia: Urodela: Plethodontidae) in Northeastern Ohio and Northwestern Pennsylvania. Copeia (1981) 371-377.

216. Karlin, A.A., Vail, V.A., Heard, W.H.: Parthenogenesis and biochemical variation in southeastern Campeloma geniculum (Gastropada: Viviparidae). Malacological Review 13 (1980) 7-15.

217. Kaufman, D.W., Selander, R.K.: Genic heterozygosity in a population of Eutamias panamintinus. J.Mamm. 54 (1973) 776-778.

218. Kawamoto, Y., Nozawa, K., Ischak, Tb.M.: Genetic variability and differentiation in local populations in Indonesian crab-eating macaque (Macaca fascicularis). Kyoto University Overseas Report of Studies on Indonesian Macaque 1 (1981) 15-39.

219. Kawamoto, Y., Shotake, T., Nozawa, K.: Genetic differentiation among three genera of family Cercopithecidae. Primates 23 (1982) 272-286.

220. Khanna, N.D., Juneja, R.K., Larsson, B., Gahne, B.: Electrophoretic studies on proteins and enzymes in the Atlantic salmon, Salmo salar, L. Swedish J. Agric.Res. 5 (1975) 185-192.

221. Kilpatrick, C.W.: Genetic structure of insular populations. In: Mammalian Population Genetics, Smith, M.H., Joule, J., (eds.), Univ. of Georgia (Athens) (1981) pp. 28-59.

222. Kilpatrick, C.W., Zimmerman, E.G.: Genetic variation and systematics of four species of mice of the Peromyscus boylii species group. Syst. Zool. 24 (1975) 143-162.

223. Kilpatrick, C.W., Zimmerman, E.G.: Biochemical variation and systematics of Peromyscus pectoralis. J.Mamm. 57 (1976) 506-522.

224. Kim, Y.J., Gorman, G.C., Huey, R.B.: Genetic variation and differentiation in two species of the fossorial African skink Typhlosaurus (Sauria: Scincidae). Herpetologica 34 (1978) 192-194.

225. Kim, Y.J., Gorman, G.C., Papenfuss, T., Roychoudhury, A.K.: Genetic relationship and genetic variation in the amphisbaenian genus Bipes. Copeia (1976) 120-124.

226. King, M.-C., Wilson, A.C.: Evolution at two levels in humans and chimpanzees. Science 188 (1975) 107-116.

227. Kirkpatrick,M., Selander,R.K.: Genetics of speciation in lake whitefishes in the Allegash Basin. Evolution 33 (1979) 478-485.

228. Kitching,I.J., (1982) Personal communication

229. Kohn,P.H., Tamarin,R.H.: Selection at electrophoretic loci for reproductive parameters in island and mainland voles. Evolution 32 (1978) 15-28.

230. Kojima,K.-i., Gillespie,J., Tobari,Y.N.: A profile of Drosophila species enzymes assayed by electrophoresis. I. Number of alleles, heterozygosities, and linkage disequilibrium in glucose-metabolizing systems and some other enzymes. Biochem.Genet. 4 (1970) 627-637.

231. Koniuszek,J.W.J., Verkleij,J.A.C.: Genetical variation in two related annual Senecio species occuring on the same habitat. Genetica 59 (1982) 133-137.

232. Kornfield,I.L.: Evolutionary genetics of endemic cichlid fishes (Pisces Cichlidae) in Lake Malawi, Africa. (1974) Personal communication.

233. Kornfield,I.L., Koehn,R.K.: Genetic variation and speciation in New World cichlids. Evolution 29 (1975) 427-437.

234. Kornfield,I.L., Nevo,E.: Likely pre-Suez occurrence of a Red Sea fish Aphanius dispar in the Mediterranean. Nature 264 (1976) 289-291.

235. Kornfield,I.L., Ritte,U., Richler,C., Wahrman,J.: Biochemical and cytological differntiation among cichlid fishes of the Sea of Galilee. Evolution 33 (1979) 1-14.

236. Kovacic,D.A., Guttman,S.I.: An electrophoretic comparison of genetic variability between eastern and western populations of the opossum (Didelphis virginiana). Am.Midl.Nat. 101 (1979) 269-277.

237. Krepp,S.R., Smith,M.H.: Genic heterozygosity in the 13-year cicada, Magicicada. Evolution 28 (1974) 396-401.

238. Kristiansson,A.C., McIntyre,J.D.: Genetic variation in chinook salmon (Oncorhynchus tshawytscha) from the Columbia River and three Oregon coastal rivers. Trans.Am.Fish.Soc. 5 (1976) 620-623.

239. Laing,C.D., Carmody,G.R., Peck,S.B.: How common are sibling species in cave-inhabiting invertebrates? Amer.Natur. 110 (1976) 184-189.

240. Lakovaara,S., Saura,A.: Genic variation in marginal populations of Drosophila subobscura. Hereditas 69 (1971) 77-82.

241. Lakovaara,S., Saura,A.: Genetic variation in natural populations of Drosophila obscura. Genetics 69 (1971) 377-384.

242. Langley,C.H., Tobari,Y.N., Kojima,K.-i.: Linkage disequilibrium in natural populations of Drosophila melanogaster. Genetics 78 (1974) 921-936.

243. Larson,A.: Paedomorphosis in relation to rates of morphological and molecular evolution in the salamander Aneides flavipunctatus (Amphibia, Plethodontidae). Evolution 34 (1980) 1-17.

244. Larson, A.: A reevaluation of the relationship between genome size and genetic variation. Amer.Natur. 118 (1981) 119-120

245. Larson, A., Highton, R.: Geographic protein variation and divergence in the salamanders of the Plethodon welleri group (Amphibia, Plethodontidae). Syst.Zool. 27 (1978) 431-448.

246. Lavee, D.: Ph.D. Thesis, The Hebrew University, Jerusalem, Israel (1981).

247. Lavie, B., Nevo, E.: Genetic diversity in marine molluscs: A test of the niche-width variation hypothesis. Marine Ecology 2 (1981) 335-342.

248. Lavigne, D.M., Bogart, J.P., Downer, R.G.H., Dawzmann, R.G., Barchard, W.W. Earle, M.: Genetic uniformity in northwest Atlantic harp seal. In: The Harp Seal, Lavigne, D.M., Ronald, K., Stewaart, R.E.A., (eds.) (1982) Dr. W.Jonk, b.v. Publishers, The Hugue, Netherland, (in press).

249. Lawson, R., Dessauer, H.C.: Biochemical genetics and systematics of garter snakes of the Thamnophis elegans-couchii-ordinoides complex. Occas. Papers 56 (1979) 1-24.

250. Leary, R., Booke, H.E.: Genetic stock analysis of yellow perch from Green Bay and Lake Michigan. Trans.Am.Fish.Soc. 111 (1982) 50-57.

251. Lester, L.J.: Population genetics of penaeid shrimp from the Gulf of Mexico. J.Hered. 70 (1979) 175-180.

252. Lester, L.J., Selander, R.K.: Population genetics of haplodiploid insects. Genetics 92 (1979) 1329-1345.

253. Levin, D.A.: Genic heterozygosity and protein polymorphism among local populations of Oenothera biennis. Genetics 79 (1975) 477-491.

254. Levin, D.A.: Interspecific hybridization, heterozygosity and gene exchange in Phlox. Evolution 29 (1975) 37-51.

255. Levin, D.A.: Genetic variation in annual phlox: Self-compatible versus self-incompatible species. Evolution 32 (1978) 245-263.

256. Levin, D.A., Crepet, W.L.: Genetic variation in Lycopodium lucidulum: A phylogenetic relic. Evolution 27 (1974) 622-632.

257. Levin, D.A., Howland, G.P., Steiner, E.: Protein polymorphism and genic heterozygosity in a population of the permanent translocation heterozygote, Oenothera biennis. Proc.Natl.Acad.Sci.USA 69 (1972) 1475-1477.

258. Levy, M., Levin, D.A.: Genic heterozygosity and variation in permanent translocation heterozygotes of the Oenothera biennis complex. Genetics 79 (1975) 493-512.

259. Lewontin, R.C., Hubby, J.L.: A molecular approach to the study of genic heterozygosity in natural populations. II. Amount of variation and degree of heterozygosity in natural populations of Drosophila pseudoobscura. Genetics 54 (1966) 595-609.

260. Lidicker, W.Z.Jr., Sage, R.D., Calkins, D.G.: Biochemical variation in northern sea lions from Alaska. In: Mammalian Population Genetics Smith, M.H., Joule, J.(Eds.), Univ. of Georgia Press, Athens 1981 pp. 231-2

261. Lokki,J., Suomalainen,E., Saura,A., Lankinen,P.: Genetic polymorphism and evolution in parthenogenetic animals. II. Diploid and polyploid Solenobia triquetrella (Lepidoptera: Psychidae). Genetics 79 (1975) 513-525.

262. Loudenslager,E.J.: Variation in the genetic structure of Peromyscus populations. I. Genetic heterozygosity - its relationship to adaptive divergence. Biochem.Genet. 16 (1978) 1165-1179.

263. Loudenslager,E.J., Gall,G.A.E.: Geographic patterns of protein variation and subspeciation in cutthroat trout, Salmo clarki. Syst.Zool. 29 (1980) 27-42.

264. Loudenslager,E.J., Kitchin,R.M.: Genetic similarity of two forms of cutthroat trout, Salmo clarki in Wyoming. Copeia (1979) 673-678.

265. Lynch,J.C., Vyse,E.R.: Genetic variability and divergence in grayling Thymallus arcticus. Genetics 92 (1979) 263-278.

266. Lynch,J.F., Yang,S.Y., Papenfuss,T.J.: Studies of neotropical salamanders of the genus Pseudoeurycea, I: Systematic status of Pseudoeurycea unguidentis. Herpetologica 33 (1977) 46-52.

267. Manchenko,G.P.: Electrophoretic study of genetic variability of protein in sea-stars. Dissertation, Institute of Marine Biology, Vladivostok, (1980) 159 pp. (In Russian).

268. Manchenko, G.P.: Allozymic variation in Araneus ventricosus (Arachnida, Aranei). Isozyme Bulletin 14 (1981) 78.

269. Manchenko,G.P.: Allozymic variation and allele frequencies in tadpoles, juveniles, and adults of Bombina orientalis (Anura). Isozyme Bulletin 14 (1981) 86.

270. Manchenko,G.P.: Allozymic and colour variation in Metridium senile (Anthozoa) from the sea of Japan. (1982), Personal communication.

271. Manchenko,G.P., Balakirev,E.S.: High level of genetic variation in mushrooms. Isozyme Bulletin 13 (1980) 96.

272. Manchenko,G.P., Balakirev,E.S.: Level of genetic variability in marine invertebrates. In: Biology of Shelf Zones of the World Ocean, Kafanov,A.I., (ed.) Vol 2, Vladivostok, (1982) pp. 91-93 (In Russian).

273. Manchenko,G.P., Pudovkin,A.I., Serov,O.L.: The level of genetic variation in sea urchins and sea stars. In: Proceeding of 4th All-Union Colloquium on Echinoderms. Gongadze,G.S., (ed.) Tbilisi (1979) pp. 133-137 (In Russian).

274. Manchenko,G.P., Pudovkin,A.I., Serov,O.L.: Allozymic variation in sea urchin Strongylocentrotus intermedius. (1982) Personal communication.

275. Manlove,M.N., Avise,J.C., Hillestad,H.O., Ramsey,P.R., Smith,M.H., Straney,D.O.: Starch gel electrophoresis for the study of population genetics in white-tailed deer. In: Proc. 29th Ann. Conf. S.E. Game and Fish Comm. Rogers,W.A. (ed.) St. Louis, Mo. 1975, pp. 392-402.

276. Manlove,M.N., Baccus,R., Pelton,M.R., Smith,M.H., Graber,D.: Biochemical variation in the black bear. In: Bears - Their Biology and Management.

C.J. Martinka, K.L. McArthur (eds.), U.S. Government Printing Office Washington, D.C. pp.37-41.

277.Manwell,C.: Enzyme variability in the protochordate Amphioxus. Nature 258 (1975) 606-608.

278.Manwell,C., Baker,C.M.A.: Molecular genetics and avian proteins. XIII. Protein polymorphism in three species of Australian passerines. Aust.J.Biol.Sci. 28 (1975) 545-557.

279.Marinković,D., Ayala,F.J., Andjelković, M.: Genetic polymorphism and phylogeny of Drosophila subobscura. Evolution 32 (1978) 164-173.

280.Marlow,R.W, Patton,J.L.: Biochemical relationships of the Galapagos giant tortoises (Geochelone elephantopus). J.Zool.,Lond. 195 (1981) 413-422.

281.Marshall,D.R., Allard,R.W.: Isozyme polymorphisms in natural populations of Avena fatua and A. barbata. Heredity 25 (1970) 373-382.

282.Mascarello,J.T.: Chromosomal, biochemical, mensural, penile, and cranial variation in desert woodrats (Neotoma lepida). J.Mamm. 59 (1978) 477-495.

283.Massey,D.R., Joule,J.: Spatial-temporal changes in genetic composition of deer mouse populations. In: Mammalian Population Genetics, Smith,M.H., Joule,J. (Eds.), Univ. of Georgia Press (Athens) 1981, pp. 180-201.

284.Mastro,E., Chow,V., Hedgecock,D.: Littorina scutulata and Littorina plena: Sibling species status of two prosobranch gastropod species confirmed by electrophoresis. The Veliger 24 (1982) 239-246.

285.Matthaeis,De E., Allegrucci,G., Caccone,A., Cesaroni,D., Cobolli Sbordoni, M., Sbordoni,V.: Genetic differentiation between Penaeus kerathurus and P. japonicus (Crustacea, Decapoda). (1982) Personal communication.

286.Matthaeis,De E., Colognola,R., Sbordoni,V., Cobolli Sbordoni,M., Pesce, G.L.: Genetic differentiation and variability in cave dwelling and brackish water populations of Mysidacea (Crustacea). (1982) Personal communication.

287.Matthews,T.C.: Biochemical polymorphism in populations of the Argentine toad Bufo arenarum. Copeia (1975) 454-465.

288.May,B., Holbrook,F.R.: Absence of genetic variability in the green peach aphid, Myzus persicae (Hemiptera: Aphididae). Ann.Entomol.Soc.Am. 71 (1978) 809-812.

289.May,B., Utter,F.M., Allendorf,F.M.: Biochemical genetic variation in pink and chum salmon. J.Hered. 66 (1975) 227-232.

290.Mayer, Von,W.: Elektrophoretische untersuchungen an europaischen arten der gattungen Lacerta und Podarcis. III. Podarcis tiliguerta - art oder unterart? Zool.Anz. 207 (1981) 151-157.

291.Mayer Von,W., Tiedemann,F.: Elektrophoretische untersuchungen an europaeischen arten der gattungen Lacerta und Podarcis. I. Die Podarcis - formen der griechischen inseln Milos und Skiros. Z.Zool.Syst.Evolut.-Forsch. 18 (1980) 147-152.

292.McClenaghan,L.R.Jr.,: The genetic structure of an isolated population of Sigmodon hispidus from the Lower Colorado River Valley. J.Mamm. 61 (1980) 304-307.

293.McClenaghan,L.R.Jr., Gaines,M.B.: Genic and morphological variability in central and marginal populations of Sigmodon hispidus. In: Mammalian Population Genetics, Smith,M.H., Joule,J., (eds.) Univ. of Georgia Press (Athens) 1981, pp. 202-213.

294.McCracken,G.F., Selander,R.K.: Self-fertilization and monogenic strains in natural populations of terrestrial slugs. Proc.Natl.Acad.Sci. USA 77 (1980) 684-688.

295.McDermid,E.M., Ananthakrishnan,R., Agar,N.S.: Electrophoretic investigation of plasma and red cell proteins and enzymes of Macquarie Island elephant seals. Anim.Blood Grps.Biochem.Genet. 3 (1972) 85-94.

296.McKinney,C.O., Selander,R.K., Johnson,W.E., Yang,A.S.: XV Genetic variation in the side-blotched lizard (Uta stansburiana). In: Studies in Genetics VII, Univ. Texas Publ. 7213, (July, 1972), pp. 307-318.

297.McLeod,M.J., Eshbaugh,W.H., Guttman,S.I.: An electrophoretic study of Capsicum (Solanaceae): The purple flowered taxa. Bulletin of the Torrey Botanical Club 106 (1979) 326-333.

298.McLeod,M.J., Eshbaugh,W.H., Guttman,S.I.: A preliminary biochemical systematic study of the genus Capsicum-Solanaceae. In: Biology and Taxonomy of the Solanaceae, Hawkes,J.G., Lester,R.N., Skelding,A.D., (eds.) Linnean Society Symposium Series No.7, (1979) pp. 701-714.

299.McWalter,D.B., Hebert,P.D.N.: Genetic variation in arctic Daphnia reproducing by obligate parthenogenesis. (1982) Personal communication.

300.Menken,S.B.J.: Allozyme polymorphism and the speciation process in small ermine moths (Lepidoptera, Yponomeutidae). Studies in Yponomeuta 2. Rijiksuniversiteit te Leiden, Holand (1980).

301.Menzies.R.A., Kushlan,J., Dessauer,H.C.: Low degree of genetic variability in the American alligator (Alligator mississippiensis). Isozyme Bulletin 12 (1979) 61.

302.Merkle,D.A., Guttman,S.I.: Geographic variation in the cave salamander Eurycea lucifuga. Herpetologica 33 (1977) 313-321.

303.Merkle,D.A., Guttman,S.I., Nickerson,M.A.: Genetic uniformity throughout the range of the hellbender Cryptobranchus alleganiensis. Copeia (1977) 549-553.

304.Merrit,R.B., Rogers,J.F., Kurz,B.J.: Genic variability in the longnose dace Rhinichthys cataractae. Evolution 32 (1978) 116-124.

305.Metcalf,R.A., Marlin,J.C., Whitt,G.S.: Low levels of genetic heterozygosity in Hymenoptera. Nature 257 (1975) 792-794.

305a.Metcalf,R.A.: Personal communication (1982).

306.Mickevich,M.F., Johnson,M.S.: Congruence between morphological and allozyme data in evolutionary inference and character evolution. Syst. Zool. 25 (1976) 260-270.

307.Millet,M.C., Britton-Davidian,J., Orsini,P.: Genetique biochimique comparee de Microtus cabrerae Thomas, 1906 et de trois autres especes d'Arvicolidae mediterraneens. Mammalia 46 (1982) 381-388.

308.Mitton,J.R., Koehn,R.K.: Genetic organization and adaptive response of allozymes to ecological variables in Fundulus heteroclitus. Genetics 79 (1975) 97-111.

309.Moffit,C.M., Leary,R., Booke,H.E.: Electrophoretic comparisons of American shad (Alosa sapidissima from the Connecticut and Hudson Rivers. (1982) Personal communication.

310.Montanucci,R.R., Axtell,R.W., Dessauer,H.C.: Evolutionary divergence among collared lizards (Crotaphytus), with comments on the status of Gambelia. Herpetologica 31 (1975) 336-347.

311.Mork,J., Reuterwall,C., Ryman,N., Stahl,G.: Genetic variation in Atlanta cod (Gadus morhua L.): A quantitative estimate from a Norwegian coastal population. Hereditas 96 (1982) 55-61.

312.Mulley,J.C.: Polymorphism in the gastropod Austrocochlea constricta (Prosobranchia). Isozyme Bulletin 14 (1981) 89.

313.Mulley,J.C., Latter,B.D.H.: Genetic variation and evolutionary relationships within a group of thirteen species of penaeid prawns. Evolution 34 (1980) 904-916.

314.Nadler,C.F., Hoffmann,R.S.: Patterns of evolution and migration in the arctic ground squirrel, Spermophilus parryii (Richardson). Can.J.Zool. 55 (1977) 748-758.

315.Nadler,C.F., Hoffmann,R.S., Vorontsov,N.N., Koeppl,J.W., Deutsch,L., Sukernik,R.I.: Evolution in ground squirrels. II. Biochemical comparisons in Holarcic populations of Spermophilus. Z.Saugetierkunde 47 (1982) 198-215.

316.Nair,P.S., Brncic,D.: Allelic variations within identical chromosomal inversions. Amer.Natur. 105 (1971) 291-294.

317.Nair,P.S., Brncic,D., Kojima,K.: Isozyme variation and evolutionary relationships in the mesophragmatica species group of Drosophila. Univ. Texas Publ. 7103 (1971) pp. 17-28.

318.Nascetti,G., Grappelli,C., Bullini,L., Montalenti,S.G.: Ricerche sul differenziamento genetico di Ascaris lumbricoides e Ascaris suum. Academia Nazionale dei Lincei. Serie VIII, Vol.LXVII (1979) 457-465.

319.Nascetti,G., Tizi,L., Bullini,L.: Differenziazione biochimica e variabilita genetica in due popolazioni simpatriche di Apodemus sylvaticus (L., 1758) e Apodemus flavicollis (Melchior, 1834) (Rodentia, Muridae). Lincei - Rend.Sc.Fis.Mat. e Nat. - Vol. LXVII (1979) 131-136.

320.Nei,M., Roychoudhury,A.K.: Genetic relationship and evolution of human races. Evol.Biol. 14 (1981) 1-59.

321.Nelson,K., Hedgecock,D.: Enzyme polymorphism and adaptive strategy in the decapod Crustacea. Amer.Natur. 116 (1980) 238-280.

322. Nemeth,S.T., Tracey,L.T.: Allozyme variability and relatedness in six crayfish species. J.Hered. 70 (1979) 37-43.

323. Nevo,E.: Adaptive strategies of genetic systems in constant and varying environments. In: Population Genetics and Ecology, Karlin,S., Nevo,E., (eds.) Academic Press, New York (1976) pp. 141-158.

324. Nevo,E.: Genetic variation in constant environments. Experientia 32 (1976) 858-859.

325. Nevo,E.: Genetic variation in natural populations: Patterns and theory. Theor.Popul.Biol. 13 (1978) 121-177.

326. Nevo,E.: Genetic variation and climatic selection in the lizard Agama stellio in Israel and Sinai. Theor.Appl.Genet. 60 (1981) 369-380.

327. Nevo,E.: Genetic differentiation and speciation in spiny mice, Acomys. Act.Zool.Fenn. (1983) In press.

327a. Nevo,E.: Genetic structure and differentiation during speciation in fossorial gerbil rodents. Mammalia, 46 (1982) 523-530.

328. Nevo,E.: Genetic summary of 2 species and 4 populations of Patella. (unpublished data).

328a. Nevo,E., Alkalay, C., Blondheim, S. Unpublished data.

329. Nevo,E., Bar,Z.: Natural selection of genetic polymorphisms along climatic gradients. In: Population Genetics and Ecology, Karlin,S., Nevo,E., (eds.), Academic Press New York, (1975) pp. 159-184.

330. Nevo,E., Bar-El,C., Bar,Z.: Genetic diversity, climatic selection and speciation of Sphincterochila landsnails in Israel. Biol.J.Linn.Soc. (1983) In press.

331. Nevo,E., Bar-El,C., Bar,Z., Beiles,A.: Genetic structure and climatic correlates of desert landsnails. Oecologia 48 (1981) 199-208.

332. Nevo,E., Golenberg,E., Beiles,A., Brown,A.H.D., Zohary,D.: Genetic diversity and environmental associations of wild wheat, Triticum dicoccoides, in Israel. Theor.Appl.Genet. 62 (1982) 241-254.

333. Nevo,E., Kim,Y.J., Shaw,C.R., Thaeler,C.S.Jr.: Genetic variation, selection and speciation in Thomomys talpoides pocket gophers. Evolution 28 (1974) 1-23.

334. Nevo,E., Shaw,C.R.: Genetic variation in a subterranean mammal, Spalax ehrenbergi. Biochem.Genet. 7 (1972) 235-241.

335. Nevo,E., Shimony,T., Libni,M.: Pollution selection of allozyme polymorphisms in barnacles. Experientia 34 (1978) 1562-1564.

336. Nevo,E., Yang,S.Y.: Genetic diversity and climatic determinants of tree frogs in Israel. Oecologia (Berl.) 41 (1979) 47-63.

337. Nevo,E., Zohary,D., Brown,A.H.D., Haber,M.: Genetic diversity and environmental associations of wild barley, Hordeum spontaneum, in Israel. Evolution 33 (1979) 815-833.

338. Nottebohm,F., Selander,R.K.: Vocal dialects and gene frequencies in the chingolo sparrow (Zonotricia capensis). Condor 74 (1972) 137-143.

339. Nozawa,K., Shotake,T., Kawamoto,Y.: Population genetics of Japanese monkeys: II. Blood protein polymorphisms and population structure. Primates 23 (1982) 252-271.

340. Nozawa,K., Shotake,T., Okura,Y.: Blood protein polymorphisms and population structure of the Japanese macaque, Macaca fuscata fuscata. Isozymes: Genetics and Evolution, Yale University, 1975. pp.225-241.

341. Nozawa,K., Shotake,T., Ohkura,Y., Kitajima,M., Tanabe,Y.: Genetic variations within and between troops of Macaca fuscata fuscata. In: Contemporary Primatology, Kondo,S., Kawai,M., Inuyama,A.E. (Eds.) 5th Int.Cong.Primat., Nagoya 1974, pp. 75-89, Karger, Basel 1975.

342. Nozawa,K., Shotake,T., Ohkura,Y., Tanabe,Y.: Gentic variations within and between species of Asian macaques. Japan.J.Genetics 52 (1977) 15-30.

343. Nygren,J.: Allozyme variation in natural populations of field vole (Microtus agrestis L.). II. Survey of an isolated island population. Hereditas 93 (1980) 107-114.

344. Nygren,J., Rasmuson,M.: Allozyme variation in natural populations of field vole (Microtus agrestis L.). Hereditas 92 (1980) 65-72

345. O'Brien,S.J., Gail,M.H., Levin,D.L.: Correlative genetic variation in natural populations of cats, mice and men. Nature 288 (1980) 580-583.

346. Ochman,H., Stille,B., Niklasson,M., Selander,R.K., Templeton,A.R.: Evolution of clonal diversity in the parthenogenetic fly Lonchoptera dubia. Evolution 34 (1980) 539-547.

347. Palmour,R.M., Cronin,J.E., Childs,A., Grunbaum,B.W.: Studies of primate protein variation and evolution: microelectrophoretic detection. Biochem.Genet. 18 (1980) 793-808.

348. Pamilo,P., Varvio-Aho,S.-L., Pekkarinen,A.: Low enzyme gene variability in Hymenoptera as a consequence of haplodiploidy. Hereditas 88 (1978) 93-99.

349. Parker,E.D.Jr., Selander,R.K.: The organization of genetic diversity in the parthenogenetic lizard Cnemidophorus tesselatus. Genetics 84 (1976) 791-805.

350. Pashtan,A.: Genetic variability and colonization success in two species of the genus Cerithium (Mollusca: Gastropoda). A thesis submitted for MSc. Degree, Department of Genetics, The Hebrew University, Jerusalem, Israel, (1978).

351. Pasteur,G., Pasteur,N., Orsini,J.-P.G.: On genetic variability in a population of the widespread gecko Hemidactylus brooki. Experientia 34 (1978) 1557-1558.

352. Pasteur,N., Worms,J., Tohari,M., Iskandar,D.: Genetic differentiation in Indonesian and French rats of the subgenus Rattus. Biochem.Syst.Ecol. 10 (1982) (in press).

353. Patton,J.L., Feder,J.H.: Genetic divergence between populations of the pocket gopher, Thomomys umbrinus (Richardson). Z.f.Saugetiere Knude Bd. 43 (1978) 17-30.

354. Patton, J.L., MacArthur, H., Yang, S.Y.: Systematic relationships of the four-toed populations of Dipodomys heermanni. J.Mamm. 57 (1976) 159-163.

355. Patton, J.L., Selander, R.K., Smith, M.H.: Genic variation in hybridizing populations of gophers (Genus Thomomys). Syst.Zool. 21 (1972) 263-270.

356. Patton, J.L., Sherwood, S.W., Yang, S.Y.: Biochemical systematics of chaetodipine pocket mice, genus Perognathus. J.Mamm. 62 (1981) 477-492.

357. Patton, J.L., Yang, S.Y.: Genetic variation in Thomomys bottae pocket gophers. Macrogeographic patterns. Evolution 31 (1977) 697-720.

358. Patton, J.L., Yang, S.Y., Myers, P.: Genetic and morphologic divergence among introduced rat populations (Rattus rattus) of the Galapagos Archipelago, Ecuador. Syst.Zool. 24 (1975) 296-310.

359. Peck, C.T., Biggers, C.J.: Electrophoretic analysis of plasma proteins of Mississippi Peromyscus. J.Hered. 66 (1975) 237-241.

360. Penney, D.F., Zimmerman, E.G.: Genic divergence and local population differentiation by random drift in the pocket gopher genus Geomys. Evolution 30 (1976) 473-483.

361. Phelps, S.R., Allendorf, F.W.: Genetic comparison of upper Missouri cutthroat trout to other Salmo clarki lewisi populations. Proc.Mont. Acad.Sci. 41 (1982) 14-22.

362. Phelps, S.R., Allendorf, F.W.: Genetic identity of pallid and shovelnose sturgeon (Scaphirhynchus albus and S. platorynchus). (1982), Personal communication.

363. Pinsker, W., Lankinen, P., Sperlich, D.: Allozyme and inversion polymorphism in a central European population of Drosophila subobscura. Genetica 48 (1978) 207-214.

364. Pope, M.H., Highton,, R.: Geographic genetic variation in the Sacramento Mountain salamander, Aneides hardii. Journal of Herpetology 14 (1980) 343-346.

365. Prakash, S.: Genic variation in a natural population of Drosophila persimilis. Proc.Natl.Acad.Sci. 62 (1969) 778-784.

366. Prakash, S.: Patterns of gene variation in central and marginal populations of Drosophila robusta. Genetics 75 (1973) 347-369.

367. Prakash, S.: Gene polymorphism in natural populations of Drosophila persimilis. Genetics 85 (1977) 513-520.

368. Prakash, S.: Genetic divergence in closely related sibling species Drosophila pseudoobscura, Drosophila persimilis and Drosophila miranda. Evolution 31 (1977) 14-23.

369. Prakash, S.: Further studies of gene polymorphism in the mainbody and geographically isolated populations of Drosophila pseudoobscura. Genetics 85 (1977) 713-719.

370. Prakash,S., Lewontin,R.C., Hubby,J.L.: A molecular approach to the study of genic heterozygosity in natural populations IV. Patterns of genic variation in central, marginal and isolated populations of Drosophila pseudoobscura. Genetics 61 (1969) 841-858.

371. Ralin,D.B., Kloek,G.P.: Genic variability in some Kentucky populations of seventeen-year periodical cicadas (Homoptera:Magicicada) Trans.Ky. Acad.Sci. 39 (1978) 111-116.

372. Ralin,D.B., Selander,R.K.: Evolutionary genetics of diploid-tetraploid species of treefrogs of the genus Hyla. Evolution 33 (1979) 595-608.

373. Ramsey,P.R., Avise,J.C., Smith,M.H., Urbston,D.F.: Biochemical variation and genetic heterogeneity in South Carolina deer populations. J.Wildl.Manage. 43 (1979) 136-142.

374. Rice,M.C., Garner,M.B., O'Brien,S.J.: Genetic diversity in leukemia-prone feral house mice infected with murine leukemia virus. Biochem. Genet. 18 (1980) 915-928.

375. Rice,M.C., O'Brien,S.J.: Genetic variance of laboratory outbred Swiss mice. Nature 283 (1980) 157-161.

376. Richardson,B.J., Rogers,P.M., Hewitt,G.M.: Ecological genetics of the wild rabbit in Australia. II. Protein variation in British, French and Australian rabbits and the geographical distribution of the variation in Australia. Aust.J.Biol.Sci. 33 (1980) 371-383.

377. Richmond,R.C.: Enzyme variability in the Drosophila willistoni group. III. Amounts of variability in the superspecies, D. paulistorum. Genetics 70 (1972) 87-112.

378. Richmond,R.C., Sabath,M.D., Jones,J.M.: Patterns of allozyme polymorphism in three species of the Drosophila affinis subgroup. (1977) Personal communication.

379. Rick,C.M., Fobes,J.F.: Allozymes of Galapagos tomatoes: Polymorphism, geographic distribution, and affinities. Evolution 29 (1975) 443-457.

380. Ritte,U., Pashtan,A.: Extreme levels of genetic variability in two Red Sea Cerithium species (Gastropoda: Cerithidae). Evolution 36 (1982) 403-407.

381. Rodino,E., Comparini,A.: Genetic variability in the European eel, Anguilla anguilla L. In: Marine Organisms: Genetic, Ecology and Evolution, Battaglia,B., Beardmore,J., (eds.), Plenum Press New York, (1978) pp. 389-423.

382. Rodino,E., Comparini,A.: Biochemical polymorphism in teleosts. The eel problem. Boll.Zool. 45 (1978) 47-61.

383. Rose,R.K., Gaines,M.S.: Relationships of genotype, reproduction, and wounding in Kansas prairie voles. In: Mammalian Population Genetics, Smith,M.H., Joule,J., (eds.), Univ. of Georgia Press (Athens) (1981). pp.160-179.

384. Ruddle,F.H., Roderick,T.H., Shows,T.B., Weigl,P.G., Chipman,R.K., Anderson,P.K.: Measurment of genetic heterogeneity by means of enzyme polymorphisms in wild populations of the mouse. J.Hered. 60 (1969) 321-322.

385. Ryman,N., Beckman,G., Bruun-Petersen,G., Reuterwall,C.: Variability of red cell enzymes and genetic implications of management policies in Scandinavian moose (Alces alces). Hereditas 85 (1977) 157-162.

386. Ryman,N., Reuterwall,C., Nygren,K., Nygren,T.: Genetic variation and differentiation in Scandinavian moose (Alces alces): are large mammals monomorphic? Evolution 34 (1980) 1037-1049.

387. Ryman,N., Stahl,G.: Genetic perspectives of the identification and conservation of Scandinavian stocks of fish. Can.J.Fish.Aquat.Sci. 38 (1981) 1562-1575.

388. Sage,R.D., Selander,R.K.: Trophic radiation through polymorphism in cichlid fishes. Proc.Nat.Acad.Sci.USA 72 (1975) 4669-4673.

389. Salmon,M., Ferris,S.D., Johnston,D., Hyatt,G., Whitt,G.S.: Behavioral and biochemical evidence for species distinctiveness in the fiddler crabs, Uca speciosa and U. spinicarpa. Evolution 33 (1979) 182-191.

390. Sattler,P.W.: Genetic relationships among selected species of North American Scaphiopus. Copeia (1980) 605-610.

391. Sattler,P.W., Guttman,S.I.: An electrophoretic analysis of Thamnophis sirtalis from western Ohio. Copeia (1976) 352-356.

392. Saura,A.: Genic variation in Scandinavian populations of Drosophila bifasciata. Hereditas 76 (1974) 161-172.

393. Saura,A., Halkka,O., Lokki,J.: Enzyme gene heterozygosity in small island populations of Philaenus spumarius (L.) (Homoptera). Genetica 44 (1973) 459-473.

394. Saura,A., Lakovaara,S., Lokki,J., Lankinen,P.: Genic variation in central and marginal populations of Drosophila subobscura. Hereditas 75 (1973) 33-46.

395. Sbordoni,V.: Advances in speciation of cave animals. Personal communication.

396. Sbordoni,V., Allegrucci,G., Caccone,A., Cesaroni,D., Cobolli-Sbordoni,M., Matthaeis, De E.: Genetic variability and divergence in cave populations of Troglophilus cavicola and T. andreinii (Orthoptera, Rhaphidophoridae). Evolution 35 (1981) 226-233.

397. Sbordoni,V., Caccone,A., Matthaeis,De E., Cobolli Sbordoni,M.: Biochemical divergence between cavernicolous and marine Sphaeromidae and the Mediterranean salinity crisis. Personal communication.

398. Sbordoni,V., Cobolli Sbordoni,M., Matthaeis,De E.: Divergenza genetica tra popolazioni e specie ipogee ed epigee di Niphargus (Crustacea, Amphipoda). Estratto de "Lavori della Societa Italiana di Biogeografia" nuova serie, vol.IV. Sienna 1976.

399. Schaal,B.A.: Population structure and local differentiation in Liatris cylindracea. Amer.Natur. 109 (1975) 511-528.

400. Schaumar,N., Rojas-Rousse,D., Pasteur,N.: Allozyme polymorphism in the parasitic hymenoptera Diadromus pulchellus WSM. (Ichneumonidae). Genet.Res.,Camb. 32 (1978) 47-54.

401. Schmidly, D.J., Zimmerman, E.G.: Genetic and chromosomal affinities of *Peromyscus hooperi*. In Preparation (1982).

402. Schmitt, L.H.: Genetic variation in isolated populations of the Australian bush-rat, *Rattus fuscipes*. Evolution 32 (1978) 1-14.

403. Schoen, D.J.: Genetic variation and the breeding system of *Gilia achilleifolia*. Evolution 36 (1982) 361-370.

404. Schopf, T.J.M., Murphy, L.S.: Protein polymorphism of the hybridizing seastars *Asterias forbesi* and *Asterias vulgaris* and implications for their evolution. Biol.Bull. 145 (1973) 589-597.

405. Schwartz, O.A. Armitage, K.B.: Social substructure and dispersion in yellow-bellied marmot (*Marmota flaviventris*), In: Mammalian Population Genetics, Smith, M.N., Joule, J. (eds.) Univ. of Georgia Press (Athens) (1981) pp. 139-159.

406. Seidel, M.E., Lucchino, R.V.: Allozymic and morphological variation among the musk turtles *Sternotherus carinatus*, *S. depressus* and *S. minor* (*Kinosternidae*). Copeia (1981) 119-128.

407. Seidel, M.E., Reynolds, S.L., Lucchino, R.V.: Phylogenetic relationships among musk turtles (genus *Sternotherus*) and genic variation in *Sternotherus odoratus*. Herpetologica 37 (1981) 161-165.

408. Selander, R.K.: Biochemical polymorphism in populations of the house mouse and old-field mouse. Symp.Zool.Soc.Lond. 26 (1970) 73-91.

409. Selander, R.K., Hudson, R.O.: Animal population structure under close inbreeding: The land snail *Rumina* in Southern France. Amer.Natur. 110 (1976) 695-718.

410. Selander, R.K., Hunt, W.G., Yang, S.Y.: Protein polymorphism and genic heterozygosity in two European subspecies of the house mouse. Evolution 23 (1969) 379-390.

411. Selander, R.K., Johnson, W.E.: Genetic variation among vertebrate species, in "Proc. XVII Int. Cong. Zool.," (1972) pp. 1-31; and Ann.Rev.Ecol. Syst. 4 (1973) 75-91.

412. Selander, R.K., Kaufman, D.W.: Self-fertilization and genetic population structure in a colonizing land snail. Proc.Nat.Acad.Sci.USA 70 (1973) 1186-1190.

413. Selander, R.K., Kaufman, D.W.: Genetic structure of populations of the brown snail (*Helix aspersa*). I. Microgeographic variation. 29 (1975) 385-401.

414. Selander, R.K., Kaufman, D.W., Baker, R.J., Williams, S.L.: Genic and chromosomal differentiation in pocket gophers of the *Goemys bursarius* group. Evolution 28 (1975) 557-564.

414a. Selander, R.K., Levin, B.R.: Genetic diversity and structure in *Escherichia coli* populations. Science 210 (1980) 545-547.

415. Selander, R.K., Smith, M.H., Yang, S.Y., Johnson, W.E., Gentry, J.B.: IV. Biochemical polymorphism and systematics in the genus *Peromyscus*.

I. Variation in the old field mouse (Peromyscus polionotus). Studies in Genetics VI Univ. Texas Publ. 7103 (1971). pp. 49-90.

416. Selander,R.K., Yang,S.Y.: Protein polymorphism and genic heterozygosity in a wild population of the house mouse (Mus musculus). Genetics 63 (1969) 653-667.

417. Selander,R.K., Yang,S.Y., Lewontin,R.C., Johnson,W.E.: Genetic variation in the horseshoe crab (Limulus polyphemus), a phylogenetic "relic". Evolution 24 (1970) 402-414.

418. Sene.F.M., Carson,H.L.: Genetic variation in Hawaiian Drosophila. IV. Allozymic similarity between D. silvestris and D. Heyteroneura from the island of Hawaii. Genetics 86 (1977) 187-198.

419. Shaklee,J.B., Tamaru,C.S.: Biochemical and morphological evolution of Hawaiian bonefishes (Albula). Syst.Zool. 30 (1981) 125-146.

420. Shami,S.A., Beardmore,J.A.: Genetic studies of enzyme variation in the guppy Poecilia reticulata (Peters). Genetica 48 (1978) 67-73.

421. Shotake,T.: Serum albumin and erythrocyte adenosine deaminase polymorphisms in Asian macaques with special reference to taxonomic relationships among Macaca assamensis, M .radiata, and M. mulatta. Primates 20 (1979) 443-451.

422. Shotake,T.: Genetic variability within and between herds of gelada baboons in Central Ethiopian Highland. Antropol. Conteporanea 3 (1980) 270.

423. Shotake,T.: Population genetical study of natural hybridization between Papio anubis and P.hamadryas. Primates 22 (1981) 285-308.

424. Shotake,T., Nozawa,K., Tanabe,Y.: Blood protein variations in baboons. I. Gene exchange and genetic distance between Papio anubis, Papio hamadryas and their hybrid. Japan.J.Genetics 52 (1977) 223-237.

425. Sibley,C.G., Corbin,K.W.: Ornithological field studies in the great plains and Nova Scotia. Discovery 6 (1970) 3-6.

426. Siebenaller,J.F.: Genetic variability in deep-sea fishes of the genus Sebastolobus (Scorpaenidae). In: Marine Organisms: Genetics, Ecology and Evolution. Battaglia,B., Beardmore,J., (eds.) Plenum Press New York, (1978) pp. 95-122.

427. Siebenaller,J.F.: Genetic variation in deep-sea invertebrate populations: The bathyal gastropod Bathybembix bairdii. Marine Biology 47 (1978) 265-275.

428. Simon,C.M.: Evolution of periodical cicadas: Phylogenetic inferences based on allozymic data. Syst.Zool. 28 (1979) 22-39.

429. Simonsen,V.: Electrophoretic variation in large mammals. II. The red fox, Vulpes vulpes, the stoat, Mustela erminea, the weasel, Mustela nivalis, the pole cat, Mustela putorius, the pine marten, Martes martes, the beech marten, Martes foina, and the badger, Meles meles. Hereditas 96 (1982) 299-305.

430. Simonsen,V., Allendorf,F.W., Eanes,W.F., Kapel,F.O.: Electrophoretic variation in large mammals. III. The ringed seal, Pusa hispida, the

harp seal, Pagophilus groenlandicus, and the hooded seal, Cystophora cristata. Hereditas 97 (1982) 87-90.

431. Simonsen,V., Born,E.W., Kristensen,T.: Electrophoretic variation in large mammals. IV. The Atlantic walrus, Odobenus rosmarus (L.). Hereditas 97 (1982) 91-94.

432. Simonsen,V., Kapel,F., Larsen,F.: Electrophoretic variation in the minke whale, Balaenoptera acutorostrata Lacepede. Rep.Int.Whal.Commn. 32 (1982) 275-278.

433. Singh,R.S., Coulthart,M.B.: Genic variation in abundant soluble proteins of Drosophila melanogaster and Drosophila pseudoobscura. Genetics (1982) In press.

434. Singh,R.S., Hickey,D.A., David,J.: Genetic differentiation between geographically distant populations of Drosophila melanogaster. Genetics (1982) In press.

435. Skibinsky,D.O.F., McNee,A.R., Beardmore,J.A.: Protein variation in the marine bivalve Scrobicularia plana. Anim.Blood Grps bioch.Genet.9 (1978) 223-228.

436. Sluss,T.P., Graham,H.M.: Allozyme variation in natural populations of Heliothis virescens. Ann.Entomol.Soc.Am. 72 (1979) 317-322.

437. Smith,J.K., Zimmerman,E.G.: Biochemical genetics and evolution of North American blackbirds, family Icteridae. Comp.Biochem.Physiol. 53B (1976) 319-324.

438. Smith,M.F.: Geographic variation in genic and morphological characters in Peromyscus californicus. J.Mamm. 60 (1979) 705-722.

439. Smith,M.F.: Relationships between genetic variability and niche dimensions among coexisting species of Peromyscus. J.Mamm. 62 (1981) 273-285.

440. Smith,M.F., Patton,J.L.: Relationships of pocket gopher (Thomomys bottae) populations of the lower Colorado River. J.Mamm. 61 (1980) 681-696.

441. Smith,M.H., Britton,J., Burke,P., Chesser,R.K., Smith,M.W., Hagen,J.: Genetic variability in Corbicula, an invading species. Proceedings, First International Corbicula Symposium. Texas Christian University (1977) 243-248.

442. Smith,M.H., Hillestad,H.O., Manlove,M.N., Straney,D.O., Dean,J.M.: Management implications of genetic variability in loggerhead and green sea turtles. XIIIth Congress of Game Biologists. pp. 302-312.

443. Smith,M.H., Manlove,M.N., Joule,J.: Genetic organization in space and time: Spatial and temporal dynamics of the genetic organization of small mammal population. In: Population of Small Mammals Under Natural Conditions. Snyder,D.P., (ed.), (1978) pp. 99-113. Univ. of Pittsburgh, Pittsburgh.

444. Smith,M.H., Selander,R.K..,Johnson,W.E.: Biochemical polymorphism and systematics in the genus Peromyscus. III. Variation in the Florida deer mouse (Peromyscus floridanus) a Pleistocene relict. J.Mamm. 54 (1973) 1-13.

445. Smith,M.W., Smith,M.H., Brisbin,I.L.Jr.: Genetic variability and domestication in swine. J.Mamm. 61 (1980) 39-45.

446. Smith,P.J., McKoy,J.: Low levels of genetic heterozygosity in Jasus edwardsii. Isozyme Bulletin 11 (1978) 68.

447. Snyder,T.P.: Lack of allozymic variability in three bee species. Evolution 28 (1974)687-689.

448. Snyder,T.P.: Electrophoretic characterizations of black flies in the Simulium venustum and verecundum species complexes (Diptera: Simulidae). Can.Ent. 114 (1982) 503-507.

449. Snyder,T.P., Linton,M.C.: Electrophoretic and morphological separation of Prosimilium fuscum and P.mixtum larvae (Diptera: Simulidae). Can.Ent. In press.

450. Somero,G.N., Soule,M.: Genetic variation in marine fishes as a test of the niche-variation hypothesis. Nature 249 (1974) 670-672.

451. Soule,M., Yang,S.Y.: Genetic variation in side-blotched lizards from islands in the Gulf of California. Evolution 27 (1973) 539-600.

452. Stahl,G.: Genetic differentiation among natural populations of Atlantic salmon (Salmo salar) in northern Sweden. Ecol.Bull. (Stockholm) (1980).

453. Steiner,W.W.M.: Enzyme variability in exotic and endemic Drosophila in Hawaii. Genetics s78 (1975).

454. Steiner,W.W.M.: Genetic variation in Hawaiian Drosophila. V. Allozyme variation and events in D. mimica and D. engyochracea. (1975) Personal communication.

455. Steiner,W.W.M., Kitzmiller,J.B., Osterbur,D.L.: Gene differentiation in chromosome races of Anopheles nuneztovari (Gabaldon). Mosquito Systematics 12 (1980) 306-319.

456. Steiner,W.W.M., Kitzmiller,J.B., Osterbur,D.L.: Genetic identity and evolution in Anopheles aquasalis. In: Cytogenetics and Genetics of Vectors (Proc.of a Symp. of the XVI Int. Cong. of Entomology, Japan), Pol,R., Kitzmiller,J.B., Kanda,T., (eds.) Kodansh Scientific, 21-12-2, Otoba, Bunkyo, Tokyo, 112 Japan (1981).

457. Steiner,W.W.M., Sung,K.C., Paik,Y.K.: Electrophoretic variability in island populations of Drosophila simulans and Drosophila immagrans. Biochem.Genet. 14 (1976) 495-506.

458. Stoneking,M., Wagner,D.J., Hildebrand,A.C.: Genetic evidence suggesting subspecific differences between northern and southern populations of brook trout (Salvelinus fontinalis). Copeia (1981) 810-819.

459. Straney,D.O., O'Farreli,M.J., Smith,H.: Biochemical genetics of Myotis californicus and Pipistrellus hesperus from southern Nevada. Mammalia 40 (1976) 344-347.

460. Straney,D.O., Smith,M.H., Greenbaum,I.F., Baker,R.J.: Biochemical genetics In: Biology of the Bats of the New World Family Phyllostomatidae. Part III. Baker R.J., Jones,J.K.Jr., Carter,D.C., (eds.) Spec.Publ. 16 (1979) pp.157-176, Texas Tech Press, Lubbock.

461. Suomalainen, E., Saura, A.: Genetic polymorphism and evolution in parthenogenetic animals. I. Polyploid curculionidae. Genetics 74 (1973) 489-508.

462. Swofford, D.L., Branson, B.A., Sievert, G.A.: Genetic differentiation of cavefish populations (Amblyopsidae). Isozyme Bulletin 13 (1980) 109-110.

463. Taylor, C.E., Gorman, G.C.: Population genetics of a "colonising" lizard: Natural selection for allozyme morphs in Anolis grahami. Heredity 35 (1975) 241-247.

464. Tegelstrom, H., (1982) Personal communication.

465. Tegelstrom, H., Jansson, H., Ryttman, H.: Isozyme differences among three related Larus species (Aves). Hereditas 92 (1980) 117-122.

466. Tegelstrom, H., Larsen, T., (1982) Personal communication.

467. Thaler, L., Bonhomme, F., Britton-Davidian, J., Hamar, M.: The house mouse complex of species: Sympatric occurrence of biochemical groups Mus 2 and Mus 4 in Rumania. Z. Saugetierkunde 46 (1981) 169-173.

468. Tilley, S.G., Merritt, R.B., Wu, B., Highton, R.: Genetic differentiation in salamanders of the Desmognathus ochrophaeus complex (Plethodontidae). Evolution 32 (1978) 93-115

469. Tilley, S.G., Schwerdtfeger, P.M.: Electrophoretic variation in Appalachian populations of the Desmognathus fuscus complex (Amphibia: Plethodontidae) Copeia (1981) 109-119.

470. Tinkle, D.W., Selander, R.K.: Age-dependent allozymic variation in a natural population of lizards. Biochem. Genet. 8 (1973) 231-237.

471. Tracey, M.L., Nelson, K.B.: Allozyme variation in the American lobster, Homarus americanus. Genetics 80 (1975) s81.

472. Tracey, M.L., Nelson, K., Hedgecock, D., Shleser, R.A., Pressick, M.L.: Biochemical genetics of lobsters: Genetic variation and the structure of American lobster (Homarus americanus) populations. J. Fish. Res. Board Can. 32 (1975) 2091-2101.

473. Turner, B.J.: Genetic divergence of death valley pupfish species: biochemical versus morphological evidence. Evolution 28 (1974) 281-294.

474. Tuttle, R.D., Lindahl, R.G.: Gene-enzyme variation in three sympatric morphs of the Othilia (Echinaster) species complex (Echinodermata; Asteroidea). Isozyme Bulletin 12 (1979) 65.

475. Utter, F.M., Allendorf, F.W., Hodgins, H.O.: Genetic variability and relationships in Pacific salmon and related trout based on protein variations. Syst. Zool. 22 (1973) 257-270.

476. Utter, F., Folmar, L.: Protein systems of grass carp: Allelic variants and their application to management of introduced populations. Trans. Am. Fish. Soc. 107 (1978) 129-134.

477. Valentine, J.W., Ayala, F.J.: Genetic variation in Frieleia halli, a deep-sea brachiopod. Deep-Sea Research 22 (1975) 37-44.

478. Valentine,J.W., Ayala,F.J.: Genetic variability in krill. Proc.Nat.Acad. Sci. USA 73 (1976) 658-660.

479. Varvio-Aho,S.-L., Jarvinen,O., Vepsalainen,K.: Enzyme gene variation in three species of water-striders (Gerris) (Heteroptera, Gerridae). Ann. Ent.Fenn. 44 (1978) 87-94.

479a. Varvio-Aho,S.-L., Pamilo,P.: Genetic differentiation of Gerris lacustris populations. Hereditas 90 (1979) 237- 249.

479b. Varvio-Aho,S.-L., Pamilo,P.: Genetic differentiation of northern Finnish water-strider (Gerris) populations. Hereditas 92 (1980) 363-371.

480. Verkleij,J.A.C.: PhD. Thesis, Vrije Universiteit te Amsterdam, (1980).

481. Vrijenhoek,R.C.: Genetics of a sexually reproducing fish in a highly fluctuating environment. Amer.Natur. 113 (1979) 17-29.

482. Vrijenhoek,R.C., Allendorf,F.W.: Protein polymorphism and inheritance in the fish Poecilia mexicana (Poeciliidae). Isozyme Bulletin 13 (1980) 92

483. Vrijenhoek,R.C., Angus,R.A., Schultz,R.J.: Variation and heterozygosity in sexually vs. clonally reproducing populations of Poeciliopsis. Evolution 31 (1977) 767-781.

484. Vuorinen,J., Himberg,M.K.-J., Lankinen,P.: Genetic differentiation in Coregonus albula (L.) (Salmonidae) populations in Finland. Hereditas 94 (1981) 113-121.

485. Wake,D.B., Maxson,L.R., Wurst,G.Z.: Genetic differentiation, albumin evolution, and their biogeographic implications in plethodontid salamanders of California and Southern Europe. Evolution 32 (1978) 529-539.

486. Ward,P.S.: Genetic variation and population differentiation in the Rhytidoponera impressa group, a species complex of ponerine ants (Hymenoptera: Formicidae). Evolution 34 (1980) 1060-1076.

487. Ward,R.D., Beardmore,J.A.: Protein variation in the plaice, Pleuronectes platessa L. Genet.Res.,Camb. 30 (1977) 45-62.

488. Ward,R.D., Galleguillos,R.A.: Protein variation in the plaice, dab and flounder and their genetic relationships. In: Marine Organisms: Genetics, Ecology and Evolution, Battaglia,B., Beardmore,J.A., (eds.) Plenum Press New York (1978) pp. 71-93.

489. Ward,R.D., Galleguilos,R.A.: Systematic Association (Book) (1982) In press.

490. Ward,R.D., McAndrew,B.J., Wallis,G.P.: Enzyme variation in the brook lamprey, Lampetra planeri (Bloch), a member of the vertebrate group Agnatha. Genetica 55 (1981) 67-73.

491. Ward,R.D., Pasteur,N., Rioux,J.A.: Electrophoretic studies on genetic polymorphism and differentiation of phlebotomine sandflies (Diptera: Psychodidae) from France and Tunisia. Annals of Tropical Medicine and Parasitology 75 (1981) 235-245.

492. Webster.T.P., Selander,R.K., Yang,S.Y.: Genetic variability and similarity in the Anolis lizards of Bimini. Evolution 26 (1973) 523-535.

493.Werner Von,M.: Electrophoretic investigations on European species of the genus Lacerta and Podarcis. III. Podarcis tiliguerta - species or subspecies? Zool.Anz.,Jena 207 (1981) 151-157.

494.Wheeler,L.L., Selander,R.K.: XIII. Genetic variation in populations of the house mouse, Mus musculus, in the Hawaiian Islands. Studies in Genetics VII. Univ.Texas Publ. 7213 (1972) pp. 269-296.

495.White,M.M., Turner,B.J.: Genetic differentiation in Goodea, a lake dwelling goodeid fish. Isozyme Bulletin 15 (1982) 131

496.Williams,E.E., Case,S.M., (1982), Personal communication.

497.Winans,G.A.: Geographic variation in the milkfish Chanos chanos. I. Biochemical evidence. Evolution 34 (1980) 558-574.

498.Woodfield,D.G., Scragg,R.F.R., Blake,N.M., Kirk,R.L., McDermid,E.M.: Distribution of blood, serum protein and enzyme groups among the Fuyuge speakers of the Goilala sub-district. Human heredity 24 (1974) 507-519.

499.Woodruff,D.S.: Allozyme variation and genic heterozygosity in the Bahaman pulmonate snail Cerion bendalli. Malacological Review 8 (1975) 47-55.

500.Wooten,M.C., Zimmerman,E.G.: Ecological correlates to genetic variation in minnows of the genus Notropis. (1982) In preparation.

501.Worms,J., Pasteur,N.: Polymorphisme biochimique de la palourde, Venerupis decussata de l'etang du Prevost (France). Oceanologica Acta 5 (1982) 395-397.

502.Yanev,K.P., Wake,D.B.: Genetic differentiation in a relict desert salamander, Batrachoseps campi. Herpetologica 37 (1981) 16-28.

503.Yang,S.Y., Patton,J.L.: Genic variability and differentiation in the Galapagos finches. The Auk 98 (1981) 230-242.

504.Yang,S.Y., Soulè,M.,Gorman,G.C.: Anolis lizards of the eastern Caribbean: a case study in evolution. I. Genetic relationships, phylogeny, and colonization sequence of the roquet group. Syst.Zool. 23 (1974) 387-399.

505.Yang,S.Y., Wheeler,L.L., Bock,I.R.: IX. Isozyme variations and phylogenetic relationships in the Drosophila bipectinata species complex. Studies in Genetics VII. Univ. of Texas Publ. 7213 (1972) pp.213-227.

506.Zera,A.J.: Genetic structure of two species of waterstriders (Gerridae: Hemiptera) with differing degrees of winglessness. Evolution 35 (1981) 218-225.

507.Zimmerman,E.G., Gayden,N.A.: Analysis of genic heterogeneity among local populations of the pocket gopher, Geomys bursarius. In: Mammalian Population Genetics. Smith,M.H., Joule,J., (eds.) Univ. of Georgia Press (Athens) (1981) 272-287.

508.Zimmerman,E.G., Merritt,R.L., Wooten,M.C.: Genetic variation and ecology of stoneroller minnows. Biochem.Syst.Ecol. 8 (1980) 447-453.

509.Zimmerman,E.G., Nejtek,M.E.: Genetics and speciation of three semispecies of Neotoma. J.Mamm. 58 (1977) 391-402.

510. Zouros, E.: Genic differentiation associated with the early stages of speciation in the mulleri subgroup of Drosophila. Evolution 27 (1974) 601-621.

511. Zouros, E., Krimbas, C.B., Tsakas, S., Loukas, M.: Genic versus chromosomal variation in natural populations of Drosophila subobscura. Genetics 78 (1974) 1223-1244.

512. (--) : Geographic variation in venom and blood proteins of Crotalus viridis. 57th Annual Meeting of the American Society of Icthyologists and Herpetologists (1977) Univ. of Florida, Gainesville.

REFERENCES OF THEORETICAL DISCUSSION

Ayala,F.J. 1968. Genotype, environment, and population numbers. Science, 162: 1453-1459.

Ayala,F.J. 1977. Protein evolution in different species: Is it a random process? In: Molecular Evolution And Polymorphism. Kimura,M. (ed.), pp. 73-102. Proceedings of the Second Taniguchi International Symposium on Biophysics, National Institute of Genetics, Mishima, Japan.

Baker,H.G. 1965. Characteristics and modes of origin of weeds. In: The Genetics of Colonizing Species. Baker,H.G., Stebbins,G.L. (eds.), pp. 147-168. Academic Press, New York.

Bargellio,T. and Grossfield,J. 1979. Biochemical polymorphisms: The unit of selection and hypothesis of conditional neutrality. BioSystems, 11: 183-192.

Bell,G. 1982. The Masterpiece of Nature: The Evolution and Genetics of Sexuality. Univ. California Press, Berkeley, California.

Bodmer,W.F. and Felsenstein,J. 1967. Linkage and selection: Theoretical analysis of the deterministic two-locus random mating model. Genetics, 57: 237-265.

Brown,A.J.L. and Langley,C.H. 1979. Correlation between heterozygosity and subunit molecular weight. Nature, 277: 649-651.

Bryant,E.H. 1974. On the adaptive significance of enzyme polymorphisms in relation to environmental variability. Amer.Natur. 108: 1-19.

Bryant,E.H. 1976. A comment on the role of environment variation in maintaining polymorphisms in natural populations. Evolution, 30: 188-190.

Burton,R.S. and Feldman,M.W. 1983. Physiological and fitness effects of an allozyme polymorphism: Glutamate-pyruvate Transaminase and response to hyperosmotic stress in the copepod _Tigriopus californicus_. Biochem.Genet. 20, In press.

Carson,H.L. 1982. Speciation as a major reorganization of polygenic balances. In: Mechanisms of Speciation. Barigozzi,C. (ed.), pp. 411-433, Alan R. Liss, Inc., New York.

Clarke,B. 1979. The evolution of genetic diversity. Proceedings of the Royal Society of London, B 205: 453-474.

Dempster,E. 1955. Maintenance of genetic heterogeneity. Cold Spring Harbor Symposium of Quantitative Biology, 20: 25-32.

Ewens,W.J. 1979. Mathematical Population Genetics. Springer Verlag, Berlin.

Ewens,W.J. and Feldman,M.W. 1976. The theoretical assessment of selective neutrality. In: Population Genetics and Ecology. Karlin,S. and Nevo,E. (eds.), pp. 303-337. Academic Press, New York.

Feldman,M.W., Franklin.I. and Thomson,G.J. 1974. Selection in complex genetic systems. I. The symmetric equilibria of the three-locus symmetric viability model. Genetics, 76: 135-162.

Felsenstein,J. 1976. The theoretical population genetics of variable selection and migration. Ann.Rev.Genet. 10: 253-280.

Gillespie,J.H. 1974a. Polymorphism in patchy environments. Amer.Natur. 108: 145-151.

Gillespie,J.H. 1974b. The role of environmental grain in the maintenance of genetic variation. Amer.Natur. 108: 831-836.

Gillespie,J.H. 1977a. A general model to account for enzyme variation in natural populations. III. Multiple alleles. Evolution, 31: 85-90.

Gillespie,J.H. 1978. A general model to account for enzyme variation in natural populations. V. The SAS-CFF model. Theor.Pop.Biol. 14: 1-45.

Gillespie,J.H. and Langley,C.H. 1974. A general model to account for enzyme variation in natural populations. Genetics, 76: 837-848.

Glesener,R.R. and Tilman,D. 1978. Sexuality and the components of environmental uncertainty: Clues from geographic parthenogenesis in terrestrial animals. Amer.Natur. 112: 659-673.

Guttman,L. 1968. A general nonmetric technique for finding the smallest coordinate space for a configuration of points. Psychometrika, 33: 469-506.

Haldane,J.B.S. and Jayakar,S.D. 1963. Polymorphism due to selection of varying direction. J.Genet. 58: 237-242.

Hamilton,W.D. 1982. Unraveling the riddle of nature's masterpiece. Bioscience, 32: 745-746.

Hamilton,W.D., Henderson,P. and Moran,N. 1979. Fluctuation of environment and coevolved antagonist polymorphism as factors in the maintenance of sex. In: Natural Selection and Social Behavior. Alexander,R.D. and Tinkle,D. (eds.).

Hamrick,J.L., Linhart,Y.B. and Mitton,J.B. 1979. Relationships between life history characteristics and electrophoretically detectable genetic variation in plants. Ann.Rev.Ecol.Syst. 10: 173-200.

Hartl,D.L. 1980. Principles of Population Genetics. Sinauer Assoc. Inc. Sunderland, Mass.

Hartl,D.L. and Cook,R.D. 1973. Balanced polymorphisms of quasineutral alleles. Theor.Pop.Biol. 4: 163-172.

Hartl,D.L. and Dykhuizen,D.E. 1981. Potential for selection among nearly neutral allozymes of 6-phosphogluconate dehydrogenase in _Escherichia coli_. Proc.Nat.Acad.Sci.USA, 78: 6344-6348.

Hedgecock,D., Tracey,M.L. and Nelson,K. 1982. In: The Biology of Crustacea. Genetics. Vol. 2, pp. 283-403. Academic Press, New York.

Hedrick,P.W., Ginevan,M.E. and Ewing,E.P. 1976. Genetic polymorphism in heterogeneous environments. Ann.Rev.Ecol.Syst. 7: 1-32.

Hedrick,P.W., Jain,S. and Holden,L. 1978. Multilocus systems in evolution. Evol.Biol. 11: 101-185.

Hochachka,P.W. and Somero,G.N. 1973. Strategies of Biochemical Adaptation. W.B. Saunders Co. Philadelphia.

Hull,C.H. and Nie,N.H. 1981. SPSS - update 7-9. McGraw Hill Co. New York.

Jaenike,J. and Selander,R.K. 1979. Evolution and ecology of parthenogenesis in earthworms. Amer.Zool. 19: 729-737.

Johnson,G.B. 1979. Genetic polymorphism among enzyme loci. In: Physiological Genetics. Scandalios,J.G. (ed.) pp. 239-273. Academic Press, New York.

Karlin,S. 1976. Population subdivision and selection-migration interactions. In: Population Genetics and Ecology. Karlin,S. and Nevo,E. (eds.), pp. 617-657. Academic Press, New York.

Karlin,S. 1977a. Gene frequency patterns in Levene subdivided population model. Theor.Pop.Biol. 11: 356-385.

Karlin,S. 1977b. Protection of recessive and dominant traits in a subdivided population with general migration structure. Amer.Natur. 11: 1145-1162.

Karlin,S. 1979a. Principles of polymorphism and epistasis for multilocus systems. Proc.Nat.Acad.Sci.USA, 76: 541-545.

Karlin,S. 1979b. Models of multifactorial inheritance. Theor.Pop.Biol. 15: 308-438.

Karlin,S. 1981. Some natural viability systems for multiallelic locus: A theoretical study. Genetics, 97: 457-473.

Karlin,S. 1982a. Theoretical studies in multilocus systems. III. A generalized symmetric heterotic regime in a presence of tight linkage. In press.

Karlin,S. 1982b. Classifications of selection migration structures and conditions for a protected polymorphism. Evol.Biol. 14: 61-204.

Karlin,S. and Avni,H. 1981. Analysis of central equilibria in multilocus systems: A generalized symmetric viability regimes. Theor.Pop.Biol. 20: 241-280.

Karlin,S. and Campbell,R.B. 1978. Analysis of central equilibrium configurations for certain multilocus systems in subdivided populations. Genet.Res.,Camb. 32: 151-168.

Karlin,S. and Campbell,R.B. 1979. Selection-migration regimes characterized by a globally stable equilibrium. Genetics, 94: 1064-1084.

Karlin,S. and Campbell,R.B. 1980. Polymorphism in subdivided populations characterized by a major and subordinate demes. Heredity, 44: 151-168.

Karlin,S. and Feldman,M.W. 1970. Linkage and selection: Two locus symmetric viability models. Theor.Pop.Biol. 1: 39-71.

Karlin,S. and Feldman,M.W. 1981. A theoretical and numerical assessment of genetic variability. Genetics, 97: 475-493.

Karlin,S. and Levikson,B. 1974. Temporal fluctuations in selection intensities case of small population size. Theor.Pop.Biol. 6: 383-412.

Karlin,S. and Liberman,U. 1974. Random temporal variation in selection intensities: Case of large population size. Theor.Pop.Biol. 6: 355-382.

Karlin,S. and Liberman,U. 1979a. Central equilibria in multilocus systems. I. Generalized nonepistatic selection regimes. Genetics, 91: 777-798.

Karlin,S. and Liberman,U. 1979b. Central equilibria in multilocus systems II. Bisexual generalized nonepistatic selection models. Genetics, 91: 799-816.

Karlin,S. and McGregor,J.L. 1972a. Application of method of small parameters to multiniche population genetic models. Theor.Pop.Biol. 3: 186-209.

Karlin,S. and McGregor,J.L. 1972b. Polymorphisms for genetics and ecological systems with weak coupling. Theor.Pop.Biol. 3: 210-238.

Karlin,S. and Richter-Dyn,N. 1976. Some theoretical analyses of migration selection interaction in a cline: A generalized two range environment. In: Population Genetics and Ecology. Karlin,S. and Nevo,E. (eds.), pp. 659-706. Academic Press, New York.

Kimura,M. 1968. Evolutionary rate of the molecular level. Nature, 217: 624-626.

Kimura,M. 1969. The number of heterozygous nucleotide sites maintained in a finite population due to steady flux of mutation. Genetics, 61:893.

Kimura,M. 1979a. Model of effectively neutral mutations in which selective constraint is incorporated. Proc.Nat.Acad.Sci.USA, 76: 3440-3444.

Kimura,M. 1979b. The neutral theory of molecular evolution. Scientific American, 241: 98-126.

Kimura,M. 1981. Possibility of extensive neutral evolution under stabilizing selection with special reference to nonrandom usage of synonymous codons. Proc.Nat.Acad.Sci.USA, 78: 5773-5777.

Kimura,M. and Crow,J. 1964. The number of alleles that can be maintained in a finite population. Genetics, 49: 725-738.

Kimura,M. and Ohta,T. 1971. Theoretical Aspects of Poulation Genetics. Princeton University Press, Princeton, New Jersey.

King,J.L. and Ohta,T.H. 1975. Polyallelic mutational equilibria. Genetics, 79: 681-691.

Koehn,R.K. 1978. Physiology and biochemistry of enzyme variation: The interface of ecology and population genetics. In: Ecological Genetics: The Interface. Brussard,P.F. (ed.), pp. 51-72. Springer Verlag, New York.

Koehn,R.K. and Immerman,F.W. 1981. Biochemical studies of aminopeptidase polymorphism in Mytillus edulis. I. Dependence of enzyme activity on season, tissue, and genotype. Biochem.Genet. 19: 1115-1142.

Kojima,K. and Lewontin,R.C. 1970. Evolutionary significance of linkage and epistasis. In: Topics in Mathematical Genetics. Kojima,K.(ed.), Springer Verlag, New York.

Levene,H. 1953. Genetic equilibrium when more than one ecological niche is available. Amer.Natur. 87: 331-333.

Levin,D.A. 1975. Pest pressures and recombination systems in plants. Amer.Natur. 109: 437-451.

Levins,R. 1968. Evolution in Changing Environments. Princeton University Press, Princeton, New Jersey.

Levins,R. and Lewontin,R. 1980. Dialectics and reductionism in ecology. Synthese, 43: 47-78.

Levy,S. 1981. Lawful roles of facets in social theories. In: Multidimensional Data Representation: When and Why. Borg,I. (ed.), Mathesis Press, Ann Arbor,Michigan.

Lewontin,R.C. 1974. The Genetic Basis of Evolutionary Change. Columbia University Press, New York.

Lewontin,R.C.,Ginzburg,L.R. and Tuljapurkar,S.D. 1978. Heterosis as an explanation for large amounts of genic polymorphism. Genetics, 88: 149-169.

Lewontin,R.C. and Kojima,K. 1960. The evolutionary dynamics of complex polymorphism. Evolution, 14: 458-472.

Maynard Smith,J. 1978. The Evolution of Sex. Cambridge University Press, Cambridge.

McDonald,J.F. and Ayala,F.J. 1974. Genetic response to environmental heterogeneity. Nature, 250: 572-574.

Milkman,R. 1978. The maintenance of polymorphisms by natural selection. In: Marine Organisms, Genetics, Ecology and Evolution. Battaglia,B. and Beardmore,J.A. (eds.), pp. 3-22. Plenum Press, New York.

Moran,P.A.P. 1976. A selective model for electrophoretic profiles in protein polymorphisms. Genet.Res.,Camb. 28: 47-53.

Mukai,T.T.K., Watanabe,E. and Yamaguchi,O. 1974. The genetic structure of natural populations of Drosophila melanogaster XII., Genetics, 77: 771-793.

Nei,M. 1975. Molecular Population Genetics and Evolution. North Holland Publication Co., Amsterdam.

Nei,M. 1980. Stochastic theory of population genetics and evolution. In: Vito Volterra Symposium on Mathematical Models in Biology. Barigozzi,C. (ed.), pp. 17-47. Springer Verlag, Berlin.

Nei,M.,Fuerst,P.A. and Chakraborty,R. 1976. Testing the neutral mutation hypothesis by distribution of single loci heterozygosity. Nature, 262: 491-493.

Nei,M. and Graur,D. 1982. Relationship between gene diversity and population size and the mechanism of protein polymorphism. Personal communication.

Nelson,K. and Hedgecock,D. 1980. Enzyme polymorphism and adaptive strategy in the decapod crustacea. Amer.Natur. 116: 238-280.

Nevo,E. 1978. Genetic variation in natural populatins: Patterns and theory. Theor.Pop.Biol. 13: 121-177.

Nevo,E. 1983a. Population genetics and ecology: The interface. In: Evolution from Molecules to Man. Bendall,D. (ed.), Cambridge University Press, Cambridge.

Nevo,E. 1983b. Adaptive significance of protein variation. In: Adaptation and Taxonomic Significanceof Protein Variation. Oxford,G. and Rollinson,D. (eds.), In press, Academic Press, London.

Nevo,E. 1983c. Adaptive allozyme polymorphisms in mammals. Acta, Zool.Fenn. In press.

Nevo,E., Lavie,B. and Ben-Shlomo,R. 1983. Selection of allozyme polymorphisms in marine organisms: Pattern theory and application. In: Isozymes: Current Topics in Biological and Medical Research. Ratazzi,M.C., Scandalios,J.G. and Whitt,G.S. (eds.), Liss Publ. New York. In press.

Ohta,T. and Kimura,M. 1973. A model of mutation appropriate to estimate the number of electrophoretically detectable alleles in a finite population. Genet.Res.,Camb. 22: 201-204.

Orgel,L.E. and Crick,F.H.C. 1980. Selfish DNA: The ultimate parasite. Nature, 284: 604-607.

Parker,E.D.,Jr. and Selander,R.K. 1976. The organization of genetic diversity in the parthenogenetic lizard Cnemidophorus tesselatus. Genetics, 84: 791-805.

Platt,J.R. 1964. Strong inference. Science, 146: 347-353.

Powell,J.R. 1971. Genetic polymorphism in varied environment. Science, 174: 1035-1036.

Powell,J.R. 1975. Protein variation in natural populations of animals. Evol.Biol. 8: 79-119.

Powell,J.R. and Taylor,C.E. 1979. Genetic variation in ecologically diverse environments. Amer.Sci. 67: 590-596.

Powell,J.R. and Wistrand,H. 1978. The effect of heterogeneous environments and a competitor on genetic variation in Drosophila. Amer.Natur. 112: 935-947.

Ramshaw,J.A.M., Coyne,L.A. and Lewontin,R.C. 1979. The sensitivity of gel electrophoresis as a detector of genetic variation. Genetics, 93: 1019-1037.

Roskam,E. and Lingoes,J.C. 1970. MiniSSA-I: A fortran iv (g) program for the smallest space analysis of square symmetric matrices. Behav.Sci. 15: 204-205.

Schnell,D.G. and Selander,R.K. 1981. Environmental and morphological correlates of genetic variation in mammals. In: Mammalian Population Genetics. Smith,M.H. and Joule,J. (eds.), pp. 60-99. Athens, The University of Georgia Press.

Selander,R.K. 1975. Stochastic factors in population structure. In: Proceeding of the Eighth International Conference on Numerical Taxonomy. Estabrook,G.H. (ed.), pp. 284-332, W.H. Freeman, San Francisco.

Selander,R.K. 1976. Genic variation in natural populations. In: Molecular Evolution. Ayala,F.J. (ed.), pp. 21-45. Sinauer Press Assoc. Sunderland, Mass.

Simmons,M.J., Sheldon,W. and Crow,J.F. 1978. Heterozygous effects on fitness of EMS-treated chromosomes in Drosophila melanogaster. Genetics, 88: 575-590.

Simonsen,V. 1982. Electrophoretic variation in large mammals. II. The red fox, Vulpes vulpes, the stoat, Mustela erminea, the weasel, Mustela nivalis, the pole cat, Mustela putorius, the pine marten, Martes martes, the beech marten, Martes foina, and the badger, Meles meles. Hereditas, 96: 299-305.

Soule,M. 1976. Allozyme variation: Its determinants in space and time. In: Molecular Evolution. Ayala,F.J. (ed.), pp. 60-70. Sinauer Assoc. Sunderland, Mass.

Turelli,M. 1977. Random environments and stochastic calculus. Theor.Pop.Biol. 12: 140-178.

Turelli,M. and Ginzburg,L.R. 1983. Should individual fitness increase with heterozygosity? Genetics, 104: 191-209.

Van Valen,L. 1965. Morphological variation and width of ecological niche. Amer.Natur. XCIC: 377-390.

Vrijenhoek,R.C., Angus,R. and Schultz,R.J. 1976. Variation and clonal structure in a unisexsual fish. Isozyme Bull. 9: 60.

Wallace,B. 1958. The role of heterozygosity in Drosophila populations. Proc. Tenth Cong.Genet. 1: 408-419.

Wills,C. 1981. Genetic Variability. Claredon Press, Oxford.

Wilton,A.N. and Sved J.A. 1979. X-chromosomal heterosis in Drosophila melanogaster Genet.Res.,Camb. 34: 303-315.

Wright,S. 1970. Random drift and the shifting balance theory of evolution. In: Mathematical Topics in Population Genetics. Kojima,K. (ed.), pp. 1-31. Springer Verlag, New York.

Wright,S. 1982. The shifting balance theory and macroevolution. Ann. Rev. Genet. 16: 1-19.

Yamazaki,T. and Maruyama,T. 1975. Isozyme polymorphism maintenance mechanisms viewed from standpoint of population genetics. In: Isozymes IV Genetics and Evolution. Markert,C.L, (ed.) Academic Press New York.

Zouros,E., Singh,S.N. and Miles,H.N. 1982. Growth rate in oysters: An overdominnant phenotype and its possible explanations. Evolution, In press.

Appendix I. The data set: All species ,their indices of genetic diversity (polymorphism, P, and heterozygosity,H) and their biotic profile (ecological, demographic and life history characteristic), accompanied by their references.

Species	No. of:			Genetic indices		Biotic Profile @																						References (Listed according to references of the data)
	populations	individuals	loci	H	P	Lz	Gr	Ht	Hr	Ar	Ps	Ss	Bs	Mo	Gf	Lo	Fe	Gl	Or	Tr	So	Pl	Ch	Ds	Ma	Rp	Po	
V E R T E B R A T A $																												
MAMMALIA																												
Acomys cahirinus 2n=36	2	39	35	0.026	0.164	2	1	2	1	1	3	2	1	4	3	2	1	1	1	2	2	2	36	1		2		327
Acomys cahirinus 2n=38	5	133	35	0.028	0.172	2	3	2	1	4	3	2	1	4	3	2	1	1	2	2	2	2	38	1		2		327
Acomys russatus	3	51	33	0.005	0.115	2	2	2	1	1	2	2	1	4	3	2	1	1	1	2	2	2	66	1		2		327
Alces alces	21	794	23	0.018	0.095	9	4		2		3	2	3	4	3	4	1	3	1	1	3	2	68	1		2		385,386
Anoura geoffroyi	1	30	17	0.017	0.234																							460
Apodemus agrarius	1	12	11	0.007	0.129	2	4	2	2	4	3	4	1	3	2	1	2	1	1	1	3	2	48	1		2		157
Apodemus flavicollis	2	53	26	0.025	0.250	2	4	2	2	4	3	4	1	3	2	1	2	1	1	1	3	2	48	1		2		157,319
Apodemus sylvaticus	6	134	21	0.033	0.200	2	4	2	2	4	3	4	1	3	2	1	2	1	1	1	3	2	48	1		2		157,319
Artibeus jamaicensis	3	57	17	0.080	0.317																							460
Arvicola sapidus	2	12	19	0.035	0.125																							173
Arvicola terrestris	7	66	20	0.014	0.453																							173
Balaenoptera acutorostara	1	95	15	0.043	0.095	2	4	62	2	7	3	1	1	4	1	4	1	3	2	1	1	2	44	4		2		432
Brachyphylla cavernum	1	10	17	--	0.0																							49
Calomys laucha	1	80	18	0.158	0.797	2	3	3	2	3	2	3	1			1	1	1			3	2	64	1				152
Calomys musculinus	1	92	15	0.184	0.666	2	3	3	2	3	2	3	1	4		1	1	1			3	2	38	1				153
Canis lupus	1	11	53	0.025	0.155																							147
Carollia perspicillata	3	13	17	0.037	0.317																							460
Cercopithecus aethiops	2	353	25	0.046	0.302	3	4	3	2	2	2	3	2	4	1	4	1	3	2	1	2	2	60	1		2		219,347
Cervus canadensis	1	113	24	0.012	0.040	2	3	3	2	2	2	3	3	4	2	4	1	3	2	2	4	2		1		2		93
Cervus elaphus	22	110	34	0.021	0.092	2	4	3	2		2	2	3	4	3	4	1	3	1	1	3			1		2		185
Chionomys nivalis	14	190	22	0.025	0.305																							173
Cystophora cristata	1	10	21	0.009	0.076	1	2	62	1	7	3	2	3	5	2	4	1	3		1	3	2	34	4		2		430
Didelphis virginiana	5	79	31	0.113	0.384	4	4	4	2	4	3	3	1	4	2	3	2	2	2	2				1				236
Dipodomys agilis	1	14	18	0.040	0.151	2	1	2	1	2	2	1	1	3	1	3	2	1	1	1	1			1		2		210
Dipodomys compactus	3	34	18	0.023	0.048	2	1	2	1	1	1	1	1	3	1	3	2	1	1	1	1			2		2		210
Dipodomys deserti	4	22	18	0.010	0.086	2	1	2	1	1	2	1	1	3	1	4	2	1	1	1	1			1		2		210
Dipodomys elator	1	23	18	0.002	0.0	2	1	2	1		1	1	1	3	1	3	2	1	1	1	1			1		2		210
Dipodomys heermanni	8	121	21	0.026	0.119	2	1	2	1	2	2	1	1	3	1	3	2	1	1	1	1			1		2		210,354

Appendix I. (contin.)

Species	No. of: populations	No. of: indi.	No. of: loci	Genetic indices: H	Genetic indices: P	Biotic Profile @: Lz	Gr	Ht	Hr	Ar	Ps	Ss	Bs	Mo	Gf	Lo	Fe	Gl	Or	Tr	So	Pl	Ch	Ds	Ma	Rp	Po	References (according to ref. of the data)
Dipodomys merriami	11	251	18	0.051	0.220	2	3	2	1	2	2	2	1	3	1	3	2	1	1	1	1			1		2		210
Dipodomys microps	7	103	18	0.007	0.048	2	2	2	1	2	2	2	1	3	1	3	2	1	1	1	1			1		2		210
Dipodomys nitratoides	2	21	21	0.058	0.221	2	1	2	1	2	2	1	1	3	1	3	2	1	1	1	1			1		2		210
Dipodomys ordii	12	405	18	0.008	0.048	2	3	2	1	1	2	2	1	3	1	4	2	1	1	1	1			1		2		210
Dipodomys panamintinus	2	10	18	0.0	0.0	2	1	2	1	2	2	1	1	3	1	3	2	1	1	1	1			1		2		210
Dipodomys spectabilis	2	48	18	0.008	0.086	2	1	2	1	1	2	1	1	3	1	4	2	1	1	1	1			1		2		210
Erophylla sezekorni	1	12	17	--	0.059																							49
Eumetopias jubatus	5	157	32	0.018	0.214	9	4	62	2	7		2	3	5	3	4	1	3	2	1	3			4	3	2		260
Eutamias panamintinus	1	15	36	0.061	0.250																							217
Felis domesticus	2	117	40	0.131	0.454																							345
Geomys arenarius	5	68	23	0.054	0.207	2	2	1	1	1	1	2	1	1	1	3	1	1	1	1	1	2	70	1		2		360,414
Geomys breviceps	1	15	21	0.039	--	2	2	1	2	4	2	2	1	2	1	3	1	1	1	1	1	2	74	1	3	2		71
Geomys bursarius	46	638	22	0.038	0.333	2	3	1	1	3	2	3	1	2	1	3	1	1	1	1	1			1	3	2		360,414,507
Geomys bursarius (2n=72)	19	173	21	0.061	0.224																							71
Geomys bursarius (2n=74)	17	152	21	0.039	0.136																							71
Geomys personatus	10	154	23	0.046	0.219	2	2	1	1	2	2	2	1	1	1	3	1	1	1	1	1	2	70	2		2		360,414
Geomys pinetis	4	30	22	0.044	0.207	2	5	1	1	4	2	1	1	1	1	3	1	1	1	1	1			1		2		18,360
Geomys tropicalis	1	30	34	0.0	0.0	3	1	1	1	1	1	1	1	2	1	3	1	1	1	1	1	2	38	1				414
Gerbillus allenbyi	3	103	29	0.019	0.149	2	1	2	1	3	2	2	1	4	3	2	1	1	1	2	2	2	40	1		2		327b
Gerbillus pyramidum c.p.	1	27	28	0.004	0.071	2	1	2	1	3	2	2	1	4	3	2	1	1	2	2	2	2	52	1		2		327b
Glossophaga soricina	1	19	17	0.018	0.086																							460
Gorilla gorilla	1	10	22	0.036	0.129																							80
Homo sapiens	9	7349	107	0.125	0.470	5	4	3	2	6	3	4	3	4	2	4	1	3	1	1	4	2	46	1				80,93,191,226, 320,345,408,498
Macaca assamensis	1	28	30	--	0.267	4	2	3	1	4	2	2	2	4	1	4	1	3	2	1	2	2	42	1		2		421
Macaca cyclopis	1	50	29	0.041	0.241	2	2	3	1	3	2	1	2	4	1	4	1	3	1	1	4	2	42	2		2		342
Macaca fascicularis	20	359	32	0.040	0.162	3	4	3	1	3	2	3	2	4	1	4	1	3	1	1	4	2	42	2		2		218,219,342
Macaca fuscata	57	2709	31	0.013	0.096	2	3	3	1	3	2	1	2	4	1	4	1	3	1	1	4	2	42	2		2		219,339,340,341, 342
Macaca mulatta	4	205	29	0.075	0.284	4	4	3	1	3	2	3	2	4	1	4	1	3	1	1	4	2	42	1		2		219,342,421
Macaca nemestrina	1	350	27	--	0.259	3	4	3	1	3	2	2	2	4	1	4	1	3	1	1	4	2	42	1		2		9,342
Macrotus californicus	5	115	21	0.034	0.347	2	3	9	2	1	3	2	1	4	3	3	1	2	1		4	2	40	1	3	2		177
Macrotus waterhausii	3	69	21	0.028	0.200	4	3	9	2	1	3	3	1	4	3	3	1	2	1		4	2	46	1	3	2		177
Marmota flaviventris	6	762	26	0.082	0.322	2	3	2	1	4	2	3	2	3	3	3	2	3	2	1	4			1		2		405
Martes foina	1	121	21	0.0	0.0		4	4	2	4	3	2	2	4	1	4	1	2	2	1	1	2	38	3		2		429
Microdipodops megacephalus	1	20	22	0.064	0.217	2	1	3	1	1	2	1	1	3	1	1	1	1	1	1	1	2	40	1				186
Microdipodops pallidus	1	20	22	0.064	0.217	2	1	3	1	1	2	1	1	3	1	1	1	1	1	1	1	2	42	1				186
Microtus agretis	9	146	18	0.132	0.353	2	4	3	2	2	3	3	1	4	3	3	1	1	1	1	4	2	50	1		2		307,343,344

Appendix I. (contin.)

Species	No. of: populations	indi.	loci	Genetic indices: H	P	Biotic Profile @: Lz	Gr	Ht	Hr	Ar	Ps	Ss	Bs	Mo	Gf	Lo	Fe	Gl	Or	Tr	So	Pl	Ch	Ds	Ma	Rp	Po	References (according to ref. of the data)
Microtus arvalis	10	128	19	0.046	0.440																							173,307
Microtus breweri	1	15	15	--	0.096	2	1	3	1	4	1	1	1	3	2	1	1	1	1	1	1	2	46	2		2		229
Microtus cabrerae	4	22	20	--	0.220																							307
Microtus californicus	1	105	30	--	0.575																							74
Microtus mariae	2	28	23	0.093	0.394																							173
Microtus (Pitymys) multiplex	6	53	25	0.040	0.408																							173,174
Microtus ochrogaster	1	640	26	--	0.269	2	3	3	2	2	3	3	1	3	2	1	1	1	1	1	2	2		1		2		383
Microtus pennsylvanicus	9	135	10	0.089	0.300	2	4	3	2	4	2	3	1	3	2	1	2	1	2	1	1	2	46	2		2		221,229
Mirounga angustirostris	5	97	24	0.0	0.0	2	2	62	2	5	3	1	3	4	3	4	2	3	2	1	3	2	34	2		2		73
Mirounga leonina	3	99	19	0.022	0.297																							295
Molossus molossus	1	30	17	0.015	0.086																							460
Monophyllus plethodon	1	10	17	--	0.084																							49
Monophyllus redmani	1	25	17	--	0.118																							49
Mus hispanicus (spretus)	1	15	56	0.034	0.237																							375
Mus musculus	95	2748	33	0.066	0.273	5	4	2	2	3	3	2	1	4	3	2	2	1	1	2	3	2	40	1		2		60,61,62,63,64, 65,72,76,77,374, 375,384,408,410, 416,467
Mus spicilegus	1	15	27	--	0.106																							467
Mus spretus	4	69	23	0.024	0.123																							76
Mustela erminea	1	39	21	0.0	0.0	9	4	4	2	4	3	2	2	4	1	4	1	2	2	1	1	2	44	3		2		429
Mustela nivalis	1	13	21	0.0	0.0	5	4	4	2	4	3	2	1	4	1	4	1	2	2	1	1	4	42	3		2		429
Mustela putorius	1	24	21	0.0	0.0		4	4	2	4	3	2	2	4	1	4	1	2	2	1	1	2	40	3		2		429
Myotis californicus	1	23	21	0.126	0.429																							443,459
Myotis velfer	3	15	17	0.144	0.656																							443
Neotoma albigula	9	107	21	0.073	0.325																							282,509
Neotoma floridana	5	35	20	0.078	0.385																							509
Neotoma lepida (baja type)	3	15	27	--	0.296																							282
Neotoma lepida (eastern type)	4	20	27	--	0.157																							282
Neotoma lepida (western type)	7	35	27	--	0.254																							282
Neotoma micropus	6	37	20	0.084	0.293																							509
Nitalus sp.	1	30	17	0.034	0.165																							460
Ochotoma princeps	5	197	26	0.011	0.121																							159
Odobenus rosmarus	1	102	32	0.033	0.094	1	3	62	2	7	3	2	3	4	1	4	1	3	2	1	4	2	32	4		2		431
Odocoileus virginianus	7	830	26	0.108	0.313	4	4	3	2	6	3	3	3	3	2	4	1	2	1	1	3	2	70	1		2		201,275,373,443
Oryctolagus cuniculus	25	1622	25	0.052	0.148	2	4	2	2	6	2	3	2	3	2	3	2	1	2	1	4	2		1	3	2		376
Pagophilus groenlandicus	2	21	27	0.017	0.076	1	2	62	1	7	3	2	3	5	2	4	2	3	1	1	3							248,430
Pan troglodytes	2	182	42	0.010	0.079	3	3	3	2	4	2	1	3	4	1	4	1	3	1	1	3	2	48	1	3	2		80,226

Appendix I. (contin.)

Species	No. of:			Genetic indices		Biotic Profile @																						References (according to ref. of the data)
	populations	indi.	loci	H	P	Lz	Gr	Ht	Hr	Ar	Ps	Ss	Bs	Mo	Gf	Lo	Fe	Gl	Or	Tr	So	Pl	Ch	Ds	Ma	Rp	Po	
Papio anubis	4	73	35	0.025	0.112	4	4	3	2	2	2	3	3	4	1	4	1	3	2	1	2	2	42	1		2		219,423,424
Papio humadryas	5	434	35	0.044	0.217	3	2	3	1	1	2	3	3	4	1	4	1	3	1	1	3	2	42	1		2		219,423,424
Pappogeomys fumosus	1	14	22	0.0	0.0	3	5	1	1	1	2	1	1	1	2	3	1	2	1	1	1	2	40	1	3	2		200
Pappogeomys merriami	3	10	22	0.0	0.072	3	5	1	1	4	2	1	1	1	2	3	1	2	1	1	1	2	36	1	3	2		200
Pappogeomys zinseri	1	13	22	0.0	0.0	3	5	1	1	1	2	1	1	1	2	3	1	2	1	1	1	2	40	1	3	2		200
Perognathus arenarius	1	18	26	0.041	0.137																							356
Perognathus artus	2	12	26	0.072	0.228																							356
Perognathus baileyi	21	289	26	0.049	0.242																							356
Perognathus californicus	2	21	26	0.096	0.137																							356
Perognathus fallax	5	16	27	0.059	0.125																							356
Perognathus formousus	7	49	26	0.022	0.096																							356
Perognathus goldmani	19	256	26	0.027	0.320																							356
Perognathus intermedius	6	47	24	0.054	0.190																							356
Perognathus nelsoni	2	35	28	0.022	0.170																							356
Perognathus penicillatus	22	179	26	0.047	0.198																							356
Perognathus pernix	6	40	26	0.087	0.290																							356
Peromyscus attwateri	4	125	17	0.005	0.063	2	2	3	1	2	2	3	1	3	2	2	1	1	1	1	1	2	48	1	3	2		26
Peromyscus boylii	26	465	26	0.026	0.130	2	4	3	1	2	2	4	1	3	2	2	1	2	1	1	2	2	48	1	3	2		27,222,439
Peromyscus californicus	13	224	30	0.074	0.327																							28,438,439
Peromyscus caniceps	1	15	23	0.011	0.068																							28
Peromyscus comanche	1	21	25	0.070	0.396																							204
Peromyscus dickey	1	15	23	0.0	0.0																							28
Peromyscus difficilis	3	57	24	0.045	0.229	2	3			2		3	1	4	3	3	1	2		1	2	2	48	1		2		26,204
Peromyscus eremicus	9	45	23	0.040	0.159																							28
Peromyscus eva	1	15	23	0.0	0.0																							28
Peromyscus floridanus	4	71	39	0.053	0.214	2	2	3	1	3	1	2	1	3	2	1	1	1	2	1	1	2	48	1	3	2		444
Peromyscus gossypinus	1	15	10	0.051	--	2	3	3	2	4	2	2	1	3	2	1	1	1	1	1	1	2	48	1	3	2		414
Peromyscus guardia	1	15	23	0.014	0.068																							28
Peromyscus hooperi	1	15	17	0.031	--	2	1	3	1	1	2	1	1	3	2	2	1	1	2	2	1	2	48	1	3	2		401
Peromyscus interparietalis	1	15	23	0.0	0.0																							28
Peromyscus leucopus	3	45	10	0.068	0.196	4	4	3	2	6	2	3	1	3	2	1	1	1	2	1	1	2	48	2	3	2		79,414
Peromyscus maniculatus	84	1958	22	0.117	0.441	2	4	3	2	6	3	4	1	4	3	2	2	2	2	1	2	2	48	2	3	2		12,42,158,262, 283,439,443
Peromyscus melanotis	1	11	21	0.035	0.136																							26
Peromyscus merriami	1	15	23	0.016	0.068																							28
Peromyscus pectoralis	16	150	17	0.042	0.178	2	3	3	1	2	2	2	1	3	3	2	1	2	1	1	2	2	48	1	3	2		27,222,223
Peromyscus polionotus	35	800	32	0.064	0.256	2	3	3	2	3	2	2	1	3	2	1	1	1	1	1	2	2	48	1	3	2		154,359,408,415
Peromyscus sejugis	2	15	23	0.017	0.068																							28

Appendix I. (contin.)

Species	No. of: populations	indi.	loci	Genetic indices H	P	Lz	Gr	Ht	Hr	Ar	Ps	Ss	Bs	Mo	Gf	Lo	Fe	Gl	Or	Tr	So	Pl	Ch	Ds	Ma	Rp	Po	References (according to ref. of the data)
Peromyscus stephani	1	15	23	0.0	0.0																							28
Peromyscus truei	6	228	29	0.029	0.120	2	4			2		3	1	4	3	3	1	2		1	2	2	48	1		2		26,204,439
Phyllonycteris aphylla	1	17	17	--	0.176																							49
Pipistrellus hesperus	1	23	20	0.026	0.143																							459
Pitymys duodencimostatus	6	43	20	--	0.255																							307
Proechimys guairae 2n=46	2	31	22	0.077	0.098	3	3	3	2	3	2	2	1	3	2	2	1	1	2		4	2	46	1				57
Proechimys guairae 2n=48	1	23	22	0.065	0.136	3	3	3	2	3	2	2	1	3	2	2	1	1	2		4	2	48	1				57
Proechimys guairae 2n=50	1	14	22	0.153	0.454	3	3	3	2	3	2	2	1	3	2	2	1	1	2		4	2	50	1				57
Proechimys guairae 2n=62	1	16	22	0.069	0.136	3	3	3	2	3	2	2	1	3	2	2	1	1	2		4	2	62	1				57
Proechimys urichi	1	22	22	0.080	0.182	3	1	3	2	3	2	1	1	3	2	2	1	1	2		4	2	62	1				57
Pusa hispida	1	82	21	0.009	0.095	1	2	62	1	7	3	2	3	5	2	4	1	3		1	3	2	32	4		2		430
Rattus argentiventer	1	16	28	0.050	0.143	3	3	3		4			1	4	3	3	2	1	1	1	2	2	42	2				352
Rattus diardi	1	13	28	0.089	0.286	3	3	3		4			1	4	3	3	2	1	1	1	2	2	42	2				352
Rattus exulanus	1	15	28	0.027	0.107	3	3	3		4			1	4	3	3	2	1	1	1	2	2	42	2				352
Rattus fuscipes	11	417	16	0.010	0.054	4	2	2	1	4	2	2	1	1	1	1	1	1	1	1	1	2	38	2				402
Rattus norvegicus	3	95	25	0.064	0.319	5	4	3	2	6	2	4	1	4	3	3	2	1	1	1	4	2	42	1				140,352
Rattus rattus	7	130	36	0.032	0.157	5	4	4	2	4		4	1	4	3	3	2	1	1	1	2		44	2				352,358
Rattus tiomanicus	1	15	28	0.026	0.143	3	3	3		4			1	4	3	3	2	1	1	1	2	2	42	2				352
Sigmodon arizonae	1	50	23	0.029	0.129	2	1	2	1	3	2		1	3	1	3	2	1		1	2					2		211
Sigmodon hispidus	38	1570	23	0.040	0.185	4	3	2	2	3	2	2	1	4	1	3	2	1		1	2			1		2		211,292,293
Sorex alpinus	1	11	22	0.011	0.273																							101
Sorex araneus	5	31	22	0.021	0.127																							101
Sorex coronatus	2	14	22	0.026	0.068																							101
Spalax ehrenbergi 2n=52	1	73	25	0.035	0.120	2	1	1	1	3	3	2	1	3	2	2	1	1	2	1	2	2	52	1		2		334
Spalax ehrenbergi 2n=54	1	118	25	0.016	0.240	2	1	1	1	3	3	2	1	3	2	2	1	1	2	1	2	2	54	1		2		334
Spalax ehrenbergi 2n=58	1	88	25	0.037	0.160	2	1	1	1	3	3	2	1	3	2	2	1	1	2	1	2	2	58	1		2		334
Spalax ehrenbergi 2n=60	1	130	25	0.069	0.280	2	1	1	1	3	3	2	1	3	2	2	1	1	2	1	2	2	60	1		2		334
Spermophilus armatus	1	62	18	0.082	--	2	3	2	1	2	2	2	2	3	2	3	2	2	1	1	3	2	34	1		2		315
Spermophilus beldingi	1	27	18	0.028	--	2	3	2	1	2	2	3	2	3	2	3	2	2	1	1	3	2	30	1		2		315
Spermophilus columbianus	1	179	18	0.104	--	9	3	2	2	2	2	3	2	3	2	3	2	2	1	1	3	2	32	1		2		315
Spermophilus elegans	1	120	18	0.044	--	2	3	2	1	1	2	2	2	3	2	3	2	2	1	1	3	2	34	1		2		315
Spermophilus erythrogenys	1	24	18	0.0	--	2	3	2	1	2	2	3	2	3	2	3	2	2	1	1	3	2	36	1		2		315
Spermophilus major	1	23	18	0.035	--	2	2	2	1	1	2	2	2	3	2	3	2	2	1	1	3	2	36	1		2		315
Spermophilus mexicanus	5	105	29	0.079	0.300	1	3	2	2	2	2	3	1	3	2	3	1	1	1	1	1	2	34	1		2		113,114
Spermophilus mollis	3	92	18	0.036	--	2	2	2	1	1	2	2	2	3	2	3	2	2	1	1	3	2	38	1		2		315
Spermophilus musicus	1	56	18	0.0	--	2	5	2	1	2	2	1	2	3	2	3	2	2	1	1	3	2	36	1		2		315
Spermophilus parryii	6	287	18	0.053	0.172	1	3	2	1	2	2	3	2	3	2	3	2	2	1	1	3	2	34	1		2		314
Spermophilus pygmaeus	1	25	18	0.015	--	2	4	2	1	2	2	3	2	3	2	3	2	2	2	1	3	2	36	1		2		315

Biotic Profile @

Appendix I. (contin.)

Species	No. of: populations	indi.	loci	Genetic indices H	P	Biotic Profile @ Lz	Gr	Ht	Hr	Ar	Ps	Ss	Bs	Mo	Gf	Lo	Fe	Gl	Or	Tr	So	Pl	Ch	Ds	Ma	Rp	Po	References (according to ref. of the data)
Spermophilus relictus	1	25	18	0.019	--	2	5	2	1	2	2	1	2	3	2	3	2	2	2	1	3	2	36	1		2		315
Spermophilus richardsonii	1	105	18	0.050	--	2	3	2	1	2	2	3	2	3	2	3	2	2	1	1	3	2	36	1		2		315
Spermophilus spilosoma	3	77	28	0.087	0.312	1	3	2	2	2	2	3	1	3	2	3	1	1	1	1	1	2	32	1		2		114
Spermophilus suslicus	1	12	18	0.024	--	2	3	2	1	2	2	2	2	3	2	3	2	2	2	1	3	2	36	1		2		315
Spermophilus townsendi	1	42	18	0.072	--	2	2	2	1	1	2	1	2	3	2	3	2	2	1	1	3	2	36	1		2		315
Spermophilus tridecemlineatus	4	100	30	0.050	0.234	1	4	2	2	4	2	3	1	3	2	3	1	1	1	1	1	2	34	1		2		113, 114
Spermophilus undulatus	2	11	18	0.027	--	9	4	2	2	2	2	3	2	3	2	3	2	2	1	1	3	2	32	1		2		315
Spermophilus vigilis	1	59	18	0.035	--	2	2	2	1	1	2	1	2	3	2	3	2	2	1	1	3	2	46	1		2		315
Spermophilus washingtoni	1	31	18	0.032	--	2	1	2	1	1	2	1	2	3	2	3	2	2	2	1	3	2	36	1		2		315
Sus scrofa	2	114	20	0.049	0.186	4	4	3	2	6	2	3	3	4	2	4	1	1	1	1	1	2	37	1		2		445
Theopithecus gelada	4	213	34	0.001	0.088	9	5	3	1	4	2	3	3	4	1	4	1	3	2	1	3	2	42	1		2		422
Thomomys bottae	43	1007	23	0.094	0.402	2	3	1	1	6	2	3	1	2	2	3	1	2	1	1	1	2	76	1				355, 357, 440
Thomomys talpoides 2n=40	2	54	31	0.022	0.130	2	2	1	1	2	2	2	1	2	2	3	1	2	1	1	1	2	40	1				333
Thomomys talpoides 2n=44	1	26	31	0.042	0.320	2	2	1	1	2	2	2	1	2	2	3	1	2	1	1	1	2	44	1				333
Thomomys talpoides 2n=46	1	30	31	0.061	0.290	2	2	1	1	2	2	2	1	2	2	3	1	2	1	1	1	2	46	1				333
Thomomys talpoides 2n=48 a	3	87	31	0.061	0.300	2	2	1	1	2	2	2	1	2	2	3	1	2	1	1	1	2	48	1				333
Thomomys talpoides 2n=48 b	1	30	31	0.045	0.130	2	3	1	1	2	2	2	1	2	2	3	1	2	1	1	1	2	48	1				333
Thomomys talpoides 2n=60	2	49	31	0.047	0.230	2	2	1	1	2	2	2	1	2	2	3	1	2	1	1	1	2	60	1				333
Thomomys umbrinus	24	362	23	0.056	0.333	4	3	1	1	1	2	2	1	2	2	3	1	2	1	1	1	2	78	1	3	2		200, 353, 355
Uroderma bilobatum 2n=38	4	110	22	0.023	0.127	3	3	9	2	4	3	3	1	4	3	3	1	1	1		3	2	38	1	3	2		176
Uroderma bilobatum 2n=44	1	54	22	0.008	0.090	3	3	9	2	4	3	3	1	4	3	3	1	1	1		3	2	44	1	3	2		176
Ursus americanus	4	166	24	0.016	0.120	2	4	3	2	3	2	2	3	4	2	4	1	2	2	1	1	2	74	1		2		93, 276
Ursus (Thalarctos) maritimus	4	106	29	0.0	0.021	1	2	7	1	7	2	1	3	4	1	4	1	3	1	2	1			1				5, 466
Vulpes vulpes	1	282	21	0.0	0.0	5	4	4	2	4	3	2	2	4	1	4	1	2	1	1	3	2	36	3		2		429
Zapus hudsonius	2	25	21	0.006	0.049																							187
Zapus princeps	2	19	21	0.020	0.088																							187
AVES																												
Agelaius phoeniceus	5	30	15	--	0.533																							437
Ammodramus sandwichensis	1	10	20	0.049	0.330	2	4	9				3		5	3	3	1	2		1	2			1		2		20
Ammodramus savannarum	1	10	20	0.045	0.275	2	4	9				3		5	3	3	1	2		1	2			1		2		20
Aplonis cantorodies	4	108	18	0.008	0.151	3	2	4	1	5	2	2	2	2	2	3	1	2	1	2	4			3		2		109
Aplonis metallica	1	354	18	0.047	0.234	3	2	4	1	5	2	2	2	2	2	3	1	2	1	2	4			3		2		109
Ardea herodias	2	46	28	0.007	0.030	4	4	4	2	4	3	3	3	4	3	4	1	3	2	1	2			1				181
Camarhynchus parvulus	4	21	27	0.030	0.296																							503
Cassidix mexicanus	5	30	15	--	0.400																							437

Appendix I. (contin.)

Species	No. of: populations	indi.	loci	Genetic indices H	P	Lz	Gr	Ht	Hr	Ar	Ps	Ss	Bs	Mo	Gf	Lo	Fe	Gl	Or	Tr	So	Pl	Ch	Ds	Ma	Rp	Po	References (according to ref. of the data)
						Biotic Profile @																						
Catharus guttatus	1	13	28	0.047	0.250	2	4	9				3		5	3	3	1	2		1	2			1		2		19
Catharus ustulatus	1	17	28	0.053	0.286	2	4	9				3		5	3	3	1	2		1	2			1		2		19
Cherthidae diracea	4	15	27	0.027	0.296																							503
Coturnix coturnix	1	104	24	--	0.582																							43
Demdrpoca castanea	1	12	21	0.024	--	2	4	9				3		5	3	3	1	2		1	2			1		2		21
Dendroica coronata	3	72	30	0.034	0.331	2	4	4	2	4	3	3	1	5	2	3	1	2	1	1	2			1		2		21,52,53
Dendroica magnolia	2	26	27	0.004	0.409	2	3	4	1	4	3	2	1	5	3	3	1	2	1	1	2			1		2		21,53
Dendroica palmarum	2	45	25	0.037	0.409	2	3	3	1	3	3	2	1	5	3	3	1	2	1	1	2			1		2		21,53
Dendroica pensylvanica	1	10	21	0.014	--	2	4	9				3		5	3	3	1	2		1	2			1		2		21
Euphgus cyanocephalus	5	30	15	--	0.333																							437
Geospiza crassirostris	2	21	27	0.056	0.259																							503
Geospiza fortis	8	53	27	0.057	0.444																							503
Geospiza fuliginosa	10	80	27	0.062	0.333																							503
Geospiza magnirostris	4	17	27	0.035	0.259																							503
Geospiza scandens	3	26	27	0.058	0.407																							503
Geothlypis formosa	1	16	21	0.056	--	2	4	9				3		5	3	3	1	2		1	2			1		2		21
Geothlypis trichas	2	48	25	0.037	0.308	2	4	3	1	4	2	2	1	5	3	3	1	2	1	1	2			1		2		21,53
Hirundo tahitica	1	19	15	0.078	0.275																							278
Icterus gabula	8	429	19	0.056	0.105	2	4	4	2	3	2	2	2	2	2	2	1	2	1	1	4			1		2		108
Junco hyemalis	5	333	20	0.060	0.236	2	4	9	2	4	3	3	1	5	2	3	1	2	1	1	2	2	76	1		2		20,47,52
Lagopus lagopus	1	269	23	0.082	0.302	1	4	3	2		3	2	2	5	3	4	1	2	1	1	2			1		2		184
Lagopus mutus	1	45	23	0.044	0.202	1	4	3	2		3	2	2	5	3	4	1	2	1	1	2			1		2		184
Larus argentatus	1	28	40	--	0.121	2	4	9	2	3	3	2	2	4	2	3	1	3	1	1	3	2	72	2		2		465
Larus fuscus	1	26	40	--	0.178	2	3	9	1	3	2	2	2	5	2	3	1	3	1	1	3	2	72	2		2		465
Larus marinus	1	12	40	--	0.065	2	3	9	1	3	2	2	2	5	2	3	1	3	1	1	3	2	72	2		2		465
Molothrus ater	5	30	15	--	0.533																							437
Motacilla flava ssp.	1	15	16	0.116	0.453	4	4	9	1	6	2	4	1	5	2	2	1	2	1	1	3			1	3	2		120
Oceanites oceanicus	1	12	16	0.035	0.223	1	4	4	2	5	2	2	2	5	1	4	1	3	2	1	2			2		2		52
Olor buccinator	3	175	19	0.008	0.137	9	3	8	1		2	2	3	5	3	4	2	3	2	1	2	2		1		2		51
Pagodroma nivea	1	11	16	0.080	0.172	1	3	4	2	5	2	2	1	5	1	4	1	3	2	1	2			2		2		52
Passer domesticus	1	19	15	0.098	0.417	5	4	9			3	3	1	4	3	3	2	2	2	2				1		2		278
Petrochelidon ariel	1	30	15	0.065	0.275																							278
Phasianus colchicus	1	15	44	--	0.508																							44
Pheucticus ludovicianus	1	15	16	0.0	0.0	2	4	4	2	4	2	3	1	5	2	3	1	2	2	1	2			1		2		425
Pipilio ocai	2	30	27	0.150	--	4	2	4	2	3	3	2	2	4	2	3	1	2	1	1	2			1	3	2		128
Pipilo erythrophthalmus	1	15	18	--	0.202	2	4	3	2	4	2	3	1	2	1	3	1	2	2	1	2			3		2		425
Pomastostomus temporalis	1	80	20	0.094	0.453	4	4	9	2	2	3	2	2	3	2	4	1	3		1	4			1	3	2		208
Quisculus quiscula	6	88	17	--	0.392																							48,437

Appendix 1. (contin.)

Species	No. of:			Genetic indices		Biotic Profile @																					References (according to ref. of the data)
	populations	indi.	loci	H	P	Lz	Gr	Ht	Hr	Ar	Ps	Ss	Bs	Mo	Gf	Lo	Fe	Gl	Or	Tr	So	Pl	Ch	Ds	Ma	Rp	Po	
Regulus calendula	1	10	26	0.048	0.212	2	4	9				3		5	3	3	1	2		1	2			1		2		19
Seiurus aurocapillus	2	39	24	0.061	0.383	2	4	4	1	4	3	2	1	5	3	3	1	2	1	1	2			1		2		21,53
Seriurus noveboracensis	1	24	31	0.054	0.452	2	3	3	1	4	3	2	1	2	2	2	1	2	1	1	2			1		2		53
Setophaga ruticilla	2	28	25	0.031	0.282	2	4	4	1	4	3	2	1	5	3	3	1	2	1	1	2			1		2		21,53
Spizella passerina	1	11	21	0.065	0.197	2	4	9				3		5	3	3	1	2		1	2			1		2		20
Sturnella magna	5	30	15	--	0.600	2	3	9	2	3	3	3	1	5	3	3	1	1	1	1	1			1	3	2		437
Sturnella neglecta	5	30	15	--	0.400	2	3	9	2	3	3	3	1	5	3	3	1	1	1	1	1			1	3	2		437
Turdus iliacus	2	40	40	--	0.290	2	4	9	2	3	2	2	1	5	2	3	1	2	1	1	3	2	80	1		2		464
Vermivora celata	1	12	21	0.040	--	2	4	9				3		5	3	3	1	2		1	2			1		2		21
Vermivora peregrina	2	37	21	0.069	0.544	2	3	4	2	4	3	2	1	5	3	3	1	2	1	1	2			1		2		21,53
Vermivora ruficapilla	1	22	31	0.042	0.387	2	3	4	1	4	3	2	1	2	2	2	1	2	1	1	2			1		2		53
Zonotrichia albicollis	1	10	21	0.079	0.421	2	4	9				3		5	3	3	1	2		1	2			1		2		20
Zonotrichia capensis	4	154	24	0.035	0.052		4	4	2	6	3	3	1	5	1	3	1	2	1	1	2			3		2		338
Zonotrichia georgiana	1	10	21	0.051	0.076	2	4	9				3		5	3	3	1	2		1	2			1		2		20
Zonotrichia leucophrys	5	78	19	--	0.217	2	3	3	2	4	2	2	2	5	1	3	1	2	1	1	2			1		2		46
REPTILIA																												
Acanthodactylus boskianus	1	18	22	0.180	0.590	4	3	3	2	2	2	2	1	1	2	3	1	2	2	1		2		1	3	2		70
Acanthodactylus inornatus	1	27	22	0.140	0.500	4	3	3	1	1	2	3	1	1	2	3	1	2	2	1		2		1	3	2		70
Acanthodactylus pardalis	3	61	15	0.183	0.611	4	3	3	2	2	2	3	1	1	2	3	1	2	2	1		2		1	3	2		70
Acanthodactylus savignyi blanci	1	22	22	0.250	0.680	2	3	3	1	3	2	2	1	1	2	3	1	2	2	1		2		1	3	2		70
Agama stellio	9	242	25	0.066	0.333	2	3	3	2	4	3	2	1	4	2	3	2	3	1	1	1	2	36	1		2		326
Alligator mississippiensis	5	188	44	0.016	0.069	2	3	61	1	4	2	1	3	3	2	4	2	3	2	1	2	2	32	1	3	2		2,155,301
Anniella geronimensis	1	22	27	0.017	0.148	2	1	2	1	2	2	1	1	1	1	4	1	3	2	1	1	2	36	1		2		67
Anniella pulchra nigra	2	33	27	0.003	0.047	2	1	2	1	2	2	1	1	1	1	4	1	3	2	1	1	2	22	1		2		67
Anolis actus	1	10	22	0.069	0.443	3	1	4	2	4	3	3	1	3	2	2	2	2	2	1	1	2	31	2		2		166
Anolis angusticeps	1	38	25	0.0	0.0																							492
Anolis armouri	1	10	15	0.113	0.275																							496
Anolis bimaculatus	2	52	22	0.059	0.337	3	1	4	2	4	3	3	1	3	2	2	2	2	2	1	1	2	29	2		2		166
Anolis carolinensis	4	193	28	0.053	0.204																							492
Anolis christophei	1	27	15	0.026	0.179	3	2	3	1	4	2	2	1	3		3	1	2	2	1	3	2	36	2				496
Anolis coelestinus	1	15	15	0.031	0.179	3	3	3	2	4	3	2	1	4		3	1	2	2	1	3	2	36	2				496
Anolis cooki	1	20	19	--	0.316	3	1	4	2	4	3	3	1	3	2	2	2	2	2	1	1	2	29	2		2		164
Anolis cristellus	14	371	19	--	0.363	3	1	4	2	4	2	3	1	3	2	2	2	2	2	1	1	2	27	2		2		164
Anolis cybotes	2	55	15	0.105	0.328	3	3	3	2	4	3	3	1	4		3	1	2	2	1	3	2	36	2				496
Anolis disticus	1	55	27	0.043	0.206																							492

Appendix I. (contin.)

Species	No. of: populations	indi.	loci	Genetic indices H	P	Lz	Gr	Ht	Hr	Ar	Ps	Ss	Bs	Mo	Gf	Lo	Fe	Gl	Or	Tr	So	Pl	Ch	Ds	Ma	Rp	Po	References (according to ref. of the data)
						Biotic Profile @																						
Anolis ethridgei	1	27	15	0.034	0.179	3	2	3	1	4	3	2	1	3		3	1	2	2	1	3	2	36	2				496
Anolis eugenegrahami	1	14	15	0.0	0.100	3	1	3	1	4	1	1	1	3		3	1	2	2	1	3	2	28	2				496
Anolis ferreus	1	16	22	0.092	0.390	3	1	4	2	4	3	3	1	3	2	2	2	2	2	1	1	2	29	2		2		166
Anolis gingivinus	1	29	22	0.113	0.550	3	1	4	2	4	3	3	1	3	2	2	2	2	2	1	1	2	29	2		2		166
Anolis grahami	2	81	24	0.071	0.389																							463
Anolis insolitus	1	10	15	0.014	0.100	3	2	3	1	4	2	2	1	3		3	1	2	2	1	3	2	44	2				496
Anolis leachi	2	54	22	0.069	0.355	3	1	4	2	4	3	3	1	3	2	2	2	2	2	1	1	2	29	2		2		166
Anolis lividus	1	21	22	0.041	0.260	3	1	4	2	4	3	3	1	3	2	2	2	2	2	1	1	2	29	2		2		166
Anolis luciae	3	52	26	0.069	0.309	3	1	4	2	4	3	3	1	3	2	2	2	2	2	1	1	2	29	2		2		166
Anolis marmoratus	1	20	22	0.061	0.430	3	1	4	2	4	3	3	1	3	2	2	2	2	2	1	1	2	36	2		2		165
Anolis monensis	1	23	19	--	0.263	3	1	4	2	4	3	3	1	3	2	2	2	2	2	1	1	2	29	2		2		166
Anolis nubilus	1	17	22	0.035	0.180	3	1	4	2	4	3	3	1	3	2	2	2	2	2	1	1	2	29	2		2		164
Anolis oculatus	2	34	22	0.055	0.321	3	1	4	2	4	3	3	1	3	2	2	2	2	2	1	1	2	29	2		2		166
Anolis pogus	1	22	22	0.081	0.410	3	1	4	2	4	3	3	1	3	2	2	2	2	2	1	1	2	31	2		2		166
Anolis rimarum	1	20	15	0.070	0.356	3	2	3	1	4	3	2	1	3		3	1	2	2	1	3	2	36	2				496
Anolis roquet	13	321	23	0.067	0.166	3	1	4	2	4	3	3	1	3	2	2	2	2	2	1	1	2	34	2		2		165,504
Anolis sabanus	1	28	22	0.042	0.360	3	1	4	2	4	3	3	1	3	2	2	2	2	2	1	1	2	29	2		2		166
Anolis sagrei	2	198	26	0.020	0.064																							492
Anolis schwartzi	1	41	22	0.065	0.450	3	1	4	2	4	3	3	1	3	2	2	2	2	2	1	1	2	29	2		2		166
Anolis scriptus	1	25	19	--	0.158	3	1	4	2	4	2	3	1	3	2	2	2	2	2	1	1	2	27	2		2		164
Anolis semilineatus	1	18	15	0.010	0.275	3	3	3	2	4	3	3	1	4		3	1	2	2	1	3	2	36	2				496
Anolis shrevei	1	26	15	0.151	0.275	3	2	3	1	4	3	2	1	4		3	1	2	2	1	3	2	36	2				496
Anolis wattsi	1	21	22	0.045	0.270	3	1	4	2	4	3	3	1	3	2	2	2	2	2	1	1	2	29	2		2		166
Bipes biporus	1	20	22	0.032	0.136																							225
Bipes canaliculatus	1	28	21	0.002	0.095																							225
Caretta caretta	4	79	13	0.034	0.077	4	4	62	2	7	3	3	3	5	3	4	3	3	2	2	1	2	56	4	3	2		442
Chelonia mydas	2	34	13	0.123	0.604	4	4	62	2	7	3	3	3	5	3	4	3	3	2	2	1	2	56	4	3	2		442
Chemidophorus laredoensis (2n)	1	72	15	0.267	--																							349
Chemidophorus septemivittalus	3	62	21	0.058	0.226																							349
Chemidophorus sexlineatus	2	34	21	0.070	0.278																							349
Chemidophorus tesselatus (2n)	24	339	21	0.560	--																							349
Chemidophorus tesselatus (3n)	3	67	21	0.714	--																							349
Chemidophorus tigris	2	69	21	0.050	0.226																							349
Cnemidophorus exsangius	3	23	35	0.400	--	2	2	3	1	2	2	1	1	3	1	3	1	2	1	2	1	3		1	4	1		129
Crotalus viridis	9	10	22	--	0.635																							512
Gambelia wislizeni	1	11	27	0.018	0.206	2	3	3	1	2	3	2	1	4	2	4	1	3	2	2	3	2	36	3		2		310
Geochelone elephantopus	9	200	20	0.028	0.092																							280
Gymnophthalamus underwoodi	2	26	21	0.380	--	3	2	3	1	4	2	1	1	3	1	3	1	2	1	2	1	2		3	4	1		129

Species	No. of: populations	indi.	loci	indices H	P	Biotic Profile[e] Lz	Gr	Ht	Hr	Ar	Ps	Ss	Bs	Mo	Gf	Lo	Fe	Gl	Or	Tr	So	Pl	Ch	Ds	Ma	Rp	Po	References (according to ref. of the data)
Hemidactylus brooki	16	20	10	0.052	0.150	3	4	4	2	4	2	4	1			4	2	2		1	3	2	40	3				351
Lacerta melisellensis	12	263	22	0.039	0.170	2	3	3	2	4	2	3	1	3	1	2	2	2	2	1	1	2	38	2		2		170
Lacerta sicula (mainland)	3	90	22	0.090	0.360	2	3	3	2	4	2	3	1	3	1	2	2	2	2	1	1	2	38	1		2		170
Podarcis erbardii	1	38	18	--	0.056	2	3	3	2	4	1	3	1	3	1	3	2	3	2		4			3				291
Podarcis milensis	2	38	18	--	0.167	2	1	3	2	4	1	2	1	3	1	3	2	3	2		4			2				291
Podarcis muralis	6	80	16	--	0.090	2	4	3	2	4	2	3	1	3	1	3	2	3	2		4			3				290,291,493
Podarcis sicula	1	14	16	--	0.063	2	3	3	2	4	3	3	1	3	1	3	2	3	2		4			3				290,493
Podarcis taurica	1	20	18	--	0.111	2	3	3	2	4	2	3	1	3	1	3	2	3	2		4			3				291
Podarcis tiliguerta	3	29	16	--	0.179	2	1	3	2	4	2	2	1	3	1	3	2	3	2		4			2				290,493
Rhinophis phillipininus	1	39	26	0.040	--	2	2	2	1	4	2	1	1	2	1	3	1	2	2	2	1			2	3	2		131
Rhinopsis drummondhayi	2	12	36	0.020	0.131																							130
Sceloporus graciosus	1	149	17	0.028	0.410																							470
Sceloporus grammicus	4	118	19	0.070	0.277	4	3	4	2	3	3	3	1			3	1	2		1		2	36	1	3	2		188
Sternotherus carinatus	4	20	21	--	0.199	2	2	6	2	4	3	2	1	3	3	4	1	3			1	2		1		2		406,407
Sternotherus depressus	3	17	21	--	0.181	2	5	6	1	4	1	1	1	3	3	4	1	3			1	2		1		2		406,407
Sternotherus minor	7	32	21	--	0.299	2	2	6	2	4	3	2	1	3	3	4	1	3			1	2		1		2		406,407
Sternotherus odoratus	8	45	25	--	0.320	2	3	6	2	4	3	3	1	4	3	4	1	3			1	2		1		2		407
Thamnophis couchii	32	211	31	0.053	0.259	2	2	7	1	2	2	2	2	3	2	3	2	2	1	2	1	2	36	1	3	2		249
Thamnophis elegans	26	247	31	0.033	0.267	2	2	7	1	2	2	2	2	3	2	3	2	2	1	2	1	2	36	1	3	2		249
Thamnophis ordinoides	8	40	31	0.067	0.341																							249
Thamnophis proximus	3	20	26	0.085	0.315	2	3	3	1	4	2	2	2	3	2	3	2	2	1	2	1	2	26	1	3	2		156
Thamnophis sauritus	10	93	26	0.089	0.215																							156
Thamnophis sirtalis	8	138	17	0.044	0.185	2	4	3	2	4	3	4	2	3	1	4	2	3	1	2	1			2		2		56,391
Typhlosaurus garienpensis	2	16	34	0.007	0.084																							224
Typhlosaurus lineatus	1	10	34	0.018	0.202																							224
Uma exual	1	26	22	0.003	0.165	4	5	3	1	2	1	2	1	3		3	2	3	2	1		2	34	1		2		3
Uma inornata	1	16	18	0.003	0.111	2	5	3	1	1	1	1	1	3		3	2	3	1	1		2	34	1		2		3
Uma motata	1	19	18	0.029	0.222	2	5	3	1	1	1	2	1	3		3	2	3	1	1		2	34	1		2		3
Uma paraphygas	2	35	22	0.017	0.333	4	5	3	1	2	1	2	1	3		3	2	3	2	1		2	34	1		2		3
Uma scoparia	1	17	18	0.010	0.111	2	5	3	1	1	1	2	1	3		3	2	3	1	1		2	34	1		2		3
Uta stansburiana	34	870	19	0.052	0.254	2	3	3	2	1	3	4	1	2	1	2	1	1	1	1	1			3	3	2		296,451
Xantusia riversiana	5	47	25	0.016	0.110	2	1	3	1	2	2	1	1	1	1	4	1	3	2	1	1	2	40	2		2		66
Xantusia vigilis	1	20	30	0.026	0.140	2	3	3	1	1	2	2	1	1	1	4	1	3	2	1	1	2	40	1		2		66
AMPHIBIA																												
Acris crepitans	32	850	21	--	0.208	2	4	61	2	6	2	3	1	4	3	3	3	1	1	1	1	2	22	1		2		134

Appendix I. (contin.)

Species	No. of: populations	indi.	loci	Genetic indices H	P	@ Biotic Profile Lz	Gr	Ht	Hr	Ar	Ps	Ss	Bs	Mo	Gf	Lo	Fe	Gl	Or	Tr	So	Pl	Ch	Ds	Ma	Rp	Po	References (according to ref. of the data)
Aneides flavipunctatus	22	405	21	0.114	0.462	2	3	3	1	4	2	3	2	3	1	3	2	3	2	1	1	2	28	1	3	2		243
Aneides hardii	3	108	21	0.039	0.131																							364
Batrachoseps campi	13	106	33	--	0.210																							502
Batrachoseps wrighti	1	16	33	--	0.182																							502
Bombina orientalis	1	156	18	0.131	0.417	2	2	61	1	7	3	3	2	3	1	2	2	2	2	2					3			269
Bufo americanus	25	620	14	0.116	0.343	2	3	3	2	4	3	3	2	3	2	3	4	3	2	2	1	2	22	3				179
Bufo arenarum	15	407	16	0.163	0.442																							287
Bufo boreas	1	25	21	0.063	0.195	2	3	7	2	6	3	3	2	3	2	4	3	3	2	2	1	2	22	1		2		141
Bufo fowleri	10	50	14	0.117	0.388	2	3	3	2	4	3	3	2	3	2	3	4	3	2	2	1	2	22	3				180
Bufo punctatus	4	40	21	0.025	0.096	2	3	7	1	1	2	2	2	3	1	4	3	3	2	2	1	2	28	1		2		141
Bufo viridis	11	507	26	0.130	0.435	2	4	7	2	4	3	2	2	3	2	3	3	3	1	1	1	2	22	1		2		135
Cryptobranchus alleganiensis	12	137	24	0.006	0.015	2	3	61	1	4	2	1	3	3	2	4	2	3	2	2	1			1		2		303
Desmognathus fuscus	24	437	19	--	0.182	2	3	2		4	2	3	2	3	1	4	2	3	2	2	1	2	34	1	3	2		215,469
Desmognathus imitator	1	38	23	0.222	0.440	2	1	2	1	4	2	3	2	3		4	2	3	2	2	1			1	3	2		468
Desmognathus ochrophaeus	19	521	15	0.193	0.590	2	3	2	2	4	2	3	2	3	1	4	2	3	2	2	1	2	34	1	3	2		215,468
Eurycea lucifuga	33	766	13	0.025	0.137	2	3	3	1	4	2	2	1	3	2	4	2	3	1	2	1			1		2		302,395
Hydromantes platycephalus	1	19	18	--	0.153																							485
Hydromantes shostae	5	38	18	--	0.175																							485
Hyla andersonii	3	15	19	0.025	0.241	2	5	3	1	4	1	1	1	3	1	4	2	3	2	2	1			1	3	2		180
Hyla arborea	8	218	27	0.072	0.394	2	3	4	2	3	3	2	1	3	2	3	2	3	1	1	1	2	22	1		2		323,336
Hyla cadaverina	1	12	14	0.030	0.197	2	3	7	2	6	2	2	1	4	1	4	2	3	2	2	1	2	24	1		2		100
Hyla chrysoscelis	9	238	12	0.068	0.324	2	3	7	2	4	2	2	1	3	2	3	3	2	1		4	2		1				372
Hyla cinerea	2	15	19	0.077	0.350	2	2	3	2	4	2	3	1	3	1	4	2	3	2	2	1			1	3	2		180
Hyla eximia	1	21	12	0.051	0.452	3	3	7	2	3	2	2	1	4	1	4	2	3	2	2	1	2	24	1		2		100
Hyla regilla	17	340	13	0.049	0.381	2	3	7	2	4	3	3	1	4	1	4	2	3	2	2	1	2	24	1		2		100
Hyla versicolor	4	132	12	0.322	0.479	2	3	7	2	4	2	2	1	3	2	3	3	2	1		4	4		1				372
Leiopelma hochstetteri	5	16	33	0.048	0.166																							124
Litoria angiana	2	14	18	--	0.244	2	2	61	1	3	2	2	1	1	1	3	3	1	1	2	1	2	26	1	3	2		132
Litoria infrafrenata	5	36	19	--	0.252	3	4	61	2	3	2	2	1	1	1	3	3	1	1	2	1	2	26	1	3	2		132
Litoria iris	1	19	17	--	0.176	2	2	61	1	3	2	2	1	1	1	3	3	1	1	2	1			1	3	2		132
Litoria micromembrana	2	17	12	--	0.138	2	2	61	1	3	2	2	1	1	1	3	3	1	1	2	1			1	3	2		132
Litoria modica	2	10	13	--	0.169	2	2	61	1	3	2	2	1	1	1	3	3	1	1	2	1	2	26	1	3	2		132
Necturus lewisi	1	20	17	0.018	0.084	2	1	611	1	4	2	2	2	4	2	4	2	3	2	1	1		36	1		2		14
Necturus maculosus	3	21	17	0.006	0.012	2	4	611	2	4	2	3	3	4	2	4	2	3	2		3		36	1		2		14
Necturus punctatus	3	36	19	0.026	0.084	2	3	611		4	2	2	3	3	2	3	2	2	2	2	1		36	1		2		14,133
Pelobates syriacus	3	61	32	0.028	0.076	2	2	1	1	3	1	2	2	2	1	3	3	3	1	1	1			1		2		323,324
Plethodon caddoensis	3	86	23	0.069	0.372																							136
Plethodon cinereus	15	331	24	0.044	0.166	2	4	2	2	4	3	4	1	1		4	1	3	2	1	1	2	28	1	3	2		199

Appendix I. (contin.)

Species	No. of: populations	No. of: indi.	No. of: loci	Genetic indices H	Genetic indices P	Lz	Gr	Ht	Hr	Ar	Ps	Ss	Bs	Mo	Gf	Lo	Fe	Gl	Or	Tr	So	Pl	Ch	Ds	Ma	Rp	Po	References (according to ref. of the data)
						Biotic Profile @																						
Plethodon dorsalis	18	417	26	0.040	0.137	2	3	3	1	4	2	3	1	3	1	3	2	3	2	1	1	2	28	1	3	2		245
Plethodon dunni	3	18	24	0.009	0.065	2	2	2	1	4	2	2	2	2	1	4	2	3	2	2	1	2	28	1		2		142
Plethodon fourchesis	3	71	23	0.028	0.254																							136
Plethodon glutinosus	39	277	18	0.144	0.431																							136,244
Plethodon gordoni	2	14	24	0.018	0.065	2	5	2	1	4	1	1	2	2	1	4	2	3	2	2	1	2	28	1				142
Plethodon jordani	12	60	10	0.086	--																							244
Plethodon ouachitae	10	314	23	0.056	0.334																							136
Plethodon serratus	9	194	24	0.033	0.097	2	3	2	2	4	3	4	1	1		4	1	3	2		1	2	28	1	3	2		199
Plethodon vehiculum	1	30	24	0.089	0.330	2	3	2	1	4	2	2	2	2	1	4	2	3	2	2	1	2	28	1		2		142
Plethodon websteri	8	221	26	0.031	0.131	2	3	3	1	4	2	3	1	3	1	3	2	3	2	1	1	2	28	1	3	2		245
Plethodon welleri	3	91	26	0.043	0.140	2	5	3	1	4	1	2	1	3	1	3	2	3	2	1	1	2	28	1	3	2		245
Plethodon yonahalossee	9	214	22	0.020	0.119	2	2	3	1	4	3	2	2	3	1	3	2	3	2	1	1			1				136,182
Pseudoeurycea smithi	1	26	22	0.091	0.500	3	3	3	1	5	2	3	2	2	1	4	2	3	2	2	1	2	26	1		2		266
Pseudoeurycea unguidentis	1	21	22	0.041	0.182	3	1	3	1	5	2	2	2	2	1	4	2	3	2	2	1	2	26	1		2		266
Rana aurora	7	18	20	0.039	0.450	2	3	7	1	4	2	2	2	3	1	4	3	3	2	2	1	2	26	1				98
Rana boyleii	6	104	20	0.038	0.223	2	3	61	2	4	2	2	2	3	1	4	3	3	2	2	1	2	26	1		2		98,99
Rana cascadae	2	25	20	0.034	0.262	2	2	61	1	4	2	2	2	3	1	4		3	2	2	1	2	26	1				98
Rana catesbeiana	1	12	20	0.0	0.0	2	4	61	2	4	3	3	2	4	1	4	4	3	2	1	1	2	26	1				98
Rana muscosa	16	170	19	0.065	0.359	2	2	61	1	4	2	2	2	3	1	4	3	3	2	2	1	2	26	1		2		98,99
Rana pipiens	17	85	16	--	0.122	2	4	7	2	3	2	3	2	4	2	3	4	3	2	2	1	2	26	1		2		137
Rana pretiosa	2	22	20	0.060	0.250	2	3	61	2	4	3	2	2	3	1	4	4	3	2	2	1	2	26	1				98
Rana ridibunda	11	340	28	0.069	0.328	2	4	61	2	3	3	2	2	3	2	3	2	3	1	1	1	2	26	1		2		323,336
Rana tarahumarae	1	13	20	0.0	0.0	4	3	61	1	3	2	2	2		1	4	3	3	2	2	1	2	26	1				98
Scaphiopus bombifrons	5	130	21	0.082	0.605	2	3	2	2	1	3	3	2	2	1	3	4	2	1	2	1	2	26	1		2		180,390
Scaphiopus couchi	1	21	21	0.026	0.261	2	3	2	2	1	3	3	2	2	1	3	4	2	1	2	1	2	26	1		2		180,390
Scaphiopus holobrooki	4	63	21	0.039	0.285	2	3	2	2	4	2	3	2	2	1	3	4	2	1	2	1	2	26	2		2		180,390
Scaphiopus multiplicatus	8	137	21	0.083	0.439	2	3	2	2	1	3	2	2	2	1	3	4	2	1	2	1	2	26	1		2		180,390
Taricha granulosa	7	165	29	0.091	0.393	2	3	7	2	4	2	3	3	5	1	4	3	3	2	2	1	2	20	1		2		196,193
Taricha rivularis	16	645	36	0.067	0.228	2	2	7	2	4	2	3	3	5	1	4	3	3	2	2	1	2	20	1		2		196,194
Taricha torosa	10	325	25	0.058	0.253	2	3	7	2	4	2	3	3	5	1	4	3	3	2	2	1	2	20	1		2		196,193
Triturus alpestris	2	105	21	0.156	0.554	2	4	7	2	6	2	2	2	3	1	4	3	3	2	1	1	2	24	1				214
Triturus cristatus	4	119	30	0.035	0.241	2	4	7	2	4	2	3	3	3	1	4	3	3	2	2	1	2	24	1		2		13,214
Triturus marmoratus	1	20	50	--	0.032	2	3	7	2	4	3	3	3	3	1	3	3	3	2	2	1	2	24	1		2		13
Triturus vulgaris	3	92	16	0.090	0.401	2	4	7	2	6	2	2	2	3	1	4	3	3	2	1	1	2	24	1				214

Appendix I. (contin.)

Species	No. of: populations	indi.	loci	Genetic indices H	P	Biotic Profile @ Lz	Gr	Ht	Hr	Ar	Ps	Ss	Bs	Mo	Gf	Lo	Fe	Gl	Or	Tr	So	Pl	Ch	Ds	Ma	Rp	Po	References (according to ref. of the data)
PISCES																												
Abudefduf saxatilis	1	26	28	--	0.179	3	4	621	2	7	2	2	1	3	1	3	4	1	2	1	3					2		167
Abudefduf troschelii	2	39	25	0.050	0.260	3	4	621	2	7	2	2	1	3	2	3	4	1	2	1	3			4		2		167,450
Agrammus agrammus	1	13	14	0.060	0.197																							149
Albula glossodonta	1	90	84	0.022	0.096																							419
Albula neoguinaica	1	90	84	0.005	0.084																							419
Alosa sapidissima	1	15	16	0.005	0.063	2	4	63	2	7	3	3	2	5	2	4	4	3	2	2	1	2	48	1				309
Ambloplites rupestris	1	12	11	0.114	0.460																							29
Amblyopsis rosae	2	15	10	0.006	--																							462
Amblyopsis spela	3	15	10	0.0	--																							462
Ammodytes personatus	1	30	15	0.067	0.352																							149
Amphiprion clarkii	1	11	27	0.091	--	3	3	62	1	7	2	3	1	1	2	3	3		2	1	3			4	3	2		450
Anguilla anguilla	1	74	20	0.181	0.650	5	4	63	2	7	1	4	1	5	3	4	4	3	2	2	1	2	38		3	2		381,382
Aphanias dispar	5	150	19	0.049	0.178	4	3	621	1	7	2	2	1	4	3	2	2	1	1	2	4			4		2		234
Archoplites interruptus	1	39	11	0.038	0.0	2	4	61	2	7	2	4	2	3	3	3	4	2		1	4			1		2		29
Areliscus joyneri	1	47	20	0.146	0.544																							149
Astyanax mexicanus	9	393	17	0.086	0.373	2	3	61		7	2	3		3	3	3	3	2		1	4	2	48	1		2		22
Atherina boyeri	4	235	15	--	0.377	2	3	631	2	7	3	2	1	5	1	2	3	2		2	4					2		59
Atherinops affinis	1	15	32	--	0.160																							115
Bathygobius andrei	1	24	26	0.027	0.110	3	4	621	2	7	2	2	1	2	1	3	4	2	2	1	3					2		168
Bathygobius ramosus	1	20	26	0.056	0.305	3	4	621	2	7	2	2	1	2	1	3	4	2	2	1	3					2		168
Bathygobius soporator	1	24	26	0.069	0.212	3	4	621	2	7	2	2	1	2	1	3	4	2	2	1	3					2		168
Campostoma anomalum	17	246	18	0.070	0.351	2	3	61	2	7	2	3	2	5	2	3	4	2		1	4			1		2		88,508
Campostoma oligolepis	8	164	18	0.079	0.383	2	3	61	1	7	3	3	2	5	2	3	4	2		1	4			1		2		88,508
Campostostoma ornatum	2	60	19	0.009	0.076	2	3	61	2	7	3	3	1	5	3	3	4	3	2	1	4			1		2		88
Carpiodes carpio	1	15	31	0.089	0.449	2	3	61	2	7	2	2	2	5	2	4	4	3	2	2	1	4		1	3	2		145
Carpiodes cyprinus	1	15	31	0.083	0.409	2	3	61	2	7	2	2	2	5	2	4	4	3	2	2	1	4		1	3	2		145
Carpiodes velifer	1	15	32	0.081	0.396	2	3	61	2	7	2	2	2	5	2	4	4	3	2	2	1	4		1	3	2		145
Catostomus catostomus	1	15	29	0.038	0.236	5	4	61	2	7	2	3	3	5	2	4	4	3	2	2	1	4		1	3	2		145
Catostomus columbianus	1	15	29	0.048	0.190	2	2	61	2	7	2	2	3	5	2	4	3	3	1	2	1	4		1	3	2		145
Catostomus commersoni	2	50	27	0.024	0.153	2	4	61	2	7	3	3	3	5	2	4	4	3	1	2	1	4		1	3	2		145
Catostomus plebeius	5	53	27	0.014	0.096	2	3	61	2	7	2	2	2	4	3	3	3	3	1	2	4	4	00	1		2		144,145
Catostomus santaanae	4	200	33	0.039	0.011	2	2	61	2	7	3	2	1	4	3	3	3	3			4	4	00	1		2		90
Chanos chanos	14	609	23	0.074	0.240	3	4	621	2	7	3	4	2	4	3	3	4	3	1	2	1			4		2		497
Chelidonichtys kumu	1	28	14	0.041	0.102																							149
Chelon labrosus	1	15	16	--	0.090																							16

Appendix I. (contin.)

Species	No. of: populations	indi.	loci	Genetic indices H	P	Lz	Gr	Ht	Hr	Ar	Ps	Ss	Bs	Mo	Gf	Lo	Fe	Gl	Or	Tr	So	Pl	Ch	Ds	Ma	Rp	Po	References (according to ref. of the data)
Chologaster agassizi	10	50	10	0.026	--																							462
Chologaster cornuta	11	55	10	0.038	--																							462
Cichlosoma cyanoguttatum	4	192	26	0.0	0.187	4	1	61	1	7	2	2	2	1	2	2	4	1	1	1	2	2	70	1		2		233,388
Cichlosoma sp. (endemic Mexico)	3	118	13	0.035	0.150																							233
Cleisthenes pinetorum	1	13	23	0.107	0.345																							149
Clidoderma asperrimum	1	12	23	0.062	0.345																							149
Clupea harengus harengus	3	160	24	0.070	0.362	9	4	62	2	7	3	4	2	4	3	4	3	3	1	2	3					2		10
Clupea pallasi	1	50	17	0.058	0.310																							149
Cololabis saira	1	27	20	0.174	0.544																							149
Conger conger	1	39	19	0.076	0.316	2	4	623	2	7	2	3	3	5	2	4	4	3	2		1				3	2		382
Coregonus albula	20	695	25	0.080	0.552	2	3	612	2	7	2	4	1	4	2	3	4	3	1	2	3	2	80	1	3	2		484
Coregonus clupeaformis dwarf	2	121	18	--	0.260																							227
Coregonus clupeaformis normal	2	43	18	--	0.260																							227
Ctenopharyngdon idella	6	100	49	0.022	0.062	4		61		7			3				4	3				2	48			2		476
Cycleptus elongatus	1	15	33	0.050	0.250	2	3	61	1	7	2	1	3	5	2	4	4	3	2	2	1	4		1	3	2		145
Cyprinodon macularius	1	11	38	--	0.244																							473
Cyprinodon nevadensis	3	66	38	--	0.244																							473
Cyprinodon salinus	1	40	38	--	0.244																							473
Dascyllus reticulatus	1	10	29	0.107	--	3	3	62	2	7	2	3	1	2	2	3	3		2	1	4			4	3	2		450
Dexistes rikuzenius	1	30	15	0.058	0.183																							149
Elassoma evergladei	1	12	11	0.0	0.0	2	3	61	2	7	2	3	1	3	3	3	4	2		1	4			1		2		29
Elassoma okefenokee	1	20	11	0.013	0.129	2	3	61	2	7	2	3	1	3	2	2	4	2		1	4			1		2		29
Engraulis japonica	1	30	22	0.067	0.300																							149
Enneacanthus obesus	1	27	11	0.0	0.0	2	3	61	2	7	2	4		3	3	3	4	2		1	4			1		2		29
Eopsetta grigorjewi	1	18	22	0.015	0.130																							149
Erimyzon oblongus	1	15	29	0.043	0.236	2	2	61	2	7	2	2	1	5	2	4	3	3	2	2	1	4		1	3	2		145
Erimyzon sucetta	1	15	29	0.058	0.273	2	2	61	2	7	2	2	1	5	2	4	3	3	2	2	1	4		1	3	2		145
Erimyzon tenuis	1	15	27	0.047	0.203	2	2	61	2	7	2	2	1	5	2	4	3	3	1	2	1	4		1	3	2		145
Etheostoma fonticola	1	20	23	0.017	0.130	2	1	61	1	7	1	1	1	3	1	1	2	2	1		4			1		2		89
Etheostoma microperca	6	60	23	0.010	0.055	2	3	61	1	7	2	2	1	3	1	1	2	2	1	1	4			1		2		89
Etheostoma proeliare	3	24	23	0.033	0.140	2	3	61	2	7	2	2	1	3	1	1	2	2	1		4			1		2		89
Fundulus heteroclitus	1	65	25	0.180	0.560																							308
Gadus morhua	1	80	30	0.082	0.300	9	4	62	2	7	3	3	2	4	3	4	3	3	1	2	3					2		311
Gibbonsia metzi	1	28	28	0.043	0.247	2	3	62	1	7	2	3	1	1	2	3	3		2	2	1			4	3	2		450
Gila bicolor	1	24	24	0.059	0.277	2	3	61		7	2	3		3	3	3	3	2		1	4	2	48	1		2		17
Gillichthys mirabilis	1	30	29	0.046	0.275	2	3	62	1	7	2	3	1	1	2	3	3		2	1				4	3	2		450
Glyptocephalus cynoglossus	1	22	31	0.063	0.290																							489
Gobius niger	4	102	23	0.081	0.285	2	3	621	2	7	3	4	1	1	2	3	3	3		1	2			4				58

@ Biotic Profile

Appendix I. (contin.)

Species	No. of: populations	indi.	loci	Genetic indices: H	P	Biotic Profile @: Lz	Gr	Ht	Hr	Ar	Ps	Ss	Bs	Mo	Gf	Lo	Fe	Gl	Or	Tr	So	Pl	Ch	Ds	Ma	Rp	Po	References (according to ref. of the data)
Goodea stripinnis	5	25	34	--	0.487																							495
Halichires sp.	1	10	28	0.057	--	3	3	62		7		3	1	2	2	3			2					4	3	2		450
Haplochromis flaviijosephi	1	25	21	--	0.095																							235
Helicolenus hilgendorfi	1	15	16	0.083	0.258																							149
Hemiramphus sajori	1	12	21	0.063	0.261																							149
Hemitripterus villosus	1	12	19	0.022	0.076																							149
Hesperoleucus symmetricus	14	250	24	0.068	0.330	2	3	61		7		3		3	3	3	3	2		1	4	2	48	1		2		23
Hexagrammus otakii	1	38	21	0.003	0.0																							149
Hippoglossides platessoides	1	47	35	0.083	0.400																							489
Hypentelium etowanum	2	24	30	0.028	0.102	2	3	61	2	7	3	3	1	4	3	4	4	3			4	4	00	1		2		87
Hypentelium nigricans	4	52	19	0.016	0.127	2	3	61	2	7	2	3	2	5	2	4	4	3	2	2	4	4	00	1		2		87,145
Hypentelium roanokense	2	29	40	0.021	0.079	2	1	61	2	7	2	2	1	4	3	3	4	3			4	4	00	1		2		87
Ictiobus bubalus	1	15	32	0.103	0.395	2	3	61	2	7	2	2	3	5	2	4	4	3	2	2	1	4		1	3	2		145
Ictiobus cyprinellus	1	15	32	0.102	0.371	2	3	61	2	7	2	2	3	5	2	4	4	3	2	2	1	4		1	3	2		145
Kareius bicoloratus	1	29	22	0.099	0.402																							149
Labeotropheus fulleborni	3	166	17	0.103	0.343																							232
Lamperta planeri	1	48	30	0.076	0.300	2	4	61	1	7	2	2	1	3	1	4	3	3	1	2	4	2	46			2		490
Lateolabrax japonicus	1	15	19	0.095	0.400																							149
Lavinia exilicauda	10	300	24	0.053	0.330	2	3	61		7	2	3		3	3	3	3	2		1	4	2	48	1		2		23
Lepidotrigla microptera	1	13	18	0.042	0.153																							149
Lepomis aurilus	2	45	14	0.062	0.277	2	4	61	2	7	2	4		3	3	3	3	2		1	4	2	48	1		2		24
Lepomis gibbosus	1	29	14	0.066	0.192	2	4	61	2	7	2	4		3	3	3	3	2		1	4	2	48	1		2		24
Lepomis gulosus	2	39	14	0.030	0.151	2	4	61	2	7	2	4		3	3	3	3	2		1	4	2	48	1		2		24
Lepomis humilis	1	35	14	0.049	0.192	2	4	61	2	7	2	4		3	3	3	3	2		1	4	2	48	1		2		24
Lepomis lyanellus	1	28	14	0.074	0.192	2	4	61	2	7	2	4		3	3	3	3	2		1	4	2	48	1		2		24
Lepomis macrochirus	64	2855	16	0.041	0.149	2	4	612	2	7	2	4	2	3	2	4	3	2	1	1	4	2	48	1		2		25,143
Lepomis marginatus	1	34	14	0.069	0.192	2	4	61	2	7	2	4		3	3	3	4	2		1	4	2	48	1		2		24
Lepomis megalotis	1	30	14	0.114	0.443	2	4	61	2	7	2	4		3	3	3	3	2		1	4	2	48	1		2		24
Lepomis microlophus	3	70	14	0.037	0.114	2	4	61	2	7	2	4		3	3	3	3	2		1	4	2	48	1		2		24
Lepomis punctatus	1	29	14	0.113	0.277	2	4	61	2	7	2	4		3	3	3	3	2		1	4	2	48	1		2		24
Leuresthes tenuis	1	20	33	0.036	0.206	2	2	62	1	7	3	3	1	4	2	3	4		2	2	4			4	3	2		450
Limanda herzensteini	1	27	23	0.121	0.440																							149
Limanda limanda	2	151	34	0.069	0.231	2	4	62	2	7	3	3	2	4	2	4	4	3	2	2	4	2	48			2		488,489
Limanda yokohamae	1	21	23	0.064	0.286																							149
Liparis tanakai	1	15	22	0.0	0.0																							149
Liza aurata	1	15	16	--	0.090																							16
Liza ramanda	1	15	16	--	0.0																							16
Liza saliens	1	15	16	--	0.0																							16

Appendix I. (contin.)
contin.)

Species	No. of:			Genetic indices		Biotic Profile @																						References (according to ref. of the data)
	populations	indi.	loci	H	P	Lz	Gr	Ht	Hr	Ar	Ps	Ss	Bs	Mo	Gf	Lo	Fe	Gl	Or	Tr	So	Pl	Ch	Ds	Ma	Rp	Po	
Lophius litulor	1	15	15	0.0	0.0																							149
Lotella maximowiczi	1	15	15	0.004	0.0																							149
Menidia beryllina	8	361	24	0.042	0.100	4	3	62	2	7	3	3	1	4	2	3	3	2	1	1	1	2		3		2		205,206,306
Menidia beryllina (audens)	3	126	24	0.070	0.192	2	4	631	2	7	3	3	1	4	2	2	3	2	1	2	4			1	3	2		205,206,306
Menidia extensa	1	43	24	0.033	0.114	2	1	611	1	7	1	2	1	4	2	2	3	2	1	2	4			1	3	2		205,206,306
Menidia menidia	5	219	24	0.054	0.220	2	3	621	2	7	3	3	1	4	2	2	3	2	2	2	4			1	3	2		205,206,306
Menidia peninsulae	5	265	24	0.055	0.100	4	3	621	2	7	3	3	1	4	2	2	3	2	1		4			1	3	2		205,206,306
Microstomus achne	1	26	21	0.108	0.421																							149
Microstomus kitt	1	58	38	0.057	0.180																							489
Microterus salmoides	1	18	11	0.073	0.250	2	4	61	2	7	2	4	3	3	3	3	4	2		1	4			1		2		29
Minytrema melanops	1	15	27	0.059	0.293	2	2	61	2	7	2	2	2	5	2	4	3	3	2	2	1	4		1	3	2		145
Moxostoma cervinum	1	15	30	0.044	0.230	2	2	61	2	7	2	2	1	5	2	4	3	3	1	2	1	4		1	3	2		145
Moxostoma doquesnei	1	15	27	0.015	0.153	2	2	61	2	7	2	2	2	5	2	4	3	3	2	2	1	4		1	3	2		145
Moxostoma erythrurum	1	15	27	0.034	0.203	2	2	61	2	7	3	3	2	5	2	4	3	3	2	2	1	4		1	3	2		145
Moxostoma macrolepidotum	1	15	27	0.075	0.293	2	2	61	2	7	2	2	2	5	2	4	3	3	2	2	1	4		1	3	2		145
Mugil cephalus	2	35	26	0.071	0.196	4	4	63	2	7	3	4	2	5	3	4	4	3	2	2	4			1	3	2		16,450
Mylopharodon concephalus	1	60	24	0.006	0.064	2	3	61		7	2	3		3	3	3	3	2		1	4	2	48	1		2		17
Navodon modestus	1	30	24	0.013	0.067																							149
Nibea mitsukurii	1	15	15	0.018	0.096																							149
Notemigonus crysoleucus	1	15	24	0.068	0.277	2	3	61		7	2	4		3	3	3	3	2		1	4	2	48	1		2		17
Notropis albeolus	4	61	17	--	0.183	2	1	61	2	7	3	3	2	4	3	3	3	3		1	4	2	50	1		2		86
Notropis cerasinus	4	64	17	--	0.137	2	3	61	2	7	3	3	2	4	3	3	3	3		1	4	2	50	1		2		86
Notropis chrysocephalus	8	133	17	--	0.117	2	4	61	2	7	3	3	2	4	3	3	4	3		1	4	2	50	1		2		86
Notropis coccogenis	2	37	17	--	0.071	2	3	61	2	7	3	3	2	4	3	3	3	3		1	4	2	50	1		2		86
Notropis cornutus	2	37	17	--	0.138	2	4	61	2	7	3	3	2	4	3	3	4	3		1	4	2	50	1		2		86
Notropis lutrensis	1	15	20	0.080	--	2	3	61	2	7	3	4	1	3	2	2	3	1	1	1	1	2	50	1	3	2		500
Notropis pilsbryi	4	74	17	--	0.217	2	3	61	2	7	3	3	2	4	3	3	3	3		1	4	2	50	1		2		86
Notropis venustus	1	15	20	0.062	--	2	3	61	2	7	2	3	1	3	2	2	3	1	1	2	1	2	50	1	3	2		500
Notropis zonatus	2	37	17	--	0.0	2	3	61	2	7	3	3	2	4	3	3	3	3		1	4	2	50	1		2		86
Notropis zonistius	2	37	17	--	0.071	2	1	61	2	7	3	3	2	4	3	3	3	3		1	4	2	50	1		2		86
Oncorhynchus gorbuscha	10	910	29	0.032	0.151	4	4	63	2	7	3	3	3	5	2	3	4	3	1	2	4	2	52	3		2		7,8,15,289,475
Oncorhynchus keta	7	706	28	0.034	0.079	4	4	63	2	7	3	3	3	5	2	3	4	3	1	2	4	2	74	3		2		7,289,475
Oncorhynchus kisutch	14	1971	27	0.016	0.129	2	4	63	2	7	3	3	3	5	2	3	4	3	1	2	4	2	60	3		2		7,475
Oncorhynchus nerka	16	2067	27	0.017	0.100	2	4	63	1	7	2	3	3	5	2	3	4	3	1	2	4	2	56	3		2		7,475
Oncorhynchus tshawytscha	16	1516	28	0.028	0.196	2	4	63	2	7	3	3	3	5	2	3	4	3	1	2	4	2	68	3		2		7,238,475
Orthodon microlepidotus	1	23	24	0.015	0.064	2	3	61		7	2	2		3	3	3	3	2		1	4	2	48	1		2		17
Paralichthys olivaceus	1	12	20	0.050	0.275																							149
Perca flavescens	14	1257	19	--	0.136	2	4	61	2	7	2	3	2	3	2	3	4	3	2	2	4	2	48	1				250

Appendix I. (contin.)

Species	No. of: populations	indi.	loci	Genetic indices (ref. of the data) H	P	Biotic Profile @ Lz	Gr	Ht	Hr	Ar	Ps	Ss	Bs	Mo	Gf	Lo	Fe	Gl	Or	Tr	So	Pl	Ch	Ds	Ma	Rp	Po	References (according to species)
Petrotilapia tridentiger	1	55	14	0.069	0.192																							232
Platichthys flesus	9	370	34	0.083	0.344	2	4	62	2	7	3	3	2	4	2	4	4	3	2	2	4	2	48			2		151,488,489
Platichthys italicus	4	165	36	0.035	0.273	2	2	62	1	7	1	2	2	4	2	4	4	3	1	2	4							489
Platichthys luscus	4	26	35	0.025	0.138	2	2	62	1	7	1	2	2	4	2	4	4	3	1	2	4							489
Platichthys stellatus	1	48	22	0.047	0.700																							149
Pleurogrammus azonus	1	40	18	0.035	0.153																							149
Pleuronectes platessa lim	6	3226	45	0.102	0.480	2	4	62	2		3	3	2	4	2	4	4	3	2	2	4	2	48			2		487,488,489
Pleuronichthys cornutus	1	19	23	0.087	0.578																							149
Poecilia mexicana	1	38	24	0.022	0.101																							482
Poecilia reticulata	2	124	16	0.102	0.282																							420
Poeciliopsis monacha	8	375	25	0.052	0.163	4	2	61	1	7	1	2	1	3		2	3	1	1	2	1	2	48	1	3	2		481
Poeciliopsis occidentalis	5	202	25	0.027	0.151	4	3	61	2	7	2	3	1	3		2	3	1	1	2	1	2	48	1	3	2		483
Polyodon spathula	6	73	35	0.013	0.060	2	3	612		7	3	2	3	5	2	4	4	3	2	2	1	4	120	1		2		95,96
Pomoxis nigromaculatus	1	18	11	0.009	0.129	2	4	61	2	7	2	4		3	3	3	4	2		1	4					2		29
Pseudotropheus elegans	1	64	14	0.087	0.277																							232
Pseudotropheus livingstoni	1	99	15	0.061	0.179																							232
Pseudotropheus tropheus	1	75	15	0.062	0.179																							232
Pseudotropheus zebra	1	75	15	0.062	0.179																							232
Ptychocheilus grandis	1	28	24	0.011	0.064	2	3	61		7	2	2		3	3	3	3	2		1	4	2	48	1		2		17
Rhinichthys cataractae	13	331	21	0.057	0.215	2	4	61	1	7	2	3	1	3	2	3	3	2	2	1	1	2	54	1	3			304
Rhinoplagusia japonica	1	10	21	0.124	0.519																							149
Richardsonius egregius	1	22	24	0.030	0.114	2	3	61		7	2	2		3	3	3	3	2		1	4	2	48	1		2		17
Salmo apache	1	100	30	0.0	--																							7
Salmo clarki	44	1861	32	0.040	0.109	2	4	63	2	7	2	3	2	5	2	3	4	3	1	2	4	2	66	1		2		7,263,264,361
Salmo gairdneri	51	5041	29	0.060	0.272	2	4	63	2	7	3	3	2	5	1	3	4	3	1	1	1	2	60	1				7,84,146,475
Salmo salar	18	1434	43	0.023	0.101	9	4	632	2	7	3	3	2	5	2	4	4	3	1	2	4	2	58			2		7,119,220,387,452
Salmo trutta	22	1012	40	0.022	0.092	9	4	63	2	7	2	3	2	5	3	4	3	3	1	2	3	2	80			2		104,387
Salvelinus alpinus	9	474	37	0.006	0.016	9	4	63	2	7	2	3	2	5	3	4	3	3	1	2	3	2	80			2		387
Salvelinus fontialis	8	289	39	0.077	0.252																							458
Salvelinus namaycush	3	484	50	0.015	0.142	2	3	61	2	7	2	3	3	5	2	4	4	3	1	2	1	2	84					125
Sardinops melanostica	1	30	22	0.064	0.250																							149
Sarotherdon aureus	1	30	21	--	0.238																							235
Sarotherdon galilaeus	1	31	21	--	0.095																							235
Scaphirhynchus albus	1	15	37	0.010	0.098	2	3	61	2	7	3	1	3	4	2	4	4	3	2	2	1							362
Scaphirhynchus platorynchus	1	15	37	0.017	0.098	2	3	61	2	7	3	2	3	4	2	4	4	3	2	2	1							362
Scomber japonicus	1	39	19	0.093	0.347																							149
Sebastes alutus	3	2000	23	0.038	--	2	3	62	2	7	3	3	2	4	3	4	4	3	1		4				4	2		202,203
Sebastes caurinus	1	480	23	0.014	--	2	3	62	2	7	2	2	2	4	3	4	4	3	1		4				4	2		202,203

Appendix I. (contin.)

Species	No. of:			Genetic indices		@ Biotic Profile																						References (according to ref. of the data)
	populations	indi.	loci	H	P	Lz	Gr	Ht	Hr	Ar	Ps	Ss	Bs	Mo	Gf	Lo	Fe	Gl	Or	Tr	So	Pl	Ch	Ds	Ma	Rp	Po	
Sebastes elongatus	2	345	23	0.025	--	2	3	62	2	7	2	2	2	4	3	4	4	3	1		4			4		2		202,203
Sebastes inermis	1	30	16	0.013	0.090																							149
Sebastes thomsoni	1	16	18	0.004	0.0																							149
Sebastolobus alascanus	6	63	20	0.049	0.200	2	4	6 3	2	7	2	3	2	3	3	3	3	3								2		426
Sebastolobus altivelis	10	352	20	0.047	0.300	2	4	6 3	2	7	2	3	2	3	3	3	3	3								2		426
Sebastolobus macrochir	1	30	16	0.063	0.258																							149
Sphyraena argentea	1	10	33	--	0.166																							175
Sphyraena ensis	1	10	33	--	0.087																							175
Sphyraena lucasana	1	10	33	--	0.048																							175
Spratelloides gracilis	4	480	15	0.042	0.307	3	4	622	2	7	2	3	1	3	1	1	3	1		2	4	2		4	3	2		121
Stolephorus devisi	4	880	19	0.068	0.263	3	4	622	2	7	2	3	1	3	1	1	3	1		2	4	2		4	3	2		121
Stolephorus heterolobus	4	657	18	0.101	0.222	3	4	622	2	7	2	3	1	3	1	1	3	1		2	4	2		4	3	2		121
Tanakius kitaharai	1	10	23	0.030	0.239																							149
Theraga chalcograma	1	10	17	0.006	0.084																							149
Thoburnia atripinnis	1	15	34	0.008	0.046	2	5	61	1	7	1	1	1	4	3	3	3	3			4	4	00	1		2		85
Thoburnia hamiltoni	1	11	34	0.015	0.084	2	1	61	1	7	1	1	1	4	3	3	3	3			4	4	00	1		2		85
Thoburnia rhothoeca	2	36	34	0.055	0.242	2	2	61	1	7	2	2	1	4	3	3	3	3			4	4	00	1		2		85
Thymallus arcticus	4	60	36	0.033	0.110	9	2	61	1	7	3	3	1	4	1	3	4	3	2	1	1	2		2		2		265
Tilapia zillii	1	30	21	--	0.0																							235
Trachurus japonicus	1	15	21	0.048	0.261																							149
Trematomus bernacchii	1	30	26	0.033	0.206	1	3	62	2	7	3	3	2	3	2		3		2	2	1			4	3	2		450
Trematomus hansoni	1	26	26	0.025	0.151	1	3	62	2	7	3	3	2	3	2		3		2	2	1			4	3	2		450
Tristramella sarca	1	19	21	--	0.048																							235
Tristramella simonis	1	35	21	--	0.0																							235
Typhlichthys subterraneus	13	65	10	0.018	--																							462
Verasper variegatus	1	15	23	0.032	0.124																							149
Zoarces viviparus	2	221	32	0.089	0.290	2	3	621	2	7	3	4	2	3		4	2	3		2	1	2	50	4		2		148
CEPHALOCHORDATA																												
Brachiostoma morensis	1	44	14	0.240	0.704																							277
TUNICATA																												
Halocynthia aurantium	1	15	23	0.156	0.520	5	4	621	2	7	3	2	3	1	2	4	3	3	2	2				4				272

Appendix I. (contin.)

Species	No. of: populations	indi.	loci	Genetic indices H	P	Biotic Profile @ Lz	Gr	Ht	Hr	Ar	Ps	Ss	Bs	Mo	Gf	Lo	Fe	Gl	Or	Tr	So	Pl	Ch	Ds	Ma	Rp	Po	References (according to ref. of the data)
I N V E R T E B R A T A																												
ECHINODERMATA																												
Aphelasterias japonica	1	15	14	0.101	0.506	2	3	621	1	7	3	3	2	2	3	3	4	3	2	2								267,272
Asterias amurensis	1	15	19	0.103	0.468	2	3	621	2	7	3	3	3	3	3	3	4	3	2	2								267,272
Asterias forbesi	2	101	27	0.032	0.151																							404
Asterias vulgaris	2	96	26	0.018	0.110																							404
Benthodytes typica	1	86	14	0.021	0.357	4	2	623	1	7	2	2	2	3	2	3				2	1			4		2		112
Benthogone rosea	1	51	13	0.026	0.462	2	1	623	1	7	2	2	2	3	2	3				2	1					2		112
Lethasterias fusca	1	15	16	0.081	0.383	2	3	621	1	7	3	3	2	2	3	3	4	3	2	2								267,272
Lysastrosoma anthosticta	1	15	17	0.100	0.383	2	3	621	1	7	3	3	2	2	3	3	4	3	2	2								267,272
Nearchaster aciculosus	1	17	24	0.212	0.710	2		623	2	7	3	2	2	2	2	3		3	2	2	1			4		2		40
Ophioglypha bullata	1	25	23	0.137	0.565	1		623		7	3	2	3	4	3	3	4	2		2	1	2	43	4		2		111
Ophiomusium lymani	3	304	15	0.169	0.698	5	4	623	2	7	3	2	2	3	2	3	4	3	2	2	1	2	43	4		2		36,111
Ophiotrix fragilis	1	30	28	0.270	0.619	5	4	623	2	7	3	4	3	2	3	3	4		2	2	1	2	44	4		2		69
Othilia sp. (brown form)	1	17	16	--	0.688																							474
Othilia sp. (grey-orange form)	2	46	16	--	0.626																							474
Patria pectinifera	1	15	23	0.174	0.605	4	4	621	2	7	3	4	1	2	3	3	4	3	2	2								267,272
Scaphechinus mirabilis	1	15	30	0.196	0.589	2	3	621	1	7	3	3	2	2	3	4	4	3	2	2								272
Strongylocentrotus intermedius	1	15	18	0.246	0.671	2	3	621		7	3	3	2	2	3	4	4	3	2	2								273,274
DROSOPHILA																												
Drosophila affinis	15	1343	16	0.193	0.652	2	4	9	2	4	3	2	1	4	1	1	3	1	2	2	1	2	10	1		2		230,378
Drosophila aldrichi	1	162	16	0.121	0.481	2	3	9	1	1	2	2	1	3	3	1	2	1		2		2	12	1		2		378,510
Drosophila algoniquin	7	390	16	0.210	0.390	2	4	9	2	4	3	2	1	4	1	1	3	1	2	2	1	2	10	1		2		378
Drosophila arizonensis	2	227	17	0.126	0.383	4	3	9	1	1	2	2	1	3	3	1	2	1		2		2	12	1		2		510
Drosophila athabasca	17	890	16	0.140	0.451	2	4	9	2	4	3	2	1	4	1	1	3	1	2	2	1	2	10	1		2		230
Drosophila bifasciata	23	600	21	0.242	0.620	2	3	3	2	6	3	4	1	4	1	1	3	1	2	2	1	2	10	1				392
Drosophila bipectinata	19	196	23	--	0.493																							505
Drosophila buschii	18	90	30	0.044	--																							366
Drosophila buzzattii	52	10190	29	0.065	0.241	4	4	9	1	2	1	3	1	4	2	1	2	1	1	2	3	2	10	1		2		50

Appendix I. (contin.)

Species	No. of: populations	indi.	loci	Genetic indices H	P	Lz	Gr	Ht	Hr	Ar	Ps	Ss	Bs	Mo	Gf	Lo	Fe	Gl	Or	Tr	So	Pl	Ch	Ds	Ma	Rp	Po	References (according to ref. of the data)
Drosophila engyochracea	2	1026	20	0.120	0.300	3	1	9	1	4	1	1	1	1	1	1	2	1	1	1	3			2	3			453,454,457
Drosophila equinoxialis	5	207	31	0.165	0.624	3	4	3	2	4	3	4	1	3	1	1	3	1	2	2	1	2	6	1		2		31,33,34,35
Drosophila guanche	3	340	61	0.054	0.329	2	5	5	1	3	1	1	1			1	3	1	2			2	10	2		2		161
Drosophila heteroneura	3	541	25	0.090	0.678	4	1	9	1	5	2	1	1	3	1	2	3	3	1	1	1		9	2	3	2		418
Drosophila immigrans	1	51	17	0.115	0.710	5	4	9	2	4	2	2	1	3	1	1	2	1	2	2	3			2	3			453,454,457
Drosophila maderiensis	2	225	61	0.086	0.516	2	5	5	1	3	1	1	1			1	3	1	2			2	10	2		2		161
Drosophila malerkotliana	10	316	23	--	0.503																							505
Drosophila mauritana	2	40	55	0.045	0.182																							160
Drosophila melanogaster	29	1080	34	0.135	0.562	5	4	9	2	6	2	3	1	4	1	1	3	1	2	2	1	2	8	1		2		92,160,230,242, 433,434
Drosophila mercatorum	2	156	20	0.132	0.446																							107
Drosophila mimica	2	1225	20	0.194	0.480	3	1	9	1	4	1	2	1	1	1	1	3	1	1	2	3			2	3			453,454,457
Drosophila mojavensis	8	765	17	0.076	0.266	4	3	9	1	1	2	2	1	3	3	1	2	1		2		2	12	2		2		510
Drosophila mulleri	1	157	16	0.113	0.481	4	3	9	1	1	2	2	1	3	3	1	2	1		2		2	12	1		2		510
Drosophila nebulosa	6	151	25	0.170	0.589	3	4	9	2	6	3	4	1	3	1	1	3	1	2	2	1	2	6	2		2		33,35
Drosophila obscura	57	787	33	0.109	0.580	2	4	3	2	6	3	4	1	4	1	1	3	1	2	2	1	2	12	1				241
Drosophila ochrobasis	2	116	14	--	0.357	4	1	9	1	5	2	1	1	3	1	2	3	3	1	1	1		9	2	3	2		97
Drosophila orthofascia	1	63	13	0.025	0.310	3	1	9	1	4	2	1	1	2	1	1	2	1	1	1	3			2	3			453,454,457
Drosophila parabipectinata	12	220	23	--	0.343																							505
Drosophila paulistorum	32	1277	18	0.177	0.595	3	4	9	2	4	3	4	1	3	1	1	3	1		2	1	2	6	1		2		35,377
Drosophila pavani	14	1000	24	0.192	0.552	2	1	9	1	2	2	2	1	3	1	1	2	1	1	2	1	2	4	1	3	2		316,317
Drosophila persimilis	6	225	28	0.104	0.336																							365,367,368
Drosophila pseudoananassae	4	44	23	--	0.484																							505
Drosophila pseudoobscura	18	1792	27	0.119	0.430																							259,368,369,370, 433
Drosophila robusta	8	527	40	0.110	0.390																							366
Drosophila setosimentum	6	785	14	--	0.857	4	1	9	1	5	2	1	1	3	1	2	3	3	1	1	1		9	2	3	2		97
Drosophila silvestris	3	834	25	0.084	0.610	3	1	9	1	4	2	2	1	3	1	2	3	1	1	1	1			2		2		418
Drosophila simulans	9	327	51	0.055	0.342	5	4	9	2	4	2	3	1	4	1	1	3	1	2	2	1	2	8	2		2		92,160,230,457
Drosophila sproati	1	125	15	0.083	0.470	3	1	9	1	4	2	1	1	3	1	1	2	1	1	1	3			2	3			453,454,457
Drosophila subobscura	39	2584	23	0.145	0.610	2	4	9	2	3	2	3	1	4	1	1	3	1	2	2	1	2	12	1		2		91,240,279,363, 394,511
Drosophila tropicalis	6	185	26	0.168	0.532	3	4	9	2	4	3	4	1	3	1	1	3	1	2	2	1	2	6	2		2		31,35
Drosophila willistoni	6	284	31	0.177	0.557	3	4	3	2	4	3	4	1	3	1	1	3	1	2	2	1	2	6	1		2		31,32,33,35

(Biotic Profile columns Lz–Po are headed "@ Biotic Profile" in the original.)

Appendix I. (contin.)

Species	No. of: populations	indi.	loci	Genetic indices H	P	Lz	Gr	Ht	Hr	Ar	Ps	Ss	Bs	Mo	Gf	Lo	Fe	Gl	Or	Tr	So	Pl	Ch	Ds	Ma	Rp	Po	References (according to ref. of the data)
INSECTA EXCLUDING DROSOPHILA																												
Acraea necoda	1	29	22	0.095	0.273	3	2	9	1	1	1	2	3	3	1	1	2	1		2	1			1		2		228
Aedes mariae	1	29	26	0.070	0.350																							106
Aedes zammitii	1	30	26	0.060	0.350																							106
Anopheles aquasalis (emilianus)	3	45	18	0.142	0.660	3	4	8		4	2	2	1	3	2	1	2	1		2	3	2	6	1	3			456
Anopheles nuneztovari	2	30	22	0.130	0.432	3	3	8	1	4	2	2	1	4	3	1	2	1	2	2	3	2	6	1	3			455
Aphis pomi	1	15	13	--	0.178	5	4	9	2		2	2	1	4	3	1	2	1	2	2	1	2	8	1	4	3		212
Aphis tabae	2	30	17	--	0.079																							212
Apis mellifera	1	91	16	0.011	0.148	5	4	9	2	3	3	4	2	4	1	3	4	2	1	1	4			1	3	2		348
Augochlora pura	1	25	13	0.0	0.0	2	3	9	2	4	1	2	1	4	2	1	2	1	2	2	1			1		3		447
Bathysciola derosasi	5	250	13	0.088	0.419	2	2	1	1		1	2	1	1	1	2	1	2	2	2	1			1		2		395
Blatella germanica	1	30	19	0.012	0.211	5	4	2	1	3	2	2	2	3	1	2	2	1	1	2	2			1		2		325
Bombus americanorum	1	34	12	0.0	0.0	2	3	9	2	4	1	2	2	4	2	1	2	1	2	2	2			1		3		447
Bombus lapidarius	1	176	16	0.007	0.223	2	4	9	2	3	3	4	2	5	1	1	2	2	1	1	4			1	3	2		348
Bombus lucorum	3	119	17	0.012	0.114	2	4	9	2	3	3	4	2	5	1	1	2	2	1	1	4			1	3	2		348
Bombus pascubum	1	88	14	0.003	0.086	2	4	9	2	3	3	4	2	4	1	1	2	2	1	1	4			1	3	2		348
Bombus terrestris	1	24	15	0.035	0.158	2	4	9	2	3	3	4	2	5	1	1	2	2	1	1	4			1	3	2		348
Ceuthophilus gracilipes	7	330	26	0.024	0.131																							395
Chalybion californicum	1	51	16	0.073	0.370	2	3	9	2	4	2	2	2	4	2	1	1	1	1	2	1			1		2		305
Clepsis spectrana	1	15	18	0.086	--																							300
Danaus affinis	1	52	22	0.043	0.091	3	4	9	1	4	2	3	3	2	1	1	2	1		2	1			3		2		228
Danaus chrysippus	3	103	22	0.229	0.726	4	4	9	2	6	3	3	3	5	1	1	2	1		2	1	2	60	3		2		228
Danaus genutia	1	46	22	0.225	0.773	3	4	9	2	4	3	3	3	4	1	1	2	1		2	1			3		2		228
Danaus gilippus	1	35	22	0.199	0.723	4	4	9	2	6	3	3	3	4	1	1	2	1		2	1	2	58	3		2		228
Danaus melanippus	1	33	22	0.181	0.636	3	4	9	1	4	2	3	3	3	1	1	2	1		2	1			3		2		228
Danaus philene	1	35	22	0.171	0.682	3	4	9	1	4	2	3	3	3	1	1	2	1		2	1			3		2		228
Danaus plexippus	2	117	22	0.183	0.661	4	4	9	2	6	3	4	3	5	1	1	2	1		2	1			3		2		228
Diadromus pulchellus	1	172	22	--	0.317	2		3	1	4	2		1	4	1	1	3	1			1			1				400
Dociostaurus curvicercus	7	216	19	0.074	0.553	2	2	3	2	3	2	3	2	3	1	1	2	1	1	2	1	2	22	1		2		328a
Dociostaurus genei	3	91	18	0.057	0.641	2	3	3	2	2	2	3	1	3	1	1	2	1	1	2	1	2	22	1		2		328a
Dociostaurus sp. ("sands").	1	30	16	0.055	0.500	2	1	3	1	3	2	3	2	3	1	1	2	1	2	2	1	2	22	1		2		328a
Dolichopoda baccettii	3	102	15	0.128	0.453	2	1	1	1		1	1	2	3	1	2	2	2	2	2	1	2	31	1		2		395
Dolichopoda geniculata	9	535	15	0.160	0.571	2	3	1	1		1	2	2	3	1	2	2	2	2	2	1	2	31	3		2		395
Dolichopoda laetitiae	5	358	15	0.137	0.550	2	2	1	1		1	2	2	3	1	2	2	2	2	2	1	2	31	1		2		395
Dolichopoda schiavazzii	6	175	15	0.130	0.453	2	2	1	1		1	2	2	3	1	2	2	2	2	2	1	2	35	3		2		395
Duvalius jurececkii rasettii	1	37	20	0.124	0.397	2	2	1	1		1	1	1	2	1	2	1	2	2	1	1			1		2		395
Duvalius lepinensis	2	44	20	0.088	0.363	2	2	1	1		1	2	1	2	1	2	1	2	2	1	1			1		2		395

Biotic Profile @

Appendix I. (contin.)

Species	No. of:			Genetic indices		Biotic Profile @																						References (according to ref. of the data)
	populations	indi.	loci	H	P	Lz	Gr	Ht	Hr	Ar	Ps	Ss	Bs	Mo	Gf	Lo	Fe	Gl	Or	Tr	So	Pl	Ch	Ds	Ma	Rp	Po	
Enchenopa binotata	41	1140	15	0.113	0.411	4	4	3	1	4	2	3	1	2	1	1	2	1	1	2	4			1				183
Ephestia kuehniella	1	15	17	0.131	--																							300
Euploea mulciber	1	20	22	0.114	0.364	3	3	9	1	4	2	3	3	5	1	1	2	1		2	1	2	58	3		2		228
Gerris lacustris	6	40	22	0.164	0.413	2	3	61	1	7	1	2	2	2	1	1	2	1	1	2	1	2		1				479
Gerris lacustus	22	1018	18	--	0.439																							479a
Gerris lateralis	15	75	16	0.049	0.294	2	2	61	1	7	1	1	2	1	1	1	2	1	1	2	1	2		1				479b
Gerris odontogaster	7	35	16	0.097	0.294	2	3	61	1	7	1	2	2	3	1	1	2	1	1	2	1	2		1				479b
Gerris regimis	9	456	16	0.051	0.209	2	4	61	1	7	1	3	2	1	1	1	2	1		1					3	2		506
Grylloides hebraeus	1	34	26	0.021	0.308	2	2	2	1	3	2	2	2	3	1	2	2	1	1	1	1			1		2		325
Gryllotalpa africana	1	34	15	0.002	0.067	3	3	1	1	1	1	2	2	3	1	2	2	1	1	1	2	2	23	1		2		325
Gryllotalpa gryllotalpa 2n=19	11	414	21	0.030	0.568	2	3	1	1	3	2	3	2	3	1	2	2	1	1	1	2	2	19	1		2		325
Gryllotalpa gryllotalpa 2n=23	1	22	20	0.010	0.200	2	1	1	1	1	1	1	2	3	1	2	2	1	2	1	2	2	23	1		2		325
Gryllus bimaculatus	1	34	25	0.055	0.560	4	3	2	2	3	2	2	2	3	1	2	2	1	1	1	1			1		2		325
Gryllus integer	1	15	20	0.145	--																							325
Heliothis zea	1	15	21	0.216	--																							300
Heliothus virescens	11	327	15	0.278	0.779	5	4	3	2	4	2	3	2	4	1	1	2	1	1	1	1	2		1		2		436
Iridomyrmex purpureus	13	502	15	--	0.147	2	3	2	2	2	2	3	2	1	3	4	4	2		1	4	2	18	1		2		189
Lasioglossum zephyrum	2	75	24	0.0	0.0	2	3	9	2	4	2	2	1	4	2	1	2	1	2	2	2			1		3		447
Leptinotarsa decemlineata	1	23	21	0.099	0.381																							110
Leptodirus hohenwarti	1	22	14	0.141	0.614	2	2	1	1		1	2	2	3	1	4	1	3	2	2	1			1		2		395
Limnoporus canaliculatus	3	185	15	0.197	0.427	2	3	61	2	7	2	3	1	3	1	1	3	1							2	2		506
Lonchoptera dubia	1	1000	19	0.150	0.245																							346
Macrosiphum euphorbiae	2	36	18	0.061	0.344																							288
Macrosiphum rosae	10	60	28	0.032	0.183	5	4	9	1		2	2	2	4	3	1	2	1	2	2	1	2	10	1	4	3		212
Magicicada cassini	3	313	17	0.069	0.290	2	3	1	2	4	2	3	2	2	1	4	3	3	1	2	1			1		2		371,428
Magicicada septendecim	3	1206	17	0.062	0.411	2	3	1	2	4	2	3	2	2	1	4	3	3	1	2	1			1		2		371,428
Magicicada tredecassini	2	105	15	0.174	0.569	2	3	1	2	4	2	3	2	3	1	4	3	3	1	2	1			1		2		237
Magicicada tredecim	3	1817	18	0.071	0.389	2	3	1	2	4	2	3	2	2	1	4	3	3	1	2	1			1		2		428
Magicicada tredecula	2	37	15	0.153	0.506	2	3	1	2	4	2	2	2	3	1	4	3	3	1	2	1			1		2		237
Megachile pacifica	5	362	17	0.031	0.060	2	4	9	2	3	2	4	2	4	1	1	2	1	2	2	2			1	3	3		252
Musca autumnalis	8	251	14	0.110	0.307	2	4	9	2	3	2	3	1	4	1	1	2	1	1	2	1	2	12	1	3	2		81
Myzus persicae	9	235	17	0.0	0.0																							288
Neaphaenops tellkampfii	5	99	12	0.106	0.408																							395
Nomia melanderi (heteropoda)	1	30	15	0.070	0.130	2		9	2	3	2	3	2	4	1	1	2	1	2	2	2			1	3	2		305
Oreina cacaliae	1	28	24	0.033	0.375																							110
Ostrinia nubilalis	2	103	30	0.146	0.499																							105
Parantica melaneus	1	20	22	0.145	0.227	3	3	9	1	4	2	3	3	3	1	1	2	1		2	1	2	44	1		2		228
Parantica sita	1	20	22	0.103	0.455	3	3	9	1	4	2	3	3	3	1	1	2	1		2	1	2	94	3		2		228

Appendix I. (contin.)

Species	No. of: populations	indi.	loci	Genetic indices H	P	Lz	Gr	Ht	Hr	Ar	Ps	Ss	Bs	Mo	Gf	Lo	Fe	Gl	Or	Tr	So	Pl	Ch	Ds	Ma	Rp	Po	References (according to ref. of the data)
Periplaneta americana	1	60	20	0.016	0.400	5	4	2	1	3	2	2	2	3	1	2	2	1	1	2	2			1		2		325
Perniciosus ariasi	1	15	14	--	0.197																							491
Philaenus spumarius	7	548	21	0.087	0.450	2	4	3	2	6	3	3	2	3	1	1	3	2	2	2	1			3		2		393
Polistes annularis	1	128	15	0.046	0.130	2	4	9	2	4	2	3	2	4	1	1	2	1	2	2	4			1	3	3		252
Polistes exclamans	5	352	16	0.038	0.159	2	4	9	2	4	2	3	2	4	1	1	2	1	2	2	4			1	3	3		252
Polistes metricus	1	15	10	0.065	--	2	3	9	2	4	2	3	2	4	2	1	1	1	1	1	4			1	3	2		305a
Polistes variatus	1	15	10	0.073	--	2	3	9	2	4	2	3	2	4	2	1	1	1	1	1	4			1	3	2		305a
Pontania vesicator	1	55	18	0.020	0.198	2	4	4	1	3	2	4	2	3	1	1	2	1	1	2	1			1	3	2		348
Prosimilium fuscum	1	132	15	--	0.400	2	3	8	2	4	3	3	1	4	1	1	3	1	1	2	1	2	6	1		2		449
Prosimilium mixtum	1	371	15	--	0.400	2	3	8	2	4	3	3	1	4	1	1	3	1	1	2	1	2	6	1		2		449
Ptomaphagus hirtus	5	200	13	0.044	0.166																							395
Pyrrochoris apterus	1	23	23	0.105	0.435																							110
Rhodobium porosum	1	15	13	0.0	0.0																							212
Rhytidoponera chalybaea	10	1752	22	0.047	0.148	3	3	2	1	4	2	3	2	3	1	3	4	3	2	1	4			1	3	2		486
Rhytidoponera confusa	19	3828	22	0.034	0.141	4	3	2	1	4	2	3	2	3	1	3	4	3	2	1	4	2	42	1	3	2		486
Rhytidoponera enigmatica	2	240	22	0.033	0.134	3	2	2	1	4	2	2	2	3	1	3	4	3	2	1	4			1	3	2		486
Rhytidoponera impressa	2	96	22	0.046	0.159	3	3	2	1	4	2	3	2	3	1	3	4	3	2	1	4			1	3	2		486
Rhytidoponera purpurea	2	96	22	0.005	0.036	3	3	2	1	4	2	3	2	3	1	3	4	3	2	1	4	2	48	1	3	2		486
Savastra obliqua	1	30	16	0.038	0.250	2		9		4			3	4		1	1	1	1		1			1		2		305
Sceliphron caementarium	1	53	12	0.078	0.170	2	4	9	2	4	2	3	2	4	2	1	1	1	1	2	1			1	3	3		305
Scolia dubia	1	46	15	0.051	0.270	2		9		4			2	4		1	1	1	1		1			1		3		305
Simulinum venustum A/C	1	60	14	--	0.357	2	2	8	2	4	2	2	1	4	1	1	3	1	1	2	1	2	6	1		2		448
Simulinum venustum CC	1	60	15	--	0.467	2	3	8	2	4	3	4	1	4	1	1	3	1	1	2	1	2	6	1		2		448
Simulinum venustum CC1	1	60	15	--	0.533	2	3	8	2	4	3	3	1	4	1	1	3	1	1	2	1	2	6	1		2		448
Simulinum venustum EFG/C	1	60	15	--	0.533	5	4	8	2	4	3	4	1	4	1	1	3	1	1	2	1	2	6	2		2		448
Simulinum verecundum AA	1	60	14	--	0.357	2	2	8	2	4	2	3	1	4	1	1	3	1	1	2	1	2	6	2		2		448
Simulinum verecundum ACD	1	60	15	--	0.200	5	4	8	2	4	3	4	1	4	1	1	3	1	1	2	1	2	6	1		2		448
Solenobia triquetrella	2	49	16	0.230	0.569																							261
Speonomus delarouzeei	8	260	12	0.095	0.541	2	2	1	1		1	3	1	2	1	3	1	2	2	2	1			1		2		126
Speonomus lostiai	2	59	16	0.117	0.419	2	2	1	1		1	2	1	2	1	3	1	2	2	2	1			2		2		395
Speyeria adiaste clemencri	2	141	16	0.075	0.438	2	1	9	1	3	2	1	2	3	1	1	3	2	2	2	1	2	58	1		2		75
Speyeria atlantis irene	3	130	16	0.067	0.375	2	2	9	1	3	2	1	2	3	1	1	3	2	2	2	1	2	58	1		2		75
Speyeria callippe	6	293	16	0.140	0.653	2	2	9	1	3	2	2	2	3	1	1	3	2	2	2	1	2	60	1		2		75
Speyeria coronis	2	27	16	0.108	0.396	2	3	9	1	2	3	2	2	4	1	1	3	2	2	2	1	2	60	1		2		75
Speyeria cybele leto	1	14	16	0.089	0.188	2	2	9	1	4	1	1	2	4	1	1	3	2	2	2	1	2	58	1		2		75
Speyeria egleis egleis	3	45	16	0.133	0.500	2	3	9	1	3	2	2	2	3	1	1	3	2	2	2	1	2	58	1		2		75
Speyeria hydaspe purpurascens	5	84	16	0.118	0.625	2	3	9	1	3	2	2	2	3	1	1	3	2	2	2	1	2	58	1		2		75
Speyeria mormonia arge	1	60	16	0.137	0.467	2	2	9	1	4	1	2	2	3	1	1	2	2	2	2	1	2	58	1		2		75

@ Biotic Profile

Appendix I. (contin.)

Species	No. of: populations	indi.	loci	Genetic indices H	P	Lz	Gr	Ht	Hr	Ar	Ps	Ss	Bs	Mo	Gf	Lo	Fe	Gl	Or	Tr	So	Pl	Ch	Ds	Ma	Rp	Po	References (according to ref. of the data)
Speyeria nokomis apacheana	1	42	16	0.034	0.090	2	1	9	1	4	1	1	2	3	1	1	3	2	2	2	1			1		2		75
Speyeria zerene	7	140	16	0.122	0.491	2	3	9	1	3	2	2	2	4	1	1	3	2	2	2	1	2	58	1		2		75
Stictia carolina	1	60	17	0.056	0.350	2		9		4			3	4		1	1	1	1		1			1		2		305
Strophosomus capitatus	3	61	19	0.170	0.440																							461
Tirumala petiverana	1	20	22	0.148	0.546	3	4	9	1	4	2	3	3	4	1	1	2	1		2	1	2	88	1		2		228
Troglophilus andreinii	3	241	16	0.171	0.712	2	2	1	1		1	2	2	3	1	2	2	2	2	2	1	2	19	1		2		396
Troglophilus cavicola	2	137	16	0.065	0.482	2	2	1	1		1	3	2	3	1	2	2	2	2	2	1	2	21	1	4	3		396
Tropidotylus sp.	1	32	16	0.055	0.437																							110
Trypargilum polilum	1	34	19	0.059	0.320	2	3	9	2	4	2	2	2	4	2	1	1	1	1	2	1			1		2		305
Yponomeuta cagnagellus	1	75	51	0.106	0.490	2	4	4	1	4	3	3	2	4	1	1	2	2	1	2	1	2	62	1	3	2		300
Yponomeuta evonymellus	1	65	51	0.119	0.450	2	4	4	1	4	3	3	2	3	1	1	2	2	1	2	1	2	62	1	3	2		300
Yponomeuta irrorellus	1	48	51	0.053	0.350	2	4	4	1	4	3	3	2	3	1	1	2	2	2	2	1	2	62	1	3	2		300
Yponomeuta mahalebellus	1	75	51	0.067	0.390	2	4	4	1	4	3	3	2	3	1	1	2	2	1	2	1	2	62	1	3	2		300
Yponomeuta malinellus	1	75	51	0.061	0.330	2	4	4	1	4	3	3	2	3	1	1	2	2	1	2	1	2	62	1	3	2		300
Yponomeuta padellus	1	63	51	0.106	0.430	2	4	4	1	4	3	3	2	4	1	1	2	2	1	2	1	2	62	1	3	2		300
Yponomeuta plumbellus	1	75	51	0.083	0.410	2	4	3	1	4	3	3	2	3	1	1	2	2	2	2	1	2	62	1	3	2		300
Yponomeuta rorellus	1	64	51	0.003	0.080	2	4	4	1	4	3	3	2	3	1	1	2	2	1	2	1	2	62	1	3	2		300
Yponomeuta vigintipunctatus	1	70	51	0.058	0.270	2	4	3	1	4	2	3	2	3	1	1	2	1	2	2	1	2	62	1	3	2		300
Yponomeuta yangawanus	1	15	27	0.108	0.480																							198
Zygaena ephialtes	2	78	12	0.115	0.273																							11
Zygaena lavandulae	2	25	12	0.142	0.129																							11
Zygaena oxytropis	3	13	12	0.132	0.206																							11
Zygaena purpuralis	2	65	12	0.142	0.250																							11
Zygaena rubicundus	2	101	12	0.059	0.159																							11
CRUSTACEA																												
Artemia ?n	1	15	10	0.203	--	4	3	6	1	7	2	4	1	4	1	1	3	1	1	2	4			1	4	1		1
Artemia ?n	1	15	10	0.411	--	4	3	6	1	7	2	4	1	4	1	1	3	1	1	2	4			1	4	1		1
Artemia 2n	5	75	10	0.272	--	4	3	6	1	7	2	4	1	4	1	1	3	1	1	2	4	2	43	1	4	1		1
Artemia 4n	2	30	10	0.470	--	4	3	6	1	7	2	4	1	4	1	1	3	1	1	2	4	4	84	1	4	1		1
Balanus amphitrite	8	231	14	0.103	0.836	4	4	621	2	7	2	3	1	1	3	3	2	3	1	2	1			4		2		335
Balanus eburneus	6	128	14	0.067	0.737	4	4	621	2	7	2	2	1	1	3	3	2	3	1	2	1			4		2		325
Balanus perforatus	7	214	22	0.253	0.920	4	4	621	2	7	2	3	1	1	3	3	2	3	1	2	1			4		2		325
Calcinus obscurus	1	45	23	0.060	0.179	3	3	621	2	7	2	3	1	3	3	2	3	3	2	1	1			1		2		321
Calcinus tibicen	1	40	19	0.044	0.153	3	3	621	2	7	2	3	1	3	3	2	3	3	2	1	1			4		2		321
Callianssa californiensis	1	59	38	0.080	0.360	2	3	621	2	7	2	3	2	2	3	3	3	3	1	1	1			1		2		321

Appendix I. (contin.)

Species	No. of: populations	indi.	loci	Genetic indices H	P	Biotic Profile @ Lz	Gr	Ht	Hr	Ar	Ps	Ss	Bs	Mo	Gf	Lo	Fe	Gl	Or	Tr	So	Pl	Ch	Ds	Ma	Rp	Po	References (according to ref. of the data)
Callianssa sp.	1	24	24	0.099	0.330	3		621	2	7	2	3	2	2	3	3	3	3		1	1			1		2		321
Callinectes arcuatus	1	41	22	0.128	0.495	3	3	621	2	7	2	3	3	4	3	4	4	3		2	1			1		2		321
Callinectes sapidus	1	46	25	0.088	0.220	2	3	621	2	7	2	3	3	4	3	4	4	3	2	2	1			1		2		321
Cambarus bortonii	1	16	18	0.083	0.230	2	3	611	2	7	2	3	2	3	2	3	3	2	2	2	1	2	50	1	3	2		322
Cambarus latimanus	4	67	17	0.041	0.132	2	3	611	2	7	2	3	2	3	2	3	3	2	2	2	1	2	50	1	3	2		78,322
Cambarus robustus	2	70	19	0.039	0.148	2	3	611	2	7	2	3	2	3	2	3	3	2	2	2	1	2	50	1	3	2		322
Cancer gracilis	1	28	23	0.050	0.239	2	3	621	2	7	3	3	2	4	3	3	4	3	2	2	1			1		2		321
Cancer magister	1	91	32	0.013	0.090	2	3	621	2	7	3	3	3	4	3	4	4	3	2	2	1			1		2		321
Charybdis callianassae	1	31	23	0.026	0.124	3	3	62	2	7	3	3	2	4	3	3	4	3		2	4			4		2		321
Charybdis sp.	1	32	17	0.053	0.310	3	3	62	2	7	3	3	2	4	3	3	4	3		2	4			4		2		321
Chthamalus anisopoma	1	15	19	0.104	0.526	3	3	621	2	7	3	4	1	1	3	3	4	1	2	2	1			1		2		195
Chthamalus dalli	1	21	24	0.078	0.417	2	3	621	2	7	3	4	1	1	3	3	4	1	2	2	1			1		2		195
Chthamalus fissus	5	162	25	0.053	0.439	4	3	621	2	7	3	4	1	1	3	3	4	1	2	2	1			1		2		195
Chthamalus fragilis	1	148	16	0.090	0.607																							123
Chthamalus montagui	1	15	13	0.179	0.580																							122
Chthamalus proteus	1	124	16	0.165	0.722																							123
Chthamalus stellatus	1	49	13	0.084	0.580																							122
Clibanarius albidigitus	1	47	20	0.018	0.079	3	3	621	2	7	2	4	1	3	3	2	3	2		1	1			1		2		321
Clibanarius antillensis	1	40	25	0.037	0.165	3	3	621	2	7	2	3	1	3	3	2	3	2		1	1			4		2		321
Clibanarius panamensis	1	22	21	0.095	0.378	3	3	621	2	7	2	3	2	3	3	2	3	3		1	1			1		2		321
Coenobita compresus	2	91	18	0.080	0.293	3	3	7	2	7	3	3	2	3	3	3	3	3		1	1			1		2		321
Crangon franciscorum	1	19	30	0.057	0.352	2	3	631	2	7	2	3	2	4	3	2	3	3		2	4			1		2		321
Crangon nigricauda	1	58	30	0.053	0.230	2	3	631	2	7	2	3	2	4	3	2	3	3		2	4			1		2		321
Daphnia carinata	15	798	16	0.035	0.110	2	2	61	2	7	1	4	1	2	1	1	3	1	2	2	1			1	4	3		192
Daphnia magna	35	605	16	0.014	0.049	5	4	61	2	7	1	4	1	2	1	1	3	1	2	2	1	2	24	1	4	3		118
Daphnia middendorffiana	2	22	16	0.112	0.668	1	2	61	2	7	1	4	1	2	1	1	3	1	2	2	1	2	24	1	4	1		299
Daphnia pulex	2	22	16	0.162	0.774	5	4	61	2	7	1	4	1	2	1	1	3	1	2	2	1	2	24	1	4	1		299
Diamysis bahirensis	1	235	12	0.131	0.670																							286
Emerita analoga	1	111	22	0.125	0.495	3	3	621	2	7	2	3	2	3	3	2	3	3		2	1			1		2		321
Euphausia distinguenda	3	220	30	0.205	0.771	3	2	622	2	7	3	4	2	4	2	3		3	2	2	1			4		2		37,478
Euphausia mucronata	2	100	28	0.154	0.669	3	2	622	2	7	3	4	2	4	2	3		3	2	2	1			4		2		37,478
Euphavsia superba	2	254	36	0.057	0.361	1	2	622	2	7	3	4	2	4	2	3		3	2	2	1			4		2		37,41
Galathea californiensis	1	46	19	0.075	0.279	2	3	62	2	7	3	3	2	4	3	3	3	3		2				1		2		321
Gammarus aequicauda	1	15	31	0.042	0.258	2	3	623	1	7	2	2	2	2	1	1	3	1	2	2	1	2	52	4		2		68
Gammarus insensibilis	1	282	24	0.054	0.208	2	3	623	1	7	2	2	2	2	1	1	3	1	2	2	1	2	52	4		2		54
Gammarus minus	28	140	13	--	0.267																							395
Gecarcinus quadratus	1	36	23	0.033	0.070	3	3	621	2	7	3	3	2	4	3	3	4	3		2	1			1		2		321
Hemigrapsus oregonesis	1	39	29	0.042	0.142	2	3	621	2	7	2	4	2	4	3	3	3	3	2	2	1			1		2		321

Appendix I. (contin.)

Species	No. of: populations	No. of: indi.	No. of: loci	Genetic indices H	Genetic indices P	Lz	Gr	Ht	Hr	Ar	Ps	Ss	Bs	Mo	Gf	Lo	Fe	Gl	Or	Tr	So	Pl	Ch	Ds	Ma	Rp	Po	References (according to ref. of the data)
						Biotic Profile @																						
Hippa pacifica	1	23	19	0.008	0.0	3	3	621	2	7	2	3	2	3	3	2	3	3		2	1			4		2		321
Homarus americanus	10	453	38	0.045	0.222	2	3	62	2	7	3	3	3	4	3	4	4	3	1	1	1	2	28	1		2		321,471,472
Homarus gammarus	1	48	39	0.048	0.176	2	3	621	2	7	3	3	3	4	3	4	4	3	1	1	1	2	28	1		2		321
Idotea baltica	1	15	20	0.190	0.605	5	4	623	2	7	3	3	2	3	2	1	3	1	2	2	1			4		2		102
Jasus edwardsii	2	47	22	0.015	0.237																							446
Macrobrachium ohione	6	30	14	0.171	--																							197
Macrobrachium rosenbergii	8	378	31	0.028	0.166	3	3	61	2	7	2	3	2	4	2	3	4	3	2	1	1			3		2		197
Matuta lunaris	1	46	29	0.028	0.099	3	3	621	2	7	3	3	2	4	3	3	4	3		2	1			4		2		321
Matuta planipes	1	46	24	0.009	0.0	3	3	621	2	7	3	3	2	4	3	3	4	3		2	1			4		2		321
Meapenaeus bennettae	2	97	32	0.020	0.250	4	1	623	2	7	2	4	1	4	3	1	4	2	2	2	1			2	3	2		313
Metapenaeus eboracensis	1	44	32	0.019	0.219	3	3	623	2	7	3	4	1	5	3	1	4	2	2	2	1			2	3	2		313
Metapenaeus endeavouri	2	99	32	0.030	0.250	3	1	623	2	7	3	4	2	4	3	1	4	2	2	2	1			1	3	2		313
Metapenaeus ensis	2	67	32	0.013	0.250	3	4	623	2	7	3	4	2	4	3	1	4	2	2	2	1			1	3	2		313
Metapenaeus insolitus	1	29	32	0.010	0.125	3	3	623	2	7	3	4	1	5	3	1	4	2	2	2	1			2	3	2		313
Metapenaeus macleayi	2	150	32	0.026	0.219	2	1	623	2	7	3	4	1	5	3	1	4	2	2	2	4			2	3	2		313
Monolistra berica	1	110	14	0.257	0.726	2	1	7	1	7	1	1	1	1	1	3	1	3	2	2	1			1		2		395
Monolistra boldorii	1	51	14	0.242	0.726	2	1	7	1	7	1	1	1	1	1	3	1	3	2	2	1			1		2		395
Monolistra caeca	1	53	14	0.254	0.807	2	2	7	1	7	1	2	1	1	1	3	1	3	2	2	1			1		2		395
Munida hispida	1	35	26	0.070	0.264	2	3	62	2	7	3	3	2	4	3	3	3	3		2				1		2		321
Munidopsis hamata	2	23	29	0.079	0.250	1		623		7		2	1	3	2	3	4			2	1	2	82	4		2		111
Niphargus longicaudatus	16	1039	17	0.261	0.690	2	3	7	1	7	1	3	1	1	1	4	1	3	2	2	1			3		2		398
Niphargus romuleus	1	188	16	0.241	0.810	2	2	7	1	7	1	2	1	1	1	4	1	3	2	2	1			1		2		395
Niphargus stefanellii	1	110	15	0.217	0.703	2	2	7	1	4	1	2	1	1	1	4	1	3	2	2	1			1		2		395
Ocypode occidentalis	1	17	22	0.046	0.300	3	3	7	2	7	2	3	2	4	3	3	4	3	2	1	1			1		2		321
Ocypode quadrata	1	36	23	0.012	0.072	3	3	7	2	7	2	3	2	4	3	3	4	3	2	1	1			1		2		321
Orconectes immunis	3	170	14	0.050	0.171	2	3	611	2	7	2	3	2	3	2	3	3	2	2	2	1	2	50	1	3	2		322
Orconectes propinquus	6	233	19	0.058	0.237	2	3	611	2	7	2	3	2	3	2	3	3	2	2	2	1	2	50	1	3	2		322
Orconectes virilis	1	60	18	0.029	0.111	2	3	611	2	7	2	3	2	3	2	3	3	2	2	2	1	2	50	1	3	2		322
Pachycheles rudis	1	45	30	0.049	0.143	2	3	621	2	7	2	3	2	3	3	3	3	3		2	1			1		2		321
Pachygrapsus crassipes	1	40	23	0.023	0.179	2	3	621	2	7	2	4	2	4	3	3	3	3		2	1			1		2		321
Pachygrapsus transversus	2	114	25	0.027	0.125	3	3	621	2	7	2	3	2	4	3	3	3	2		2	1			1		2		321
Pagurus granosimanus	1	38	24	0.047	0.172	2	3	621	2	7	2	3	2	3	3	2	3	3		1	1			1		2		321
Palaemonetes pugio	10	1352	17	0.076	0.293	2	3	631	2	7	2	3	2	4	3	2	3	3		2	4			1		2		150,321
Pandalopsis ampla	1	11	15	0.072	0.569																							163
Pandalus danae	1	105	29	0.034	0.190	2	3	62	2	7	3	3	2	4	3	3	3	3	2	2	4			1		2		321
Pandalus jordani	1	15	25	0.022	--	2	3	62	2	7	3	3	2	4	3	3	3	3	2	2	4			1		2		321
Pandalus platyceros	1	15	28	0.046	--	2	3	62	2	7	3	3	2	4	3	3	3	3	2	2	4			1		2		321
Panopeus purpureus	1	39	31	0.051	0.221	3	3	621	2	7	2	3	2	3	3	3	4	3			1			1		2		321

Appendix I. (contin.)

Species	No. of: populations	indi.	loci	Genetic indices H	P	Lz	Gr	Ht	Hr	Ar	Ps	Ss	Bs	Mo	Gf	Lo	Fe	Gl	Or	Tr	So	Pl	Ch	Ds	Ma	Rp	Po	References (according to ref. of the data)
Panulirus interruptus	1	41	28	0.037	0.197	4	3	621	2	7	3	3	3	5	3	4	4	3			1			1		2		321
Panulirus longipes	1	42	29	0.061	0.236	4	3	621	2	7	3	3	3	5	3	4	4	3			1			1		2		321
Penaeus aztecus	3	15	24	0.076	0.421	4	4	62	2	7	3	4	2	4	3	2	4	1		2	1			4	3	2		251,321
Penaeus duorarum	4	20	24	0.092	0.421	4	4	62	2	7	3	4	2	4	3	2	4	1		2	1			4	3	2		251,321
Penaeus esculentus	3	93	33	0.033	0.273	3	1	623	2	7	3	4	2	4	3	1	4	2	2	2	1			1	3	2		313
Penaeus japonicus	1	25	31	0.118	0.387	4	4	621	2	7	2	4	2	4	2	2	4	2		2	1					2		285
Penaeus kerathurus	1	43	34	0.049	0.265	2	4	621	2	7	2	4	2	4	2	2	4	2		2	1					2		285
Penaeus latisulcatus	3	167	33	0.032	0.152	4	4	623	2	7	3	4	2	4	3	1	4	2	2	2	1			2	3	2		313
Penaeus longistylus	2	13	33	0.006	0.091	3	4	623	2	7	3	4	2	4	3	1	4	2	2	2	1			2	3	2		313
Penaeus merguiensis	3	501	26	0.008	0.084	3	3	62	2	7	3	4	2	4	3	3	4	2	2	2	4			1		2		313,321
Penaeus monodon	2	12	33	0.008	0.091	3	4	623	2	7	3	4	3	4	3	1	4	2	2	2	1			1	3	2		313
Penaeus plebejus	2	140	33	0.022	0.273	2	1	623	2	7	3	4	2	5	3	1	4	2	2	2	1			1	3	2		313
Penaeus semisulcatus	2	74	33	0.017	0.212	3	4	623	2	7	3	4	2	4	3	1	4	2	2	2	1			1	3	2		313
Penaeus setiferus	5	250	26	0.023	0.139	4	4	62	2	7	3	4	2	4	3	3	4	2		2	4			1		2		251,321
Petrolisthes cinctipes	1	46	28	0.052	0.197	2	3	621	2	7	2	3	2	3	3	3	3	3		2	1			1		2		321
Portunus sanguinolentis	1	72	32	0.063	0.214	3	4	62	2	7	3	3	3	4	3	3	4	3	2	2	4			4		2		321
Procambarus acutus	4	23	19	0.025	0.211																							78
Procambarus hirsutus	4	11	19	0.031	0.144																							78
Procambarus raneyi	4	18	19	0.0	0.0																							78
Procambarus robustus	4	24	19	0.015	0.105																							78
Procambarus troglodytes	4	18	19	0.012	0.144																							78
Rhithropanopeus harrisii	2	41	15	--	0.040																							162
Spelaeomysis bottazzii	7	483	12	0.063	0.424	2		7	1	7	1	3	1	2		3	1		2		1			1		2		286,395
Sphaeroma hookeri	1	50	18	0.181	0.714																							397
Sphaeroma quadridentatum	4	20	16	0.041	0.250																							127
Sphaeroma serratum	1	50	18	0.222	0.714																							395,397
Thenus orientalis	1	44	28	0.034	0.147	3	4	621	2	7	3	3	2	4	3	4	4	3		2	1			4		2		321
Tigriopus brevicornis	1	905	23	0.062	0.260	2	3	621	1	7	1	2	1	2	1	1	3	1	2	2	1	2	24	4		2		54
Tigriopus fulvus	1	1893	19	0.047	0.105	2	3	621	1	7	1	2	1	2	1	1	3	1	2	2	1	2	24	4		2		54
Tisbe biminiensis	1	592	19	0.139	0.526																							54
Tisbe clodiensis	1	590	15	0.152	0.600	2	4	623	2	7	2	3	1	3	1	1	3	1	2	2	1	2	24	4		2		54
Tisbe holothuriae	2	1382	19	0.202	0.552	5	4	623	2	7	3	3	1	3	1	1	3	1	2	2	1	2	24	4		2		54,55
Troglocaris anophtalmus	4	47	23	0.030	0.121	2	2	7	1	7	1	1	2	3	1	3	1	3	2	1	1			1		2		395
Typhlocirolana moraguesi	2	122	18	0.028	0.202	2	1	7	1	7	1	2	1	2	1	3	2		2	2	1			2		2		395
Uca musica	1	45	23	0.097	0.179	3	3	7	2	7	2	3	2	3	3	3	3	3		1	1			1		2		321
Uca princeps	1	43	23	0.029	0.124	3	3	621	2	7	2	3	2	4	3	3	3	3		1	1			1		2		321
Uca speciosa	1	23	26	--	0.230	4	3	61	2	7	2	3	1	2	3	3	4	3	2	1	4			1	3	2		389
Uca spinicarpa	1	11	26	--	0.158	4	3	61	2	7	2	3	2	2	3	3	4	3	2	1	4			1	3	2		389

Appendix I. (contin.)

Species	No. of: populations	indi.	loci	Genetic indices H	P	Biotic Profile @ Lz	Gr	Ht	Hr	Ar	Ps	Ss	Bs	Mo	Gf	Lo	Fe	Gl	Or	Tr	So	Pl	Ch	Ds	Ma	Rp	Po	References (according to ref. of the data)
Upogebia pugettensis	1	70	34	0.070	0.243	2	3	621	2	7	2	3	2	2	3	3	3	3		1	1			1		2		321
Xanthodius sternberghii	1	41	20	0.034	0.143	3	3	621	2	7	2	3	2	2	3	3	4	3			1			1		2		321
CHELIZERATA																												
Araneus ventricosus	1	50	30	0.094	0.275	2	2	3	1	3	2	2	3	3	2	1	2	2	2	1					3			268
Limulus polyphemus	4	64	25	0.057	0.250																							417
Mela menardi	1	15	15	0.025	0.121																							239,395
Nesticus eremita	13	253	18	0.106	0.347	2	3	1	1		1	3	1	2	3	2	2	2	2	1	1			1		2		103
Nesticus menozzii	1	13	18	0.090	0.222																							103
Nesticus sbordonii	1	10	19	0.109	0.400																							103
MOLLUSCA																												
Arion circumscriptus	4	312	18	0.0	0.0	2	4	2	2	4	2	4	2	3	1	2	3	1		2	1			1	1	2		294
Arion fasciatus	10	814	18	0.0	0.0	2	4	2	2	4	2	4	2	3	1	2	3	1		2	1			1	1	2		294
Arion intermedius	6	202	20	0.0	0.0	2	4	2	1	4	2	4	1	3	1	2	3	1		2	1			1	1	2		294
Arion silvaticus	3	29	17	0.0	0.0	2	4	2	1	4	2	4	1	3	1	2	3	1		2	1			1	1	2		294
Arion subfuscus	5	224	18	0.0	0.0	2	4	2	2	4	2	4	2	3	1	1	3	2		2	1			1	2	2		294
Austrochela constricta	1	15	18	0.168	0.491																							312
Bathybemix bairdii	5	479	18	0.162	0.500	2	3	6 3		7	2	3	2	3			4									2		427
Bothriembryon bulla	1	15	19	0.089	--	2	2	3	1	2	2	2	2	2	1	4	2	3	2	2	1			1	3	2		207
Bothriembryon sp.nov.	1	15	19	0.035	--	2	2	3	1	2	2	2	2	2	1	3	2	3	2	2	1			1	3	2		207
Brachidontes variabilis	10	2000	20	0.619	0.850	4	3	621	1	7	2	4	2	1	3	3	4	1	2	2	1	2		4	3	2		246
Buccinum sp.	1	22	29	0.092	0.357			623		7		1	2	2						2	1			4		2		111
Buliminus diminutus	1	20	30	0.010	0.067	2	2	3	1	1	1	1	2	2	1	3	2	2	1	2	1			1		2		325
Buliminus labrosus	4	140	31	0.040	0.273	2	3	3	1	3	3	2	2	2	1	3	2	2	1	2	1			1		2		325
Campeloma geniculum	7	111	20	0.250	0.442	2	2	611	2	7	2	4	2	2	1	4	2	3	2	2	1	2	18	1	3	3		216
Campeloma parthenum	2	19	20	0.375	0.453																							216
Cerion bendalii	15	612	19	0.056	0.198																							172,499
Cerithium caeruleum	2	400	21	0.635	1.000	3		621		7	2	3	1	2	3	4	4	3		2	1			4		2		380
Cerithium scabridum	3	600	20	0.620	1.000	4	4	621	2	7	2	3	1	2	3	3	4	3		2	1			4		2		380
Corbicula sp.	6	40	18	0.083	0.277	2	4	611	2	3	2	4	2	1	3	2	4	1	1	2	1			1		2		441
Gibula richardii	2	68	26	0.116	0.749	2	3	621	1	7	2	2	2	2	3	3	3	3	1	2	1			4		2		247
Helix aspersa	43	2218	17	0.200	--																							413
Littorina plena	5	132	10	--	0.701	2	3	621	2	7	2	3	1	3	3	3	2	3		2	1			1		2		284
Littorina scutulata	5	133	10	--	0.547	2	3	621	2	7	2	3	2	3	3	3	2	3		2	1			1		2		284

Appendix I. (contin.)

Species	No. of: populations	indi.	loci	Genetic indices H	P	Lz	Gr	Ht	Hr	Ar	Ps	Ss	Bs	Mo	Gf	Lo	Fe	Gl	Or	Tr	So	Pl	Ch	Ds	Ma	Rp	Po	References (according to ref. of the data)
						Biotic Profile @																						
Mesodon aspersa	1	15	17	0.142	--																							411
Modiolus auriculatus	4	1500	16	0.644	1.000	4	3	621	1	7	2	3	2	1	3	4	4	2	2	2	1	2		4	3	2		246
Monodonta turbiformis	3	91	26	0.108	0.571	2	3	621	1	7	2	2	2	2	3	3	3	3	1	2	1			4		2		247
Monodonta turbinata	4	121	26	0.075	0.438	2	3	621	2	7	2	3	2	2	3	3	3	3	1	2	1			4		2		247
Mytilus edulis	1	15	29	0.095	0.277																							4
Partula gibba	1	15	20	0.0	0.0	3	1	4	1	4	2	2	2	2	1	4	2	2	1	2	1	2	44	2	1	2		209
Partula mirabilis	1	23	20	0.167	0.650	3	1	4	1	4	2	2	2	2	1	4	2	2	1	2	1	2	44	2	3	2		209
Partula olympia	1	27	20	0.156	0.650	3	1	4	1	4	2	2	2	2	1	4	2	2	1	2	1	2	44	2	3	2		209
Partula otaheitana	1	10	20	0.175	0.650	3	1	4	1	4	2	2	2	2	1	4	2	2	1	2	1	2	44	2	3	2		209
Partula suturalis	1	20	20	0.167	0.800	3	1	4	1	4	2	3	2	2	1	4	2	2	1	2	1	2	44	2	3	2		209
Partula taeniata	1	40	20	0.134	0.800	3	1	4	1	4	2	3	2	2	1	4	2	2	1	2	1	2	44	2	3	2		209
Patella asperta	2	58	17	0.120	0.645	2	4	621	2	7	2	3	2	1	3	3	3	3	2	2	1			4		2		329
Patella caerolea	2	64	17	0.110	0.588	2	4	621	2	7	2	3	2	1	3	3	3	3	2	2	1			4		2		328
Rumina decollata	63	2398	26	0.0	0.309																							409,412
Scrobicularia plana	1	93	17	0.120	0.706	4	4	621	1	7	2	4	2	1	3	4	4	3	2	2	1			4	3			435
Sphincterochila aharonii	2	60	29	0.042	0.224	2	1	3	1	3	3	2	2	2	1	3	2	2	2	2	1			1		2		330
Sphincterochila cariosa	2	64	28	0.043	0.232	2	3	3	1	3	3	3	2	2	1	3	2	2	1	2	1			1		2		325,330
Sphincterochila fimbriata	3	97	27	0.104	0.423	2	3	3	1	2	2	3	2	2	1	3	2	2	1	2	1			1		2		325,330
Sphincterochila prophetarum	4	178	28	0.074	0.523	2	2	3	1	1	1	3	2	2	1	3	2	2	1	2	1			1		2		330
Sphincterochila zonata	2	110	29	0.079	0.440	2	2	3	1	1	1	3	2	2	1	3	2	2	1	2	1			1		2		330
Theba pisana	8	262	18	0.105	0.460	2	3	4	2	3	3	3	2	2	1	1	2	1	2	2	1			1		2		329
Tridacna maxima	2	104	29	0.206	0.724	3	4	621	2	7	2	3	3	1	2	4		3	2	2	1			4		2		30,94
Trochoidea (Xerocrassa) erkelii	1	29	22	0.055	0.500	2	2	4	1	1	1	2	2	2	1	3	2	2	1	2	1			1		2		331
Trochoidea (Xerocrassa) setzenii	6	194	24	0.070	0.343	2	3	4	2	2	2	3	2	2	1	3	2	2	1	2	1			1		2		325,331
Venerupis decussata	1	47	16	0.277	0.750																							501
BRACHIOPODA																												
Coptothyris grayi	1	15	26	0.202	0.670	2	3	621	1	7	3	2	2	1	2	3	3	2	2	2								272
Frieleia halli	1	45	18	0.170	0.667	2		623	2	7	3	2	2	1	2	3		3	2		1			4		2		477
Litothyrella notorcadensis	1	100	34	0.038	0.240	1	1	621	2	7	3	3	2	1	2	3		3	2	2	1			4		2		39
PHORONIDA																												
Phoronopsis viridis	3	120	39	0.088	0.482	2	2	621	2	7	2	3	2	1	2	3		3	2	2	1			4	3	2		38

Appendix I. (contin.)

Species	No. of: populations	indi.	loci	Genetic indices H	P	Lz	Gr	Ht	Hr	Ar	Ps	Ss	Bs	Mo	Gf	Lo	Fe	Gl	Or	Tr	So	Pl	Ch	Ds	Ma	Rp	Po	References (according to ref. of the data)
VERMES																												
Ascaris lumbricoides	1	15	21	0.027	0.286																							318
Ascaris suum	1	15	21	0.038	0.190																							318
Lineus torquatus	1	15	42	0.139	0.520	5	4	621	2	7	3	2	3	3	1	4	3	3	2	2						2		272
Parascaris equorum	1	127	27	0.009	0.074																							83
Parascaris univalens	1	2092	27	0.015	0.074																							83
Phascolosoma japonicum	1	15	43	0.201	0.589	5	4	621	2	7	3	4	2	3	2	4	4	3	2	2						2		272
COELENTERATA																												
Anthopleura orientalis	1	15	22	0.107	0.405	2	3	621	1	7	3	2	2	1	2	3	3	3	2	2					3	2		272
Metridium exilis	1	15	16	0.110	0.250	2	3	621	1	7	2	2	3	1	3	4			2	2	4			1		3		82
Metridium senile	1	15	18	0.143	0.417	5	4	621	1	7	3	3	3	2	2	4	3	3	2	2					3	3		270
Metridium senile clonal	3	80	16	0.129	0.585	2	4	621	2	7	2	3	2	2	3	4		2	2	1	4			1		3		82
Metridium senile solitary	2	44	16	0.210	0.750	2	4	623	2	7	2	3	3	1	3	4	4	3	2	2	1			1		2		82
PORIFERA																												
Suberites domuncula	1	15	32	0.128	0.430	5	4	621	2	7	3	3	2	3	2	4	3	3	2	2				4		2		272
P L A N T S																												
MONOCOTYLEDONEAE																												
Agrostis stolonifera	6	30	12	0.152	--		4			4				1	1	3	4					2	28		3	3	3	190
Avena barbata	83	2218	18	0.040	0.289	2	3	3	2	4	2	3	2	1	1	1	2	1	1			4	48	1	1	2	1	190,213,281
Avena canariensis	14	70	23	--	0.442	2	1	3		3	2	2	2	1	2	1	2	1				2	14	2	1	2	1	190
Avena fatua	8	720	13	--	0.540																							281
Avena hirtula	23	115	28	0.011	--	2	3			3	2	3	2	1	2	1	2	1				2	14	1	1	2	1	190
Elymus hystrix	1	15	13	--	0.094		2			4				1	2	3	4					2	28		1	2	1	190

@ Biotic Profile

Appendix I. (contin.)

Species	No. of:			Genetic indices		Biotic Profile @																						References (according to ref. of the data)
	populations	indi.	loci	H	P	Lz	Gr	Ht	Hr	Ar	Ps	Ss	Bs	Mo	Gf	Lo	Fe	Gl	Or	Tr	So	Pl	Ch	Ds	Ma	Rp	Po	
Elymus riparius	1	15	13	--	0.0		4			7				1	2	3	4					2	28		1	2	1	190
Elymus virginicus	1	15	13	--	0.351		4			4				1	2	3	4					2	28		1	2	1	190
Elymus weigandii	1	15	13	--	0.094		2			4				1	2	3	4					2	28		1	2	1	190
Hordeum spontaneum	28	1203	28	0.087	0.298	2	3	3	2	4	2	3	2	1	1	1	2	1	1			2	14	1	1	2	1	337
Lolium multiflorum	17	85	15	0.265	0.947		4			4				1	3	1	2					2	14		3	2	3	190
Triticum dicoccoides	12	457	50	0.063	0.252	2	3	3	1	3	1	3	2	1	1	1	2	1	1			4	28	1	1	2	1	332
Zea mexicana (annual)	1	15	20	0.193	0.736	3	3	3		4	2	2	2	1	3	1	3	1				2	20	1	2	2	3	190
Zea mexicana (perennial)	1	15	20	--	0.497	3	3	3		4	2	2	2	1	3	3	4	2				2	20	1	2	2	3	190
DICOTYLEDONEAE																												
Capsicum annuum	23	67	26	--	0.308																							298
Capsicum baccatum	23	67	25	--	0.280																							298
Capsicum cardenasii	39	106	24	--	0.262																							297,298
Capsicum chacoense	23	67	26	--	0.0																							298
Capsicum eximium	59	139	24	--	0.392																							297,298
Capsicum frutescens	23	67	26	--	0.192																							298
Capsicum praetermissum	23	67	26	--	0.077																							298
Capsicum sp.	23	67	23	--	0.087																							298
Capsicum tovari	1	20	25	0.017	0.110																							297,298
Chamaenerion angustifolium	5	50	16	--	0.172																							480
Colobanthus quitensis	2	15	22	0.0	0.0		2	6		7	2			1		3								1		3		190
Coreopsis cyclocarpa	8	256	20	0.076	0.512	2	2	3	1	2	2	2	1	1	1	4	2	3	2			2	24	1	3	2	2	116
Coreopsis nuecensis	7	309	15	0.083	0.730	2	1	3	1	3	2	2	2	1	1	1	2	2	1			2	14	1	3	2	2	117
Coreopsis nuecensoides	10	497	15	0.153	0.930	2	1	3	1	3	2	2	2	1	1	1	2	2	1			2	18	1	3	2	2	117
Gaura brachycarpa	1	15	12	0.060	0.290		2			4				1	1	1	3					2	14		2	2	2	190
Gaura demareei	2	15	18	0.050	0.255		1	3		4	2			1	1	1	3	1				2	14	1	3	2	2	190
Gaura longifolia	3	15	18	0.074	0.290		3	3		4	2			1	1	1	3	1				2	14	1	3	2	2	190
Gaura suffulta	1	15	12	0.030	0.374		2			4				1	3	1	3					2	14		3	2	2	190
Gaura triangulata	1	15	12	0.080	0.097		2			4				1	1	1	3					2	14		1	2	1	190
Gilia achilleifolia	7	1084	13	0.106	0.515	2	3	3	1	2	2	3	2	1	2	1	2	2	2			2	18	1	2	2		403
Isotuma petraea	1	15	25	0.009	--	2	4	3	1	2	1	2	2	1	1	1	3	2	2			2	14	1	2	2	1	207
Liatris cylindracea	1	2258	27	0.057	0.598	2	3	3		4	2	2	1	1	3	3	4	1				2	20	1	3	3	2	190,399
Limnanthes alba	6	30	13	0.159	0.595		1			7				1	3	1	3					2	10		3	2	2	190
Limnanthes floccosa	6	30	13	0.130	0.209		1			4				1	3	1	2					2	10		1	2	1	190
Lycopersicon cheesmanii	54	270	14	0.0	0.0	3	1	3		3	2	2	2	1		3	4	1				2	24	1	1	2	1	190,379
Lycopersicon chmeielewskii	8	40	14	0.046	0.159	3	1	3		4	2	2	2	1		3	4	1				2	24	1	2	2	2	190

Appendix I. (contin.)

Species	No. of: populations	indi.	loci	Genetic indices H	P	Lz	Gr	Ht	Hr	Ar	Ps	Ss	Bs	Mo	Gf	Lo	Fe	Gl	Or	Tr	So	Pl	Ch	Ds	Ma	Rp	Po	References (according to ref. of the data)
Lycopersicon esculentum	7	35	15	0.017	0.043	3	4	3		4	2	2	2	1		3	4	1				2	24	1	2	2	2	190
Lycopersicon parviflorum	8	40	14	0.0	0.0	3	1	3		4	2	2	2	1		3	4	1				2	24	1	1	2	1	190
Lycopersicon pimpinellifolium	43	215	13	0.074	0.411		2			4				1	3	3	4					2	24		2	2	2	190
Lythrum tribracteatum	7	35	18	--	0.325	2	2	6		7	2	2	2	1	1	3	3	1				2	10	1	1	2	1	190
Oenothera argillicola	10	50	20	0.080	0.200	2	3	3	2	4	2	3	2	1	1	2	4	3	1			2	14	1	3	2	2	258
Oenothera biennis	218	3343	20	0.081	0.131	2	3	3	2	4	2	4	2	1	1	2	4	3	1			2	14	1	1	2	1	190,253,257,258
Oenothera drummondii	1	15	18	--	0.333	2	1	3	1	3	2	3	2	1	1	3	4	2	1			2	14	2	2	2	2	139
Oenothera grandis	27	1055	18	0.070	0.300	2	2	3	2	3	2	3	2	1	1	1	4	1	1			2	14	1	3	2	2	138,139
Oenothera hookeri	10	50	20	0.0	0.0	2	3	3	2	2	2	3	2	1	1	1	4	3	1			2	14	1	3	2	2	258
Oenothera humifusa	1	15	18	--	0.167	2	1	3	1	3	2	3	2	1	1	3	4	2	1			2	14	2				139
Oenothera laciniata	28	1070	18	0.050	0.299	2	3	3	2	3	2	3	2	1	1	2	4	1	1			2	14	1				138,139
Oenothera mexicana	12	455	18	0.0	0.0	2	1	3	1	3	2	3	2	1	1	1	4	1	1			2	14	1	1	2	2	138,139
Oenothera parviflora	29	145	20	0.149	0.400	2	2	3	2	4	2	3	2	1	1	2	4	3	1			2	14	1	1	2	1	258
Oenothera strigosa	29	145	20	0.028	0.250	2	3	3	2	4	2	4	2	1	1	2	4	3	1			2	14	1	1	2	1	258
Phlox cuspidata	79	1735	20	0.008	0.257	2	1	3	2	2	2	3	1	1	1	1	3	1	1			2	14	1	1	2	1	190,254,255
Phlox drummondii	125	2865	20	0.051	0.267	2	1	3	2	2	2	3	1	1	1	1	3	1	1			2	14	1	2	2	2	190,254,255
Phlox roemariana	30	600	20	0.046	0.265	2	1	3	2	2	2	3	1	1	1	1	3	1	1			2	14	1	2	2	2	190,255
Senecio sylvaticus	4	400	30	0.0	0.0																							231
Senecio viscosum	3	300	28	0.0	0.0																							231
Shorea leprosula	3	15	36	--	0.552		4	3		4	2		1	1	1	4	4	3				2	14	2	3	2	2	190
Silene maritima	2	48	20	0.140	0.340																							45
Stellaria media	2	15	33	--	0.087																							480
Stellaria pallida	1	15	32	--	0.049																							480
Stephanomeria exigula	12	1095	14	0.094	0.454	2	2	3		3	2	2	1	1	1	1	2	1				2	16	1	2	2	2	171,190
Tragopogon dubius	6	141	21	0.026	0.087	2	4	3		4	2	2	1	1	3	2	3	2				2	12	1	1	2	1	190
Tragopogon mirus	8	153	42	0.003	0.021	2	1	3		4	2	2	1	1	3	2	3	2				2	24	1	1	2	1	190
Tragopogon miscelus	6	120	42	0.002	0.025	2	1	3		4	2	2	1	1	3	2	3	2				2	24	1	1	2	1	190
Tragopogon pratensis	3	67	21	0.0	0.0	2	4	3		4	2	2	1	1	3	2	3	2				2	12	1	1	2	1	190
Tragopogon porrifolius	3	84	21	0.013	0.063	2	4	3		4	2	2	1	1	3	2	3	2				2	12	1	1	2	1	190
Tripsacum dactyloides	1	15	20	--	0.178	2	4	3		4	2	2	1	1	3	3	4	1				2	54	1	2	3	3	190
Tripsacum floridinum	1	15	20	--	0.237	2	1	3		7	2	2	1	1	3	3	4	1				2	26	1	2	3	3	190
GYMNOSPERMEAE																												
Picea sitchensis	10	50	24	0.150	--	2	2	3	2	5	3	4	3	1	2	4	4	3	2			2	24	3	3	2	3	178
Pinus contorta	9	45	25	0.161	--	2	4	3	2	4	3	4	3	1	2	4	4	3	2			2	24	3	3	2	3	178
Pinus longavea	15	75	21	0.171	0.803	2	2	3		3	2	3	3	1	3	4	4	1				2	24	1	3	2	3	190

Biotic Profile columns: Lz Gr Ht Hr Ar Ps Ss Bs Mo Gf Lo Fe Gl Or Tr So Pl Ch Ds Ma Rp Po (@)

Appendix I. (contin.)

Species	No. of: populations	indi.	loci	Genetic indices H	P	Lz	Gr	Ht	Hr	Ar	Ps	Ss	Bs	Mo	Gf	Lo	Fe	Gl	Or	Tr	So	Pl	Ch	Ds	Ma	Rp	Po	References (according to ref. of the data)
Pinus ponderosa	8	50	25	0.188	0.718	2	3	3	2	2	3	4	3	1	2	4	4	3	2			2	24	1	3	2	3	6,190
Pinus pungens	3	15	15	--	0.453	2	2	3		3	2	3	3	1	3	4	4	3				2	24	1	3	2	3	190
Pinus resinosa	1	15	27	0.007	--	2	3	3	2	3	3	4	3	1	2	4	4	3	2			2	24	1	3	2	3	6
Pinus rigida	11	694	20	0.141	0.966	2	3	3	2	4	2	4	3	1	2	4	4	3	2			2	24	1	3	2	3	190
Pseudotsuga menziesii	19	95	17	0.205	0.728	2	4	3	2	4	3	4	3	1	2	4	4	3	2			2	24	3	3	2	3	190
PTERIDOPHYTA																												
Lycopodium lucidulum	16	241	18	0.060	0.100	2	3	3	2	4	2	3	2	1	1	4	3	3	2					1	2	3		256
FUNGI																												
Boletus edulis	1	30	18	0.226	0.647	5	4	3	2	3	2	4	3	1	3	1	4	3	2					3				271
B A C T E R I A																												
Escherichia coli	5	109	20	--	0.900																							414a

Biotic Profile @

This list consists of 1111 species that remained after exclusion of some species with small sample sizes.
$ Including two lower chordates.
@ For description of the variables and their abbreviations see appendix II .

APPENDIX II

Biotic - Questionnaire

Ecological Variables	Remarks
Life zone (Lz)	
1. arctic	
2. temperate	
3. tropic	
4. temperate + tropic	
5. cosmopolitan	Cosmopolitan are species ranging in several continents and climatic life zones.
Geographic range (Gr)	
1. endemic	Obviously this categorization is
2. narrow	relative to the taxonomic group
3. regional	involved.
4. widespread	Widespread on a continental scale
5. relict	
Habitat type (Ht)	
1. subterranean	Please specify all subcategories
2. fossorial	applicable.
3. above ground	
4. arboreal	
5. air	
6. aquatic (i) freshwater (a) littoral (ii) marine (b) pelagic nectonic (iii) both (c) pelagic benthonic	
Habitat range (Hr)	
1. specialist	Specialist - geographically restricted, local, rare, narrow - niched, isolated, island distribution.
2. generalist	Generalist - widespread, common, vagile, broad - niched.

Aridity index (Ar)

1. arid
2. subarid
3. subhumid
4. mesic
5. extra-mesic
6. mesic + arid

Territoriality (Tr)

1. yes
2. no

Territoriality involves either/or both sexes or only seasonal phenomen

Distribution (Ds)

1. mainland
2. island
3. both

Demographic Variables

Population structure (Ps)

1. isolated
2. patchy
3. continuous

Species size - number of individuals (Ss)

1. small (thousands)
2. medium (hundred of thousands)
3. large (millions)
4. very large (billions)

Even very subjective and tentative estimations will be helpfull.

Adult mobility (Mo)

1. sedentary
2. almost sedentary
3. low
4. high
5. migratory

Young dispersal (Gf)

1. low
2. medium
3. high

Sociality (So)

1. solitary
2. monogamic family structure
3. polygamic family structure
4. social

Life History Variables

Body size (Bs)

Relative to the taxonomic class.

1. small
2. medium
3. large

Longevity (Lo)

1. annual
2. biennial
3. short-lived perennial (3-5 years)
4. long-lived perennial (> 5 years)

Generation length (Gl)

1. < 1 yr.
2. 1 yr.
3. > 1 yr.

Time from birth to sexual maturity.

Fecundity (Fe)

1. < 10
2. 10 - 100
3. 100-1000
4. > 1000

Origin (Or)

1. recent
2. old

Recent is defined primarily as from mid- or upper-pleistocene.

Chromosome number (Pl = ploidy; Ch = chromosome no.)

Please specify exact number as well

1. haploid no. (N)
2. diploid no. (2N)
3. polyploid no (XN)

Mating system (Ms)

1. primarily selfed
2. mixed
3. primarily out-crossed
4. parthenogenetic

Mode of reproduction (Rp)

1. asexual
2. sexual
3. both

ollination mechanism (Po)

. selfed

. animal

. wind

Appendix III. Transformations of heterozygosity and polymorphism estimates.

In order to obtain a comparable set of data, we transformed all estimates of polymorphism (P) and heterozygosity (H) to a common background. For polymorphism, we transformed all other estimates (P-5% and P-unknown criteria) to an estimated value of P-1%. Likewise, we transformed He to an estimate of observed heterozygosity (H) in all cases where these two estimates were not recorded in the original studies. This procedure has been conducted for the original population or species data.

The coefficients of transformation which were used as multipliers appear in the following table, for P-5% (T-P5%), for P-unknown (T-P?) and for gene diversity, or He (expected heterozygosity, see Nei, 1975) - (T-He). These coefficients were derived empirically. We calculated the ratios P-1% / P-5% and H / He for all cases of animal species in which both indices were recorded simultaneously in the original data (cases of zero were excluded). The coefficients appearing in the table are based on about 350 cases for P and about 250 cases for He. We averaged separately each interval in the table. For P with unknown criteria we used half of the calculated correction (see Table). For selfed plants the recorded H has been ommitted and the transformed He inserted.

The transformed values and the weighting by loci x individuals were conducted by the aggregate procedure of SPSS (Hull and Nie, 1981), which generated the averages and sums for each species that appear in Appendix I. In order to conduct weighting we inserted, in cases of missing values, ten for the number of loci, one for population number and five for individuals per population (minimum ten).

Table. Coefficients of transformation.

a. For polymorphism (P).

P-5% or P-?	T-P5%	T-P?
< 0.05	1.590	1.295
0.05 - 0.10	1.430	1.215
0.10 - 0.20	1.375	1.187
0.20 - 0.30	1.320	1.160
0.30 - 0.40	1.265	1.133
0.40 - 0.50	1.210	1.105
0.50 - 0.60	1.155	1.078
0.60 - 0.70	1.100	1.050
0.70 - 0.80	1.045	1.022
> 0.80	1.000	1.000

b. For gene diversity (He).

He	T-He
< 0.05	0.938
0.05 - 0.15	0.876
0.15 - 0.25	0.840
0.25 - 0.35	0.800
> 0.35	0.750

GENETIC POLYMORPHISM AND NEOMUTATIONISM

Masatoshi Nei
Center for Demographic and Population Genetics
University of Texas at Houston, P.O. Box 20334, Houston, Texas 77225

INTRODUCTION

In the last several decades neo-Darwinism, particularly the balance theory of evolution (Dobzhansky 1955), has remained a dominant force in evolutionary biology. In this theory natural selection plays the most important role in determining the extent of genetic polymorphism and the rate of evolution. Although mutation is regarded as the ultimate source of genetic variation, its role in evolution is considered to be minor. This is because mutation occurs repeatedly at the phenotypic level and most natural populations seem to carry a sufficient amount of genetic variability, so that almost any genetic change can occur by natural selection whenever the change is needed. Namely, natural selection plays a creative role (Mayr 1963; Dobzhansky 1970). This is in sharp contrast to the mutationist's view of evolution that prevailed in the first quarter of this century (e.g., Morgan 1925, 1932). In mutationism (in Morgan's sense not de Vries') natural selection plays a less important role than mutation, and its chief role is to preserve rare useful mutations and eliminate unfit genotypes; all creative roles are given to mutation. In the 1930's and 1940's mutationism gradually became unpopular among biologists for various reasons, and by 1950 it almost disappeared.

In the 1960's and 1970's spectacular progress occurred in molecular biology, and this progress made it possible to study evolution at the most fundamental level, i.e., at the codon (amino acid) or nucleotide level. This study has led to many new discoveries in evolutionary biology. Two of the most important are (1) approximate constancy of the rate of amino acid substitution in each protein (Zuckerkandl and Pauling 1965) and (2) a large amount of genetic polymorphism at the protein level in many natural populations (Harris 1966; Lewontin and Hubby 1966). These discoveries have led a number of authors to emphasize the importance of mutation in evolution. Particularly, Kimura (1968) and King and Jukes (1969) proposed the so-called neutral mutation hypothesis in which evolution occurs mainly by random fixation of neutral or nearly neutral mutations. Ohta (1974, 1977) extended this hypothesis to the case of slightly deleterious mutations.

Nei (1975, 1980) suggested that mutation is the primary force of evolution even for morphological and physiological characters.

This new form of mutationism, particularly neutral theory, has generated a great deal of controversies among evolutionary biologists in the last decade, and many biologists are still skeptical about the validity of this new mutationism. At the same time a large amount of data have been accumulated on both polymorphism and long-term evolution at the molecular level, so that we can examine the mechanism of evolution more objectively. In the past ten years I have been involved in statistical analysis of these molecular data with the help of my colleagues. We first studied the pattern of protein polymorphism and examined whether the data available are consistent with the predictions from neutral theory or not. In recent years our statistical analysis has been extended to data on DNA polymorphism and evolution. In this paper I would like to present a summary of our recent works and discuss the implications of these works on the general theory of evolution.

PROTEIN POLYMORPHISM

When Harris (1966) and Lewontin and Hubby (1966) reported their discovery that natural populations are highly polymorphic at the protein level, many population geneticists attempted to explain the polymorphism in terms of overdominant selection or truncation selection with overdominant gene action (e.g., Sved et al. 1967; King 1967; Milkman 1967). They thought that such a high degree of polymorphism cannot be maintained without some kind of balancing selection. The following year Kimura (1968) attempted to explain the level of protein polymorphism observed by these authors in terms of the neutral mutation theory, but he was not very successful because the mathematical theory of multiallelic mutations was not well developed at that time. In the last 14 years, however, the mathematical theory has been refined extensively (Kimura and Ohta 1971; Nei 1975; Ewens 1979), and we can now examine the agreement between observed data and theoretical predictions from neutral theory and some of its alternative hypotheses.

Extent of Polymorphism and Population Size

Probably the most appropriate measure of protein polymorphism is the average heterozygosity over all loci examined. Heterozygosity at a locus is defined as $h = 1 - \Sigma p_i^2$, where p_i is the frequency of the *i*-th allele. Average heterozygosity ($\hat{H}$) is simply the average of h over all loci. In the theory of neutral mutations the expectation of $\hat{H}$ in an equilibrium population is given by

$$H = 4N_e v/(1 + 4N_e v), \quad (1)$$

where N_e and v are the effective population size and the rate of neutral mutations per generation, respectively. In neutral theory the rate of gene substitution is assumed to be equal to the mutation rate (Kimura and Ohta 1971; Nei 1975). Therefore, we can estimate v from the rate of amino acid substitution in proteins. Using this method and considering the molecular size of the proteins which are used for electrophoresis, Kimura and Ohta (1971) and Nei (1975) have estimated that the rate of neutral mutations is approximately 10^{-7} per locus per year. Therefore, we can compute v if we know the generation time. Estimation of N_e is not easy, but it is possible to have a crude estimate of actual population size (N) in certain species. N is usually larger than N_e, so that H obtained by using N instead of N_e in (1) would give the upper bound of expected heterozygosity. Therefore, if neutral theory is valid, the observed heterozygosity ($\hat{H}$) is expected to be equal to or lower than H.

With this understanding, Nei and Graur (1984) examined the relationship between $\hat{H}$ and Ng for 72 species, including Escherichia coli, Drosophila, fishes, reptiles, and mammals. Note that $Nv = Ng \times 10^{-7}$, where g is the generation time. In this study we included only those species in which an estimate of N was obtainable and there were gene frequency data for at least 20 loci studied by electrophoresis. Estimates of N were obtained by census (e.g., man, Japanese macaques) or by the multiplication of the population density by the geographical distribution. The results obtained are presented in Figure 1. It is clear that in all species except one the observed heterozygosity is lower than the expected (H; solid line) and thus the data can be accommodated with the neutral theory in most species.

One important message from Figure 1 is that although many selectionists tend to believe that the genetic variability of natural populations is too high to be neutral, actually it is too low compared with the neutral expectation under the assumption of equilibrium population. It is therefore clear that to explain the level of protein polymorphism properly we must consider the factors that reduce genetic variability. For this purpose overdominant selection or other types of balancing selection are clearly inadequate since they tend to increase heterozygosity (see Nei (1980) for other problems). The

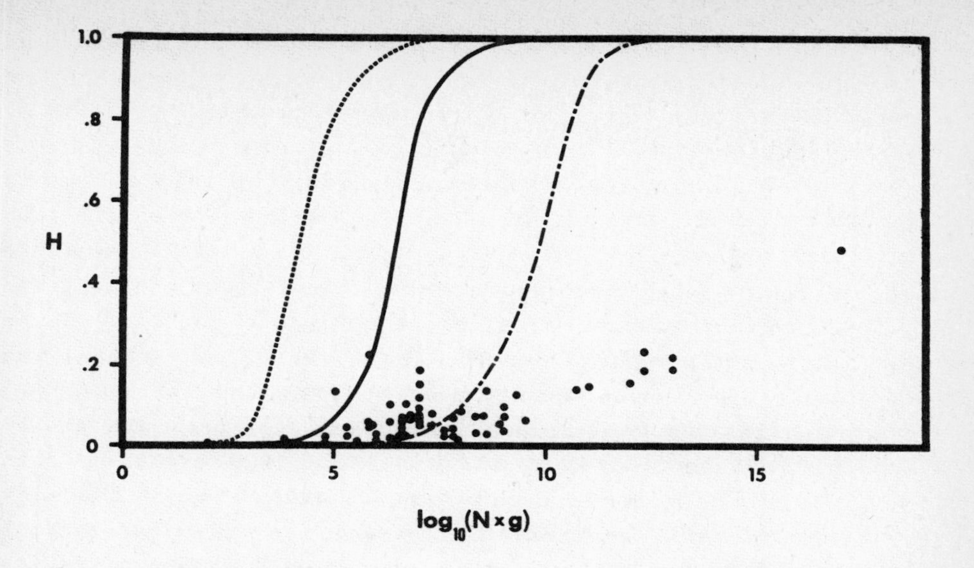

Figure 1. Relationship between the observed average heterozygosity ($\hat{H}$) and Ng for 72 species. Solid line: Expected heterozygosity for neutral alleles (H). Chain line: Expected heterozygosity for slightly deleterious alleles with a mean selection coefficient of $\bar{s}$ = 0.002. This curve was obtained by using Kimura's (1979) formulae (23) and (24) with β = 0.5. Dotted line: expected heterozygosity for overdominant alleles with s = 0.001. The curve for overdominant alleles does not change appreciably even if s varies from homozygote to homozygote as long as the mean of s remains as 0.001. For computing the expected heterozygosity, the mutation rate (v) is assumed to be 10^{-7} per locus per year.

dotted line in Figure 1 represents the expected heterozygosity when overdominant selection with a selection coefficient of s = 0.001 is operating (Li 1978; Maruyama and Nei 1981). Although selection coefficient is very small, the expected heterozygosity rises very rapidly as Nv increases and is generally much higher than that for neutral mutations. It should also be noted that the highest gene diversity (equivalent to heterozygosity) so far observed is from _E. coli_, where overdominant selection cannot occur because of haploidy.

While the observed heterozygosity is certainly lower than the expected, the difference between them is very large when Nv is large. Under the framework of neutral theory this can be explained by the following two factors. First, a high value of Nv occurs for small organisms such as Drosophila and *E. coli*, and in these organisms the effective population size is expected to be much smaller than the actual size because of frequent extinction and replacement of colonies (Maruyama and Kimura 1980). Second, average heterozygosity is affected drastically by the bottleneck effect, and this effect is expected to last for a long time in large populations – often millions of generations (Nei et al. 1975). Actually, in the Pleistocene (10^4 - 2×10^6 years ago) there were several glaciations and in these periods many organisms apparently went through bottlenecks. It is known that up to 50 percent of mammalian species became extinct and many new species appeared in the Pleistocene (Martin and Wright 1976). Therefore, the difference in $\hat{H}$ and H in Figure 1 can be explained by the assumption that the long-term effective size is much smaller than the actual size. Of course, the bottleneck effect is not the only factor that can explain the difference between $\hat{H}$ and H. There are at least two other factors, both of which are diversity-reducing selection. One is the random fluctuation of selection intensity as formulated by Nei and Yokoyama (1976) and the other is Ohta's slightly deleterious mutations. However, it is not clear at the present time how important these factors are in reality. The reader who is interested in these factors may refer to Nei (1980, 1983).

Internal Consistency of Population Parameters

It should be remembered that in the above test we have not really estimated the effective population size and thus our conclusion is only qualitative. However, there are several other tests of neutral theory and its alternatives where estimates of N_e are not required. In these tests the internal consistency of population parameters for a particular theory to be tested is examined. For example, the distribution of single-locus heterozygosity (h) for neutral genes is a function of $M \equiv 4N_ev$ only. Therefore, if we know M rather than N_e and v separately, we can compute the theoretical distribution of h and compare this with the observed distribution. An estimate ($\hat{M}$) of M can be obtained from average heterozygosity by using (1), i.e., $\hat{M} = \hat{H}/(1 - \hat{H})$. Using this method, Nei et al. (1976) and Fuerst et al. (1977) examined the agreement between the expected and observed distributions of h for 95 vertebrate and 34 invertebrate species. This study showed that the observed distribution does not deviate significantly from the expected distribution under neutral theory. On the other hand, the observed distribution was quite different from the theoretical distribution for overdominant alleles (Maruyama and Nei

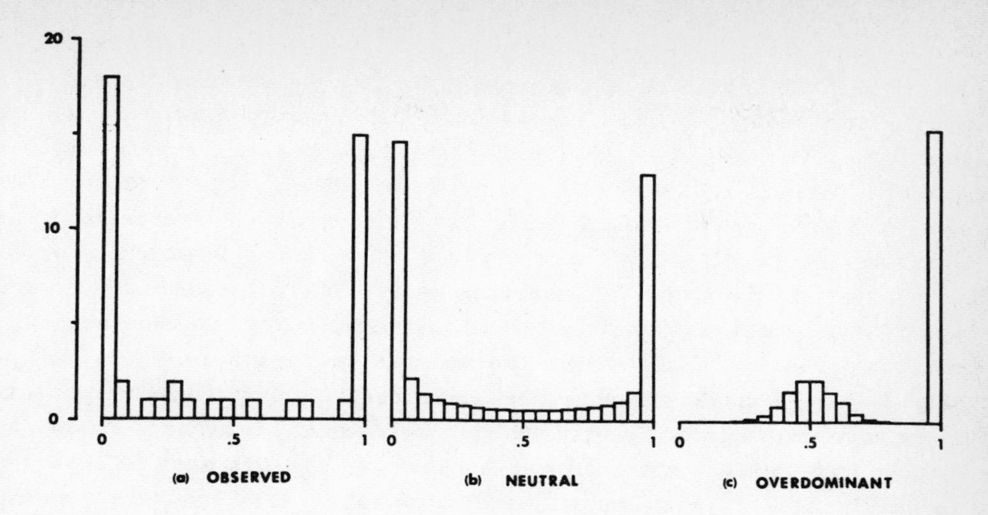

Figure 2. Observed and expected distributions of allele frequencies. (a) Observed distribution for Drosophila engyochracea ($\hat{H}$ = 0.127). (b) Expected distribution for neutral alleles with H = 0.127. (c) Expected distribution for overdominant alleles with H = 0.127.

1981). We also examined whether the relationship between the mean and variance of single-locus heterozygosity agrees with the neutral expectation or not. This test also did not reject the "null hypothesis" of neutral mutations except in some species.

Another study we conducted is to examine the observed distribution of allele frequencies or frequency spectrum. This study (Chakraborty et al. 1980) has shown that in all species examined the distribution is U-shaped and is in approximate agreement with the neutral expectation. The theoretical distribution for neutral alleles is given by

$$\Phi(x) = M(1 - x)^{M-1}x^{-1} \tag{2}$$

(Kimura and Crow 1964), where x stands for the allele frequency. Figure 2 shows one example obtained from Drosophila engyochracea. Clearly, the observed distribution (a) is very close to the neutral expectation (b) but quite different from that for overdominant alleles (c), which was obtained under the condition that the expected heterozygosity is equal to the observed value, i.e., $\hat{H}$ = 0.127.

In addition to the intrapopulational parameters we have also examined interpopulational parameters such as the distribution of genetic distance, the mean and variance of genetic distance, and the correlation of single-locus heterozygosities between related species (Chakraborty et al. 1978). The results of these studies again showed that observed data do not deviate far from neutral expectations. We can therefore conclude that the pattern of protein polymorphism within and between populations is in approximate conformity with the neutral expectations but substantially deviated from several hypotheses involving selection. Of course, the statistical methods so far used are generally very crude and would not detect small differences in prediction between alternative hypotheses. This is particularly so when selection coefficients are small. However, it now seems clear that the population dynamics of protein polymorphism is largely controlled by the stochastic factors whether some weak selection is involved or not (Nei 1980).

In the past two decades many authors have argued for various types of selection considering ecological factors. Levins' (1968) hypothesis of adaptive strategy and Nevo's (1978) specialist-generalist theory are good examples. However, these types of selection are generally ill-defined, and it is very difficult to study the corresponding mathematical model and derive any testable predictions analogous to those for neutral mutations. One common feature for these hypotheses is that genetic variability is actively maintained by selection and thus leads to a heterozygosity higher than the neutral expectation. As mentioned earlier, however, actual data indicate that the observed heterozygosity is lower than the neutral expectation in an equilibrium population. Therefore, these types of selection do not appear to be very important.

In Chapter 2 of this book Nevo has examined the relationship of genetic variability with various ecological and demographic factors and concluded that the extent of protein polymorphism is partly predictable by 3 - 4 combinations of ecological, demographic, and life-history variables. For example, he states that the organisms living in extremely arid and extra mesic environments show lower levels of genetic variation than the organisms living in a broader climatic spectrum. This type of observation is certainly interesting, but it does not really give any insight into the genetic mechanism that generated this relationship. Adaptation to a particular environment must occur by a certain set of genes, but this adaptation may increase or decrease the genetic variation of the population. Unlike neutral theory, it does not give any prediction about the extent of genetic variability unless the genetic mechanism of the adaptation is specified. It should also be noted that this type of study is largely subjective and different authors often reach different conclusions (Schnell and Selander 1981; Hedgecock et al. 1982).

Mani's Ecological Genetic Model

Among selectionists, Mani (this volume) is unique in the sense that he developed a mathematical model of dynamics of protein polymorphism, incorporating several ecological factors such as population growth, resource limitation, and environmental fluctuation of fitness. Using this model, he studied the distribution of allele frequencies, distribution of single-locus heterozygosity, relationship between average heterozygosity and population size, etc., for several different types of selection. Comparison of these theoretical expectations with observed data has led him to conclude that observed data on protein polymorphism can be explained reasonably well by some of his selection models. Careful examination of his work, however, reveals a number of problems. Here I would like to discuss a few of them.

(1) He uses a discrete time model of population growth, extending May's (1974) result to a polymorphic population. Previously, May (1974) showed that when a discrete time version of logistic growth is used, a chaotic change of population size may arise. This model is mathematically intriguing and thus has attracted attention from many theoretical ecologists. It involves two parameters, i.e., intrinsic growth rate (r) and carrying capacity (K). The chaotic change of population size (N) in this model occurs around N = K. In other words, N can be greater than K in this model. Biologically, however, this is meaningless because K is supposed to be the carrying capacity determined by the environment. Since N in the discrete-time model represents the _adult size_, it cannot be greater than K (Nei 1975, pp. 38-39). This clearly indicates that the chaos observed by May is a mathematical artifact resulting from his unrealistic model. Indeed, the chaotic status has never been observed in natural populations, as indicated by Mani. Mani's study includes this unrealistic element of mathematical model, and thus the results he obtained may not be realistic.

(2) Unlike his claim, the theoretical predictions from his ecological genetic model about the distribution (spectrum) of allele frequencies, distribution of heterozygosity, etc., do not really agree with the neutral expectations. For example, with respect to the

allele frequency distribution, he states that "as long as R and K are correlated (with $\beta \leq 0$), the results are in close agreement with the neutral model". Actually, this is not true; compared with the neutral expectation his figures often show a conspicuous excess in both the frequency classes of 1/2N - 0.1 and 0.9 - 1.0. Note that in his Figure 11 the expected number of alleles for a given gene frequency interval is not computed in the standard way (Mani, personal communication), so that they cannot be compared either with the expected number of alleles or with the observed numbers of Figure 3 of Chakraborty et al. (1980). In this case it should be noted that $\Phi(x)dx$ does not give a probability but the expected number of alleles of which the frequency is between x and x + dx. Therefore, we must consider not only the shape of the curve but also the absolute value of $\Phi(x)dx$. If we consider this factor, none of his theoretical distributions of allele frequencies is close to the neutral expectation or the observed value. The agreement for the distribution of single-locus heterozygosity (h) also is not as good as he claims. A relatively good agreement is obtained for the case of CEM; N = 5000; $\beta = 0$ in Figure 17. However, the frequency of h = 0 in this model is considerably larger than that for the neutral model. Since most of the actual data agree with the neutral expectations as mentioned earlier, these results indicate that his model does not really explain the observed data. Furthermore, unlike his claim, the results of his computer simulation about the relationship between the mean (H) and variance [V(h)] of heterozygosity also do not really agree with the observed data. Namely, V(h) is too small compared with the observed data when H is close to 0 but too large when H is between 0.1 and 0.2 (his Figure 21). It should also be noted that in this type of studies statistical tests must be carried out about the agreement between data and theory. Visual inspection is not sufficient.

(3) In Mani's models various factors such as genotype-dependent rate of population growth, genotype-dependent carrying capacity and time-dependent selection, etc., are considered, so that it is difficult to ascertain which factors are important for determining genetic variability. The primary aim of ecological genetics is to understand the mechanism of natural selection in terms of ecological factors, as in the case of industrial melanism (e.g., Kettlewell 1973; Mani 1980). In the present case, however, his study does not give much insight into the actual selection mechanism. Since his model includes many parameters, it would certainly be possible to explain observed data with a certain combination of the parameters. But this can happen by chance alone, and unless the relationship between genotype fitness and

ecological conditions is clarified, we will never know the real mechanism of maintenance of selective polymorphism.

(4) Examining Nei and Graur's (1984) data on generation times (g) and effective population sizes ($N_{e(T)}$) expected from average heterozygosities, Mani stated that "for any particular generation time, N (estimate of actual population size) could vary as much as five orders of magnitude while $N_{e(T)}$ converge to the same value within a factor of 2 - 3", and that "it is very difficult to understand how bottlenecks could produce changes in N over evolutionary time scale such that the widely varying values for any particular generation time are compressed to yield values of $N_{e(T)}$ that differ by less than a factor of three." Actually, the first part of his statement is not really correct, because when generation time is 0.5, 2 or 5, $N_{e(T)}$ varies about five orders of magnitude (Figure 1 of his paper). Nevertheless, it is true that many species with the same generation time show similar $N_{e(T)}$'s. The main reason for this seems to be that these species belong to the same genus or the same group of organisms and thus the time (t) after divergence of these species is not sufficiently long. Note that the correlation between the heterozygosities of two related species is given by $r = \exp[-(4v+1/N_e)t]$ (Li and Nei 1975; Chakraborty et al. 1978). Therefore, if t is relatively small, the related species are expected to have similar levels of heterozygosity even if N_e varies substantially. Thus, his observation is not really incompatible with neutral theory.

(5) Mani was concerned only with protein polymorphism. For a theory to be viable, it has to be able to explain all aspects of the evolutionary change of genes and proteins. It is now well established that the rate of amino acid or nucleotide substitution is higher for functionally less important genes or parts of genes than for functionally more important genes. Silent nucleotide substitution is also known to occur at a higher rate than amino acid altering nucleotide substitution. All these observations can easily be explained by neutral theory (see below), but I wonder how Mani's theory can cope with them.

DNA Polymorphism

In the study of evolution DNA sequences give much more information than protein sequences. This is because a large part of DNA sequences are not encoded into protein sequences. Thus, the genetic variation in noncoding regions of DNA (introns, leader regions, flanking regions) or in silent codons can be studied only by examining DNA sequences. Examination of DNA sequences also reveals detailed information on the mechanisms of deletion, insertion, unequal crossing over, transposition of genes, and gene conversion (Slightom et al. 1980). Unfortunately, however, the techniques for studying DNA sequences were developed only recently, so that the data on DNA polymorphism are still scanty compared with those on protein polymorphism. Nevertheless, many interesting results have already been obtained (see Nei 1983, Nei et al. 1983 for reviews). In the following I shall consider only those studies which are directly related to our problem.

Nucleotide diversity

At the DNA level the size of a gene varies extensively from gene to gene because of the varying nature of the number and size of introns and flanking regions. Furthermore, the extent of DNA polymorphism per locus is much higher than that observed at the protein level. Therefore, average heterozygosity per locus is not an appropriate measure of DNA polymorphism. Instead, DNA polymorphism is usually measured by nucleotide diversity or heterozygosity per nucleotide site (Nei and Li 1979). It is defined by

$$\pi = \Sigma \pi_{ij}/n_c$$

where π_{ij} is the proportion of different nucleotides between the *i*-th and *j*-th DNA sequences in the population and n_c is the total number of comparisons available. In other words π is the average nucleotide differences per site between two randomly chosen DNA sequences. This quantity cannot be larger than 3/4, because there are only four different types of nucleotides. When π is much smaller than 1, its expectation in an equilibrium population is $4N_e\mu$ under neutral theory, where μ is the mutation rate per nucleotide site.

Table 1 shows examples of the estimates ($\hat{\pi}$) of this quantity. Most of these estimates were obtained by an indirect method, i.e., the restriction enzyme technique (Nei and Li 1979; Nei and Tajima 1981). Table 1 indicates that nucleotide diversity is 0.002 - 0.013 in eukaryotic organisms and nearly the same for both mtDNA and nuclear genes, though the number of nuclear genes studied is very small. If we note that the coding region of structural genes usually consists of about 1000 nucleotide pairs, this result suggests that two randomly chosen structural genes from a population are rarely identical with respect to nucleotide sequence. Mitochondrial DNAs are maternally inherited and exist in the haploid form. Therefore, the effective

Table 1. Estimates of nucleotide diversity (π) or the proportion of nucleotide differences between a selected pair of DNA sequences (π_{ij}).

DNA or gene region	Organism	Method*	Sample size	No. of base pairs in region	π or π_{ij}	Source
Nucleotide diversity (π)						
mtDNA	Human	R	100	16,500	0.004	1
mtDNA	Chimpanzee	R	10	16,500	0.013	2
mtDNA	Gorilla	R	4	16,500	0.006	2
mtDNA	*Peromyscus*	R	19	16,500	0.004	3
mtDNA	Fruitfly	R	10	11,000	0.007	4
β Globin	Human	R	50	35,000	0.002	5
Adh gene	Fruitfly	R	18	12,000	0.006	6
Hemagglutinin	Influenza	S	12	320	0.510	7
Selected pair of DNA sequences (π_{ij})						
Insulin	Human	S	2	1,431	0.003	8
Adh coding	Fruitfly	S	2	765	0.009	9
Immuno-globulin Cκ	Rat	S	2	1,172	0.018	10
IgG2a	Mouse	S	2	1,114	0.100	11

*R, Restriction enzyme technique; S, Sequencing.

†Source: 1. Cann et al. (1982), 2. Ferris et al. (1981), 3. Avise et al. (1979) (Intrapopulational nucleotide diversity), 4. Shah and Langley (1979), 5. Kazazian et al. (1983), 6. Langley et al. (1982), 7. Air (1981), 8. Ullrich et al. (1980), 9. Benyajati et al. (1981), 10. Sheppard and Gutman (1981), 11. Schreier et al. (1981).

population size for mtDNAs is expected to be about 1/4 of that of nuclear genes. However, the mutation rate for mtDNA is apparently considerably higher than that for nuclear genes (Brown et al. 1979). Probably these two compensating factors make the π for mtDNA nearly equal to that of nuclear genes. Unlike eukaryotic genes, the hemagglutinin gene of influenza A virus shows an extremely high nucleotide diversity.

Although data on nucleotide diversity are rapidly increasing, they are not yet large enough to conduct a detailed statistical study similar to that for protein polymorphism data. The only case in which we can relate nucleotide diversity to its causal factors is that of the hemagglutinin gene of the influenza A virus. In this gene π is as high as 0.51. If we note that hemagglutinin is one of the two surface proteins that determine the antigenicity of this virus, it is tempting to assume that this extremely high level of nucleotide diversity is caused by some kind of balancing selection, particularly frequency-dependent selection. However, this polymorphism is not confined to the antigenic sites of the proteins but exists all over the protein (Air 1981). Actually, the real reason for this high value of nucleotide diversity seems to be an extremely high rate of mutation in RNA viruses (Holland et al. 1982). In the influenza A virus an approximate mutation rate can be obtained by using virus strains that have been kept in refrigerators. For example, the Asian flu (subtype H2) was first isolated in 1957 and its variants were isolated in 1967, 1968, 1972, 1977, etc. Therefore, comparing the nucleotide sequences of these strains one can compute the mutation rate. Applying this method to the DNA sequence data from Air (1981) and Air and Hall (1981), Nei (1983) obtained a mutation rate of about 0.01 per nucleotide site per year. This rate is about one million times higher than that for nuclear genes. If this estimate is correct, π would reach 0.5 in several hundred years, even if it is initially 0. The high mutation rate in this and other RNA viruses is believed to be due to the absence of replication-proofreading enzymes.

Silent polymorphism

From the point of view of the controversy over neutral theory, it is interesting to examine the extent of DNA polymorphism that is not expressed at the amino acid level. In neutral theory, this silent polymorphism is expected to be high compared with the polymorphism expressed at the amino acid level, because nonsilent mutations are subject to stronger purifying selection than silent ones in neutral theory (King and Jukes 1969; Dickerson 1971; Kimura and Ohta 1974). One way of testing this hypothesis is to compare the extents of polymorphism in the coding and noncoding regions of genes. Nucleotide changes in the noncoding regions usually do not affect amino acid sequences, so that we would expect a higher degree of polymorphism in the latter than in the former. This expectation is generally fulfilled in available data. Thus, between the two alleles of the human insulin gene studied by Ullrich et al. (1980) there are four nucleotide differences in the noncoding regions but none in the coding regions. A higher degree of polymorphism in the noncoding regions than in the coding regions has also been observed in the globin genes in man (Slightom et al. 1980) and the alcohol dehydrogenase (Adh) gene in Drosophila (Kreitman 1983).

Another test of this hypothesis is to examine the polymorphism at the first, second, and third positions of codons in the coding regions. In nuclear genes all nucleotide changes at the second positions lead to amino acid replacement, whereas at the third positions only about one quarter of nucleotide changes are expected to affect amino acids according to the genetic code table. In the first position about 96 percent of changes lead to amino acid changes. Therefore, if the neutral theory is valid, the extent of DNA polymorphism is expected to be higher in the third positions than in the first and second positions. Available data indicate that this is indeed the case. For example, the Adh_F and Adh_S alleles at the alcohol dehydrogenase locus in _Drosophila melanogaster_ are electrophoretically distinguishable because of the amino acid difference (threonine vs. lysine) at the 192nd position of this enzyme, which is caused by the codon change from ACG to AAG. The nucleotide sequences of approximately a dozen alleles indicate that there is no other amino acid difference in the enzymes produced but there are many third position substitutions that are silent (Benyajati et al. 1981; Kreitman 1983). Similar examples are found in other genes such as the globin genes in man (Orkin et al. 1982).

DNA length polymorphism

In addition to the polymorphism due to nucleotide substitution there are polymorphisms about the number of nucleotides (DNA length) in a given region of DNA. One important class of these DNA length polymorphisms are those due to deletion and insertion of relatively small number of nucleotides. These polymorphisms are usually observed in the noncoding regions of DNA (e.g., Langley et al. 1982), but sometimes occur also in the coding regions. The second class of DNA length polymorphisms are those with respect to duplicate genes which are apparently caused by unequal crossing over (concerted evolution in clustered genes; Arnheim 1983). In this case the number of copies of a particular gene or a particular DNA sequence may vary from individual to individual. For example, Coen et al. (1982) have shown that the number of ribosomal RNA genes varies extensively among individuals of *Drosophila melanogaster*. An extensive polymorphism of this kind was recently reported in a DNA region about 500 nucleotides upstream from the human insulin gene on chromosome 11 (Bell et al. 1982). The number of repeats of DNA sequence varies from individual to individual, and there are some variations in the sequence and the number of nucleotides involved among the repeats. Bell et al. have shown that more than 60 percent of individuals in man are heterozygous with respect to this DNA length polymorphism. The third class of DNA length polymorphisms are those which are apparently caused by gene transposition. The mechanism of this gene transposition is still under investigation (e.g., Campbell 1983), but the prevalence of repeated DNAs such as the Alu family genes in different chromosomes suggests that transposition of DNA pieces occur quite frequently in eukaryotic genomes (Singer 1982). Although this type of polymorphism has not been quantified, the amount seems to be extensive.

These data suggest that in eukaryotic genomes a large number of mutations occur in every generation with respect to DNA length. A large proportion of these mutations are presumably deleterious and quickly eliminated from the population. However, there are still a great many mutations that become frequent and fixed in the population. At the present time little is known about the evolutionary significance of these mutations, but they generally do not seem to have any noticeable effect on fitness.

NUCLEOTIDE SUBSTITUTION IN FUNCTIONAL GENES AND PSEUDOGENES

In recent years a large number of DNA sequences from many different organisms have been published, so that we can study the rate of nucleotide substitution in evolution for various genes. The rate of nucleotide substitution has a special significance for testing neutral theory. In neutral theory the rate of nucleotide substitution is expected to be higher for functionally less important genes or parts of genes than for functionally more important genes, since the latter would be subject to stronger purifying (negative) selection (Kimura and Ohta 1974). On the other hand, selectionists believe that most nucleotide substitutions are caused by positive Darwinian selection, in which case the rate of nucleotide substitution in functionally unimportant genes or parts of genes is expected to be relatively lower because mutations in these regions of DNA would not produce any significant selective advantage. Examining the pattern of nucleotide substitution in histone and globin genes, Kimura (1977) and Jukes (1978) noted that the silent nucleotide substitution occurs more often than amino acid-altering nucleotide substitution and took this as evidence for neutral theory. This was later confirmed by Miyata et al. (1980), who studied many other genes.

Recently, Kimura (1981), Li et al. (1981), and others have conducted detailed studies on the rates of nucleotide substitution for the first, second, and third nucleotide positions of codons in functional genes as well as pseudogenes. The results obtained by Li et al. (1981) and Li (1983) are presented in Table 2. It is seen that the third position rate is 3.7 times higher than the first position rate, which is in turn slightly higher than the second position rate. This pattern is clearly in accordance with the neutral expectation, since most of third position changes are silent as mentioned earlier.

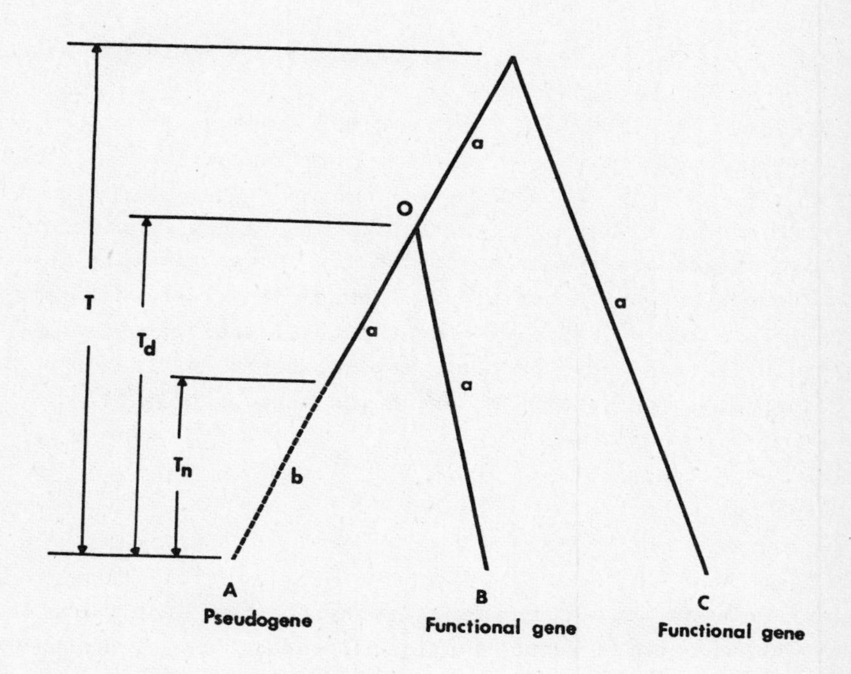

Figure 3. Possible phylogenetic tree for a pseudogene (A) and its functional counterpart (B) in the same organism and a homologous functional gene (C) in another organism. T denotes the divergence time between genes B and C, T_d the time since duplication of A and B, and T_n the time since nonfunctionalization of A. a and b denote the rates of nucleotide substitution per site per year for the functional genes and the pseudogene, respectively.

Table 2. Rates of nucleotide substitutions for the first (a_1), second (a_2), and third (a_3) nucleotide positions of codons of functional globin genes and the rate (b) for pseudogenes. T_d and T_n are the time since gene duplication and the time since nonfunctionalization, respectively (see Figure 3). From Li et al. (1981) and Li (1983).

	a_1 ($\times 10^{-9}$)	a_2 ($\times 10^{-9}$)	a_3 ($\times 10^{-9}$)	b ($\times 10^{-9}$)	T_d (Myr)	T_n (Myr)
Mouse $\psi\alpha3$	0.75	0.68	2.65	5.0±3.2	27±6	23±19
Human $\psi\alpha1$	0.75	0.68	2.65	5.1±3.3	49±8	45±37
Rabbit $\psi\beta2$	0.94	0.71	2.02	4.1±3.4	44±10	43±42
Goat $\psi\beta^X$ & $\psi\beta^Z$	0.94	0.71	2.02	4.4	46	36
Average	0.85	0.70	2.34	4.7		

a_1, a_2, and a_3 denote the rates of nucleotide substitution for the first, second, and third positions of codons in functional genes, respectively. The a_i values in the first two cases are the averages for mouse α1, human α and rabbit α and those for the last two cases are the averages for mouse β major, human β and rabbit β1.

A more direct test of neutral theory in this line is to examine the rate of nucleotide substitution in pseudogenes. A pseudogene is a DNA segment with high homology with a functional gene but containing nucleotide changes such as frameshift and nonsense mutations that prevent its expression. Pseudogenes were apparently produced by the nonfunctionalization of duplicate genes. Figure 3 shows a schematic representation of the evolution of a pseudogene (A) and its functional counterpart (B). In this figure T_d stands for the time since duplication of genes A and B, whereas T_n the time since nonfunctionalization of A. To estimate the rates of nucleotide substitution for pseudogenes and functional genes, however, we need another DNA sequence (C), of which the time of divergence from A and B is known. Li et al. (1981) devised a statistical method for estimating the rate (b) of nucleotide substitution for pseudogenes as well as the values of T_d and T_n from nucleotide sequences of A, B, and C (see Miyata and Yasunaga 1981 for a slightly different method). Using this method, Li et al. (1981) and Li (1983) estimated the rates of nucleotide substitution for globin pseudogenes from mouse, human, rabbit, and goat. The results obtained are presented in Table 2. It is clear that in all pseudogenes the rate of nucleotide substitution is even higher than the third position rate (a_3) of their functional counterparts, the average of b being about two times higher than a_3. Since about 25 percent of third position changes result in amino acid substitutions, the difference between b and a_3 can be explained by purifying selection. This result therefore supports neutral theory.

IMPLICATIONS FOR EVOLUTIONARY THEORY

Adaptive and Nonadaptive Evolution

Let us now consider the implications of the above studies on the general theory of evolution. The observations we have seen about genetic polymorphism and evolution can be summarized in the following way. (1) The extent of protein polymorphism is nearly equal to or lower than the level expected under the equilibrium theory of neutral mutations. The differences between the observed and expected levels can be explained either by the bottleneck effect or diversity-reducing selection. (2) The patterns of distributions of allele frequencies, single-locus heterozygosity, genetic distance, etc., are in rough agreement with the expectations from neutral theory but not consistent with those from several models of balancing selection. (3) Functionally important parts of genes (DNA) are generally less polymorphic than unimportant parts and evolve slower. (4) A large amount of genetic variation may be generated by mutation alone as in the case of influenza virus hemagglutinin.

The above observations all indicate the importance of mutation in the maintenance of molecular polymorphism. The studies mentioned above certainly do not rule out all sorts of natural selection, but the patterns of protein and DNA polymorphisms can be accommodated with neutral theory without much difficulty. This suggests that the dynamics of allele frequency changes in populations is dictated largely by the stochastic factors and the effect of selection, if any, is very small.

Of course, this does not mean that no allelic differences produce large fitness differences. There must be some mutations that produce a selective advantage; otherwise, no adaptive evolution can occur. Indeed, there are a number of experimental data suggesting that some allelic differences identified by electrophoresis are associated with adaptation to different environments (Koehn et al. 1983). A well-known example is a pair of alleles ($Es\text{-}I_a$ and $Es\text{-}I_b$) at an esterase locus of the freshwater fish *Catostomus clarkii* (Koehn 1969). The enzymatic activities of these alleles seem to be directly related to temperature. DiMichele and Powers (1982) also showed that the alleles B^a and B^b at the lactase dehydrogenase-B locus apparently control the adenosine triphosphate level and swimming speed in the fish *Fundulus heteroclitus*. Nevertheless, the proportion of mutations that cause significant adaptive changes seem to be generally very small at the molecular level. However, since most organisms have a large number of genes and each gene is subject to mutation in every generation, a small proportion of advantageous mutations seem to be sufficient for adaptive evolution in most organisms (Nei 1975).

We note that almost all features of molecular polymorphisms mentioned above are incompatible with neo-Darwinism. In neo-Darwinism the amount of genetic variability in populations is determined by natural selection, as mentioned earlier. Data on molecular polymorphism, however, indicates that the amount of genetic variability is largely determined by the mutation rate and effective population size. Data on the pattern of nucleotide substitution in long-term evolution also suggests that mutation rather than selection is the primary factor of evolution (Nei 1975). Indeed, as far as proteins and DNAs are concerned, the observed pattern of evolution does not seem to deviate far from the neutral pattern.

How about the evolutionary mechanism of morphological or physiological characters? All biologists are aware that morphological or physiological characters are amazingly well adapted to the environment in which the organism lives. There is no doubt that natural selection has played an important role in the evolution of these characters. Indeed, there are several observations and experimental data showing that phenotypic changes are directly caused by natural selection (Dobzhansky 1970; Ford 1974). A good example is the industrial melanism studied extensively by Kettlewell (1973) and others. On the other hand, it is also true that there are many morphological characters of which the intrapopulational variation does not seem to be directly related to fitness differences. For example, in human stature there is a great deal of genetic variation, but this variation seems to be almost irrelevant to fitness except the two extremes of the variation. It is possible that the individuals whose stature is not far from the mean are nearly identical with respect to fitness, and the genetic variation within this range is maintained largely by the balance among mutation, genetic drift, and weak selection, if any (Nei 1975, 1980). This seems to be true even if there is centripetal selection (Latter 1972; Kimura 1981; Milkman 1982). If a character is controlled by a large number of loci and subject to centripetal selection, the selection coefficient for individual genes must be very small.

Even the genetic variability between populations may not always be adaptive. The Ainu living in northern Japan are morphologically different from the Japanese, their male body being quite hairy. However, it is not clear what kind of selective advantage their hairiness confers to the Ainu. It is possible that this character was brought about by mutation in their ancestral population and the individuals with this mutant gene or genes later formed a new group, which further later became the Ainu population. In the past several decades anthropologists, evolutionists, and recently sociobiologists have attempted to explain every detail of morphological, physiological, and behavioral differences between races in terms of natural selection. However, some of these differences may well be due to non-adaptive changes.

Role of Mutation in Evolution

In neo-Darwinism only a minor role in evolution is given to mutation, as mentioned earlier. This view was formed apparently because whenever artificial selection was applied to a quantitative character a quick response was observed in many organisms. This gave the early population geneticists the impression that a random mating population contains almost any kind of genetic variation and the only force necessary for achieving a particular evolutionary change is natural selection. However, artificial selection is quite different from natural selection. The response to artificial selection is usually large in the early generations but gradually declines as generations proceed, and without further input of new genetic variability the mean value of a character under selection usually reaches a plateau in a few dozens of generations. Furthermore, it is noted that the initial responses to selection are usually greater in those characters which are remotely related to fitness than in those which are closely related (Falconer 1960). This suggests that the former characters have not been subjected to strong natural selection, so that the amount of genetic variability stored in a population is large. Therefore, if artificial selection is applied to these characters, a rapid response occurs. In the latter characters, however, natural selection (purifying selection) seems to have reduced the genetic variation and thus artificial selection is less effective. If this interpretation is right, artificial selection does not tell the picture of long-term evolution by natural selection.

Wright (1931, 1970) and Mayr (1963) have argued that the fitness of a genotype is determined by the entire set of genes in the individual and it is difficult to isolate the effect of a single gene because of the gene interaction among loci. They have then viewed adaptive evolution as a process for a population to climb up the adaptive surface in a multidimensional space. In this view the role of mutation is minor because each locus is assumed to contain many polymorphic alleles maintained by some type of balancing selection and a higher peak can be attained by natural selection and genetic drift alone. In other words, evolution is assumed to occur mainly by gene frequency shift rather than by gene substitution at each locus (Wright 1970). This shifting balance theory has been accepted by many evolutionists, but actually it lacks both theoretical and experimental proofs (Nei 1980). At the molecular level evolution occurs mainly by gene substitution rather than by gene frequency shift and the substitution at one locus is almost independent of that at another locus. It is also noted that natural populations do not have all kinds of alleles, different species usually having different sets of alleles. This raises a question about the validity of the shifting balance theory (see Nei 1980, for a more detailed discussion). Furthermore, even when epistatic gene interaction is important, evolution can occur by gene substitution rather than by gene frequency shift with changing balanced polymorphisms.

It is often said that genetic polymorphism is beneficial to the population, because in the presence of genetic variability the population can adapt easily to new environments (Dobzhansky 1970). Thus, any mechanism that increases genetic variability is selected for (Ford 1975). I am skeptical about this teleolistic explanation. In my view the genetic variability of a population at present is simply a product of evolution in the past. It may happen to be useful for future evolution as in the case of industrial melanism, but I doubt that genetic variability is stored for future use. In many cases a population may not have the genetic variability that is required for new adaptation. In this case the population may stay unevolved until new mutations occur or simply become extinct. It seems to me that in the evolution of unique characters such as camel's hump or human brain mutation was the key factor. The mutation for human-level intelligence probably occurred in the human lineage with an exceedingly small probability. It is also possible that the primary process of improvement of brain function in man was caused by only a few regulatory mutations at the molecular level.

In my view natural selection is a consequence of the existence of two or more functionally different genotypes in the same environment. For example, if one genotype is more efficient in obtaining food or more resistant to a certain disease compared with others, it will have a higher survival value. However, the functional efficiency of a genotype is determined by the genes possessed by the individual. Therefore, the most important process of adaptive evolution is the creation of better (functionally more efficient) genotypes by mutation or gene recombination. (Here mutation includes all sorts of genetic changes such as nucleotide substitution and gene duplication.) The functional efficiency of a genotype, of course, depends on the environment in which the organism lives, but it should be noted that evolution generally occurs in the direction for specialization to a particular environment.

The above argument is similar to Morgan's mutationism or Muller's (1929, 1950) classical theory but includes some new elements. This new form of mutationism (neomutationism) may be characterized as follows: (1) At the nucleotide level many mutations are deleterious but a substantial proportion of them are neutral or nearly neutral. Only a small proportion of mutations are advantageous, and that is sufficient for adaptive evolution. Under certain circumstances morphological characters may be subject to nonadaptive changes in evolution. (2) Natural selection is primarily a process to save beneficial mutations and eliminate unfit genotypes. (3) New mutations spread through the population either by selection or by genetic drift, but a large proportion of them are eliminated by chance. (4) Populations do not necessarily have the genetic variability needed for new adaptation, though the variability at the molecular level is usually very large. When there is not enough genetic variability needed, the population stays unchanged until new mutations occur or becomes extinct.

It should be noted that the above view is not incompatible with Darwin's view of evolution. At the time of Darwin the source of genetic variation was not known, so that he spent most of his discussion to the effect of natural selection. However, he was keenly aware of the importance of generation of new genetic variability. Indeed, the Morgan-Muller theory is a modern version of Darwin's theory of gradual evolution, as noted by Lewontin (1974). Lewontin coined the words _neoclassical theory_ to designate the extension of mutationism or classical theory to include the case of neutral evolution. The neomutationism I have envisioned here is somewhat different from Lewontin's definition of neoclassical theory. In my view adaptive evolution does not necessarily occur gradually. As mentioned above, some adaptive mutations seem to occur very infrequently (almost uniquely) but have a drastic phenotypic effect. Therefore, in some cases an organism may undergo a rapid phenotypic change, whereas in others it may stay unchanged for a long period of time. Eldredge and Gould (1972) have argued that in geological time scale morphological evolution occurs as a discrete process rather than a continuous process. If we accept the above neomutationism, this discontinuous evolution can be explained relatively easily. While I do not accept Goldschmidt's (1940) theory of systemic mutation (hopeful monster), it is possible that a certain advantageous mutation can be fixed in the population very rapidly in terms of geological time scale.

In conclusion, I believe that mutation is the driving force of evolution and natural selection plays a secondary role. If we accept this view, various aspects of evolution, which have been controversial in recent years, can be explained without much difficulty.

ACKNOWLEDGEMENTS

This study was supported by research grants from the U.S. National Institutes of Health and the U.S. National Science Foundation.

REFERENCES

Air, G. M. 1981. Sequence relationships among the hemagglutinin genes of 12 subtypes of influenza A virus. Proc. Natl. Acad. Sci. US 78:7639-7643.

Air, G. M. and R. M. Hall. 1981. Conservation and variation in influenza gene sequences. In: Genetic Variation Among Influenza Viruses, ICN-UCLA Symposia on Molecular and Cellular Biology. D. Nayak and C. F. Fox, eds. Academic Press, New York. Vol. 22, pp. 29-44.

Arnheim, N. 1983. Concerted evolution of multigene families. In: Evolution of Genes and Proteins. M. Nei and R. K. Koehn, eds. Sinauer Assoc., Sunderland, Mass. pp. 38-61.

Bell, G. I., M. J. Selby, and W. J. Rutter. 1982. The highly polymorphic region near the human insulin gene is composed of simple tandemly repeating sequences. Nature 295:31-35.

Benyajati, C., A. R. Place, D. A. Powers, and W. Sofer. 1981. Alcohol dehydrogenase gene of Drosophila melanogaster: Relationship of intervening sequences to functional domains in the protein. Proc. Natl. Acad. Sci. US 78:2717-2721.

Brown, W. M., M. George, Jr., and A. C. Wilson. 1979. Rapid evolution of animal mitochondrial DNA. Proc. Natl. Acad. Sci. US 76:1967-1971.

Chakraborty, R., P. A. Fuerst, and M. Nei. 1978. Statistical studies on protein polymorphism in natural populations: II. Gene differentiation between populations. Genetics 88:367-390.

Chakraborty, R., P. A. Fuerst, and M. Nei. 1980. Statistical studies on protein polymorphism in natural populations: III. Distribution of allele frequencies and the number of alleles per locus. Genetics 94:1039-1063.

Coen, E. S., J. M. Thoday, and G. Dover. 1982. Rate of turnover of structural variants in the rDNA gene family of Drosophila melanogaster. Nature 295:564-568.

Dickerson, R. E. 1971. The structure of cytochrome c and the rates of molecular evolution. J. Mol. Evol. 1:26-45.

DiMichele, L. and D. A. Powers. 1982. Physiological basis for swimming endurance differences between LDH-B genotypes of Fundulus heteroclitus. Science 216:1014-1016.

Dobzhansky, Th. 1955. A review of some fundamental concepts and problems of population genetics. Cold Spring Harbor Symp. Quant. Biol. 20:1-15.

Dobzhansky, Th. 1970. Genetics of the Evolutionary Process. Columbia Univ. Press, New York.

Eldredge, N. and S. J. Gould. 1972. Punctuated equilibria: an alternative to phyletic gradualism. In: Models in Paleo-Biology. T. J. M. Schopf, ed. Freeman, Cooper and Co., San Francisco. pp. 82-115.

Ewens, W. J. 1979. Mathematical Population Genetics. Springer-Verlag, Berlin.

Falconer, D. S. 1960. Introduction to Quantitative Genetics. Ronald Press, New York.

Ford, E. B. 1975. Ecological Genetics. 4th ed. John Wiley, New York.

Fuerst, P. A., R. Chakraborty, and M. Nei. 1977. Statistical studies on protein polymorphism in natural populations. I. Distribution of single locus heterozygosity. Genetics 86:455-483.

Goldschmidt, R. 1940. The Material Basis of Evolution. Yale Univ. Press, New Haven, Conn.

Harris, H. 1966. Enzyme polymorphisms in man. Proc. Roy. Soc. London (B) 164:298-310.

Hedgecock, D., M. L. Tracey, and K. Nelson. 1982. Genetics. In: The Biology of Crustacea, Vol. 2. L. Abele, ed. Academic Press, New York. pp. 284-403.

Holland, J., K. Spindler, F. Horodyski, E. Grabau, S. Nichol, and S. VandePol. 1982. Rapid evolution of RNA genomes. Science 215:1577-1585.

Jukes, T. H. 1978. Neutral changes during divergent evolution of hemoglobins. J. Mol. Evol. 11:267-269.

Kettlewell, H. B. D. 1973. The Evolution of Melanism. Clarendon Press, Oxford.

Kimura, M. 1968. Evolutionary rate at the molecular level. Nature 217:624-626.

Kimura, M. 1977. Preponderance of synonymous changes as evidence for the neutral theory of molecular evolution. Nature 267:275-276.

Kimura, M. 1981. Estimation of evolutionary distances between homologous nucleotide sequences. Proc. Natl. Acad. Sci. US 78:454-458.

Kimura, M. and J. F. Crow. 1964. The number of alleles that can be maintained in a finite population. Genetics 49:725-738.

Kimura, M. and T. Ohta. 1971. Theoretical Aspects of Population Genetics. Princeton Univ. Press, Princeton, New Jersey.

Kimura, M. and T. Ohta. 1974. On some principles governing molecular evolution. Proc. Natl. Acad. Sci. US 71:2848-2852.

King, J. L. 1967. Continuously distributed factors affecting fitness. Genetics 55:483-492.

King, J. L. and T. H. Jukes. 1969. Non-Darwinian evolution: Random fixation of selectively neutral mutations. Science 164:788-798.

Koehn, R. K. 1969. Esterase heterogeneity: Dynamics of a polymorphism. Science 163:943-944.

Koehn, R. K., A. J. Zera, and J. G. Hall. 1983. Enzyme polymorphism and natural selection. In: Evolution of Genes and Proteins. M. Nei and R. K. Koehn, eds. Sinauer Assoc., Sunderland, Mass. pp. 115-136.

Kreitman, M. 1983. Nucleotide polymorphism at the alcohol dehydrogenase locus of Drosophila melanogaster. Nature 304:412-417.

Langley, C. H., E. A. Montgomery, and W. F. Quattlebaum. 1982. Restriction map variation in the Adh region of Drosophila. Proc. Natl. Acad. Sci. US 79:5631-5635.

Latter, B. D. H. 1972. Selection in finite populations with multiple alleles. III. Genetic divergence with centripetal selection and mutation. Genetics 70:475-490.

Levins, R. 1968. Evolution in Changing Environments. Princeton Univ. Press, Princeton, New Jersey.

Lewontin, R. C. 1974. The Genetic Basis of Evolutionary Change. Columbia Univ. Press, New York.

Lewontin, R. C. and J. L. Hubby. 1966. A molecular approach to the study of genetic heterozygosity in natural populations. II. Amount of variation and degree of heterozygosity in natural populations of Drosophila pseudoobscura. Genetics 54:595-609.

Li, W.-H. 1978. Maintenance of genetic variability under the joint effect of mutation, selection, and random genetic drift. Genetics 90:349-382.

Li, W.-H. 1983. Evolution of duplicate genes and pseudogenes. In: Evolution of Genes and Proteins. M. Nei and R. K. Koehn, eds. Sinauer Assoc., Sunderland, Mass. pp. 14-37.

Li, W.-H. and M. Nei. 1975. Drift variances of heterozygosity and genetic distance in transient states. Genet. Res. 25:229-248.

Li, W.-H., T. Gojobori, and M. Nei. 1981. Pseudogenes as a paradigm of neutral evolution. Nature 292:237-239.

Mani, G. S. 1980. A theoretical study of morph ratio clines with special reference to melanism in moths. Proc. Roy. Soc. (London) B 210:299-316.

Martin, P. S. and H. E. Wright. 1976. Pleistocene Extinctions. Yale Univ. Press, New Haven.

Maruyama, T. and M. Kimura. 1980. Genetic variability and effective population size when local extinction and recolonization of subpopulations are frequent. Proc. Natl. Acad. Sci. US 77:6710-6714.

Maruyama, T. and M. Nei. 1981. Genetic variability maintained by mutation and overdominant selection in finite populations. Genetics 98:441-459.

May, R. M. 1974. Biological populations with nonoverlapping generations: stable points, stable cycles, and chaos. Science 186:645-647.

Mayr, E. 1963. Animal Species and Evolution. Harvard Univ. Press, Cambridge, Mass.

Milkman, R. D. 1967. Heterosis as a major cause of heterozygosity in nature. Genetics 55:493-495.

Milkman, R. D. 1982. Toward a unified selection theory. In: Perspectives on Evolution. R. D. Milkman, ed. Sinauer Assoc., Sunderland, Mass. pp. 105-118.

Miyata, T. and T. Yasunaga. 1981. Rapidly evolving mouse globin-related pseudo gene and its evolutionary history. Proc. Natl. Acad. Sci. US 78:450-453.

Miyata, T., T. Yasunaga, and T. Nishida. 1980. Nucleotide sequence divergence and functional constraint in mRNA evolution. Proc. Natl. Acad. Sci. US 77:7328-7332.

Morgan, T. H. 1925. Genetics and Evolution. Princeton Univ. Press, Princeton, New Jersey.

Morgan, T. H. 1932. The Scientific Basis of Evolution. W. W. Norton, New York.

Muller, H. J. 1929. The method of evolution. The Scientific Monthly 29:481-505.

Muller, H. J. 1950. Our load of mutations. Amer. J. Hum. Genet. 2:111-176.

Nei, M. 1975. Molecular Population Genetics and Evolution. North Holland, Amsterdam and New York.

Nei, M. 1980. Stochastic theory of population genetics and evolution. In: Vito Volterra Symposium on Mathematical Models in Biology. C. Barigozzi, ed. Springer-Verlag, Berlin. pp. 17-47.

Nei, M. 1983. Genetic polymorphism and the role of mutation in evolution. In: Evolution of Genes and Proteins. M. Nei and R. K. Koehn, eds. Sinauer Assoc., Sunderland, Mass. pp. 165-190.

Nei, M. and D. Graur. 1984. Extent of protein polymorphism and the neutral mutation theory. Evol. Biol. 17:(in press)

Nei, M. and W.-H. Li. 1979. Mathematical model for studying genetic variation in terms of restriction endonucleases. Proc. Natl. Acad. Sci. US 76:5269-5273.

Nei, M. and F. Tajima. 1981. DNA polymorphism detectable by restriction endonucleases. Genetics 97:145-163.

Nei, M. and S. Yokoyama. 1976. Effects of random fluctuations of selection intensity on genetic variability in a finite population. Japan. J. Genet. 51:355-369.

Nei, M., T. Maruyama, and R. Chakraborty. 1975. The bottleneck effect and genetic variability in populations. Evolution 29:1-10.

Nei, M., P. A. Fuerst, and R. Chakraborty. 1976. Testing the neutral mutation hypothesis by distribution of single locus heterozygosity. Nature 262:491-493.

Nei, M., F. Tajima, and T. Gojobori. 1983. Classification and measurementof DNA polymorphism. In: Methods in Human Population Genetics. A. Chakravarti, ed. Hutchinson Ross Publ., Stroudsburg, Pa. (in press)

Nevo, E. 1978. Genetic variation in natural populations: patterns and theory. Theoret. Popul. Biol. 13:121-177.

Ohta, T. 1974. Mutational pressure as the main cause of molecular evolution and polymorphism. Nature 252:351-354.

Ohta, T. 1977. Extension to the neutral mutation random drift hypothesis. In: Molecular Evolution and Polymorphism. M. Kimura, ed. Proc. 2nd Taniguchi Intl. Symp. on Biophysics. National Institute of Genetics, Mishima, Japan. pp. 148-167.

Orkin, S. H., H. H. Kazazian, S. E. Antonarakis, S. C. Goff, C. D. Boehm, J. Sexton, P. Waber, and P. V. J. Giardina. 1982. Linkage of β thalassemia mutations and β globin gene polymorphisms with DNA polymorphisms in the human β globin gene cluster. Nature 296:627-631.

Schnell, G. D. and R. K. Selander. 1981. Environmental and morphological correlates of genetic variation in mammals. In: Mammalian Population Genetics. M. H. Smith and T. Toule, eds. Univ. of Georgia Press, Athens, Georgia. pp. 60-99.

Singer, M. F. 1982. SINEs and LINEs: Highly repeated short and long interspersed sequences in mammalian genomes. Cell 28:433-434.

Slightom, J. L., A. E. Blechl, and O. Smithies. 1980. Human fetal $^{G}\gamma$- and $^{A}\gamma$-globin genes: complete nucleotide sequences suggest that DNA can be exchanged between these duplicated genes. Cell 21:627-638.

Sved, J. A., T. E. Reed, and W. F. Bodmer. 1967. The number of balanced polymorphisms that can be maintained in a natural population. Genetics 55:469-481.

Ullrich, A., T. J. Dull, A. Gray, J. Brosius, and I. Sure. 1980. Genetic variation in the human insulin gene. Science 209:612-615.

Wright, S. 1931. Evolution in Mendelian populations. Genetics 16:97-159.

Wright, S. 1970. Random drift and the shifting balance theory of evolution. In: Mathematical Topics in Population Genetics. K. Kojima, ed. Springer-Verlag, Berlin. pp. 1-31.

Zuckerkandl, E. and L. Pauling. 1965. Evolutionary divergence and convergence in proteins. In: Evolving Genes and Proteins. V. Bryson and H. J. Vogel, eds. Academic Press, New York., pp. 97-166.

A DARWINIAN THEORY OF ENZYME POLYMORPHISM

G. S. Mani
Department of Physics
Schuster Laboratory
University of Manchester
Manchester. M13 9PL

1. Introduction

The central tenet of the neutral model is that almost all of the observed polymorphism at the enzyme level is neutral or non-selective and is maintained through a balance between input through neutral mutation and loss through genetic drift in finite populations. The neutral mutations are only a fraction of all mutations that occur at a cistron; the non-neutral mutants that occur are deleterious and hence are removed through purifying selection. The model allows for a very small proportion of advantageous mutations responsible for adaptive evolution (Nei, 1975).

In the model, the definition of neutral alleles depends on the population size. What is selective in large populations becomes effectively neutral in small populations. Since disadvantageous mutations occur much more frequently than advantageous mutations, one would expect that some fraction of the observed polymorphism to be due to random fixation of slightly deleterious mutants (Mayo, 1970; Ohta and Kimura, 1971; Ohta, 1972, 1973).

In a series of papers (Nei _et al_., 1976, 1978; Fuerst _et al_., 1977; Chakraborty _et al_., 1978, 1980; Maruyama and Nei, 1981), Nei and his collaborators have shown that the predictions of the neutral model for such diverse relationships as heterozygosity, allele frequency distribution, genetic distance etc., are borne out more or less by the data on protein polymorphism. In the neutral model, the various relationships mentioned above are, apart from a few modifications, dependent on the single parameter $M = 4N_e\mu$, where N_e is the effective population size and μ the mutation rate per generation. These results have been summarized by Nei in the previous chapter.

Two questions can be asked concerning the data and the statistical tests described above: (i) is the value of M extracted from the data consistent with known values of μ and N_e? (ii) is the genetic diversity observed at the enzyme level correlated with ecological, life history and demographic parameters? If all the observed polymorphism arises through neutral mutations, the answer to the second question will be negative.

Recently Nei and Graur (1983) have attempted to answer the first question. Neutral model is based on the assumption that the mutation rate per year is a constant over the evolutionary time scale. Though there exists some evidence for the non-constancy in the amino-acid substitution rate (Goodman, 1981), it is a widely held view that most of the gene substitutions do occur more or less constantly in time. Based on the observed substitutions in protein molecules and on the estimated evolutionary time scale for various species, Kimura and Ohta (1971) estimated the neutral mutation rate per year to be of the order of 10^{-7}. This estimate is claimed by them to be an upper limit. Evidently then, the neutral mutation rate is lower than the observed value per generation of 10^{-5} - 10^{-6} for *Drosophila* and mouse and the value per generation of 10^{-4} - 10^{-6} for man, (Propping, 1972; Vogel and Rathenberg, 1975). The proponents of the neutral model would claim that almost all of the non-neutral mutations are deleterious and hence rapidly removed through purifying selection. In their paper, Nei and Graur (1983) have used the value of $\nu = 10^{-7}$ for the mutation rate per year. Defining g to be the generation time in years, we have $M = 4N_e g\nu$. The generation time for various species are taken to be the values observed at the present time. In the above equation, the values for N_e, the effective population size are more difficult to estimate. Obviously what is required is the value of N_e over the evolutionary period. The long-term effective population size is close to the smallest size (Wright, 1938) and since the populations have passed through a series of bottle-necks in the past, it is not unreasonable to assume that the present population sizes are much larger than the actual value of N_e to be used in the above equation. Secondly, even if one can estimate the present population size N, one cannot evaluate from it the effective population size. In their paper, Nei and Graur have compared the present population sizes N (assuming that N and N_e are highly correlated) with the theoretical estimate $N_{e(T)}$ obtained from the equation, given the values of g and ν and the data on genetic diversity at the enzyme level for a large number of species. They argue that the fact that $N_{e(T)}$ is always lower than N supports the neutral model, the difference being due to bottle-neck effects and due to fluctuating selection. In

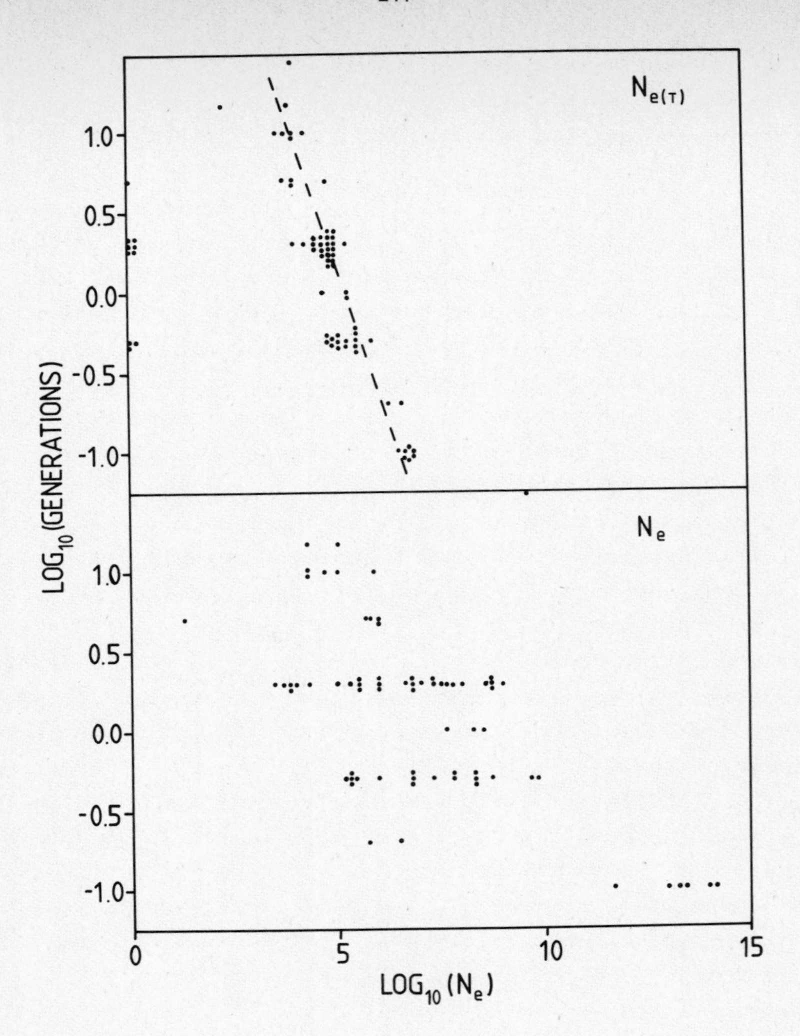

Figure 1. Relation between population size and generation time for 75 different species spanning the range from <u>Drosophila</u> to Man. The bottom curve shows the relationship between the present population size N_e and the generation time; the top curve shows the relation between $N_{e(T)}$ and generation time, where $N_{e(T)}$ is the effective population size needed to explain on the basis of the neutral model the present observed values o heterozygosity, (over at least 20 loci), in these species. The values are all taken from the compilation by Nei and Graur (1983). The broken line on the top curve represents the equation (see text for details):

$$N = 1.78 \times 10^5 \, g^{-1.25}$$

where N is the population size and g, the generation time.

Figure 1, I have plotted the relationship between the generation time g and the present population N and the population size $N_{e(T)}$ extracted from the data by Nei and Graur. All the values in the figure are taken from their compilation. As seen from the figure, while N varies over six orders of magnitude, $N_{e(T)}$ varies only by about two orders of magnitude. More interestingly, for any particular generation time, N could vary as much as five orders of magnitude while $N_{e(T)}$ converges to the same value within a factor of 2-3. It is very difficult to understand how bottle-necks could produce changes in N over evolutionary time scale such that the widely varying values for any particular generation time are compressed to yield values of $N_{e(T)}$ that differ by less than a factor of three. The fact that the effective population size N_e for the present population size N is different from N would not produce material alterations to the conclusions noted above. The other interesting feature of these results is that the transformation from N to $N_{e(T)}$ produce two distinct classes, one clustered near zero and the other clustered around 10^5.

Nevo, in Chapter 2, has attempted to answer question (ii). He concludes from a very extensive and detailed analysis of the enzyme data for a large number of species, that the genetic diversity is correlated with the ecological, life-history and demographic parameters. In particular, Nevo finds the correlation with ecological parameters to be strongest. Though one could raise objections concerning some aspects of the data or the analysis, one cannot avoid the general conclusion that the genetic diversity exhibited by diverse species is influenced more or less by ecological and environmental conditions. This extensive survey indicates that genetic polymorphism cannot rest on neutrality alone.

From the very brief survey given above, it is evident that one needs a model that can reproduce the observables such as heterozygosity, allele frequency distribution etc. as well as the neutral model does and at the same time explain the variation of heterozygosity with population size and the variation in genetic diversity with ecological factors. Such a model should also be able to account for the observed gene substitution rate. As pointed out by Nei (1980, 1983), the standard population genetic models fail to reproduce most of the above features.

In the main, population genetics has avoided the inclusion of ecological parameters in an explicit way. Most population genetic models have been concerned with very large, isolated populations, with negligible genetic drift and often with only two alleles at a single locus. Real populations are not isolated systems. The strong interactions within populations and between populations and the environment are important factors that determine the diversity present and the evolution of the population. Similarly, ecological models, by and large, have ignored genetic factors. The genetic composition of ecological systems is very important in the understanding of many ecological features such as persistence, coevolution etc. Some advance in this direction has already been made by Roughgarden (1979), May and Anderson (1983) and Hamilton (1980). But all these attempts have been based on one locus di-allelic models with mutation and genetic drift neglected.

In this chapter I present a model, the ecological-genetic model (EGM), for finite populations with mutation and genetic drift included. In the real world the various interactions at the phenotypic level discussed above and their relation to the underlying genetics are very complex. Thus the EGM is necessarily a very crude shadow of the complex biological world. As we shall see the model indicates how such interactions at the phenotypic level may have strong influence at the genetic level. My main aim in presenting this model is to emphasise the role played by ecological parameters in the evolution of genetic diversity. The results obtained are very encouraging and indicate a possible mode of reconciling the selectionists' view with the neutralists' view. It further shows that the apparently neutral phenomena at the molecular level need not necessarily imply non-Darwinian mechanisms.

2. The Model

Consider a random mating diploid population of effective size N. Let μ be the mutation rate per generation. First I consider the case with no selection. Let there be (n+1) alleles, $\{A_i\}$, $(i = 0,1,2,..,n)$ present in the population. Let $\Phi(t; \{x_i\}; \{y_i\})$ be the transition probability density for transition from the gene frequency state $\{x_i\}$, $(i = 1,2,..,n)$ to the state $\{y_i\}$, $(i = 1,2,..,n)$ in the time interval t. Then Φ satisfies the Kolmogorov backward diffusion equation

$$\partial\Phi/\partial t = \tfrac{1}{2} \sum_{i=1}^{n} \sum_{j=1}^{n} x_i(\delta_{ij} - x_j) \frac{\partial^2 \Phi}{\partial x_i \partial x_j} - 2N \sum_{i=1}^{n} \mu\, x_i\, \partial\Phi/\partial x_i \qquad (2.1)$$

where δ_{ij} is the Kronecker delta and time is measured in units of 2N generations. The Equation (2.1) can be approximated by the following difference equation, (Itoh, 1979):

$$\Delta x_i(t) = \sum_{j=0}^{n} \sigma_{ij} \{x_i(t)\, x_j(t)\}^{\frac{1}{2}} B_{ij}(\Delta t) - 2N\,\mu\, x_i(t)\Delta t$$

$$i = 1,2,..,n \qquad (2.2)$$

It was this form of the diffusion Equation (2.1) that was used by Maruyama and Nei (1981) in their study of the effects of overdominant selection in finite populations. In Equation (2.2) σ_{ij} is an anti-symmetric matrix with $\sigma_{ij} = +1$ for $i < j$. The matrix $B_{ij}(\Delta t)$ is a symmetric matrix with its off-diagonal elements following the normal distribution with mean zero and variance Δt.

Since my aim is to study the effects of various models of selection in finite populations, the method used by Maruyama and Nei (1981) is not very appropriate. Instead, I have chosen the following procedure which is less accurate than the numerical method of Maruyama and Nei. The accuracy of the present calculations is discussed in the next section.

In each generation, selection is assumed to precede mutation which precedes segregation. The gene frequency after segregation is calculated using Equation (2.2) transformed to time units of one generation and with the term involving mutation omitted. For a given population size and mutation rate μ, from the set of random numbers uniformly distributed in the interval $(1,1/\mu)$, the occurrence of mutation in any generation is decided. Note that there are 2N genes in a population of size N. Thus, instead of reducing the frequency in every generation due to possible mutations as is done through the last term in Equation (2.2), the frequency is adjusted in the generation where mutation occurs. Thus the numerical procedure used is as follows: Equation (2.2), without the last term and transformed to time units of one generation, is used to calculate the gene frequencies at the beginning of the following generation in terms of the frequencies at the time of segregation in the present generation. The gene frequencies are then combined to yield genotype frequencies on which selection acts. Thus we obtain the gene frequencies at the end of the new generation. If mutation occurs in the generation, the mutant is included in the gene pool and the frequencies of all the genes adjusted before segregation. The process is then repeated.

The programme can handle a variety of selection models. In this chapte I shall be concerned only with density-dependent models. Most natural populations exhibit a high degree of persistence. Further, most popula tions are thought to be regulated through density dependent interaction These concepts are often embedded in population ecology models describ- ing the evolution and behaviour of populations. It is thus worthwhile to ask to what extent such regulation occurs, if at all, in nature. Tanner (1966) has studied the average relation between the population density and change in the density per year for 64 populations which in- clude species of invertebrates, fish, birds and mammals. If there exist a negative correlation between change in density and actual density, then this could be interpreted as evidence for density dependent regula tion. Of the 64 populations studied, only two showed positive correla- tion, and only one of the two, the world's human population, was statistically significant. Tanner's results are summarized by Pimm (1982, Table 1). Of the remaining 62 populations, 26% had negative correlations that were statistically not significant, 66% showed negati correlation at around 5% level and the rest (8%) exhibited negative correlation at or above the 10% level. It could be argued that Tanner' data may have a large bias in the choice of the species since one would more readily chose a species that does not vary in its population size from year to year than a species whose population size fluctuates withi wide limits. To avoid such criticisms, Pimm (1982) examined 45 species of birds in English farmlands and woodlands. These species were chosen for their abundance and not for their stability in population size with time. Again his results showed that only 2 species showed positive correlations, neither of which was statistically significant; of the rest 40% showed a slight tendency towards negative correlation, 40% had negative correlation at the 5% level and the remaining 20% at the 10% level. These two sets of results indicate that one cannot avoid the conclusion that the majority of species experience strong density dep- endent regulation.

The exact nature of such population regulation in natural populations i very complex, involving a variety of interactions and behaviour. The question that one can hope to answer at present is, given a descriptio of density dependent regulation, what effect it has on the underlying genetic diversity in the populations. If a simple model is capable of reproducing observations on genetic diversity, then one can justifiably add more realistic details into the model.

The simplest model that we consider is the single species model, with non-overlapping generations, in which the population regulation is maintained through resource limitation. Here "resource" is defined in a broad sense. The population growth is then represented by

$$X_{t+1} = X_t \; F(X_t;K;R_o) \tag{2.3}$$

where F(.) is a function of the present population density X_t, the parameter K that defines the maximum population that can be supported through available resources and the parameter R_o that determines the growth rate at very low densities. The parameter K is usually referred to as the carrying capacity and R_o as the intrinsic growth rate. Various forms of F(.) have been proposed in the literature, (see May, 1981). I choose two of the forms for this discussion, to illustrate the effect the form of the function has on the results.

$$F(X_t;K;R_o) = 1 + R_o(1 - X_t/K) \tag{2.4}$$

$$F(X_t;K;R_o) = \exp\{\, R_o\,(1 - X_t/K)\} \tag{2.5}$$

The requirement that $X_t \geq 0 \;\forall\; t$ limits R_o in (2.4) to the interval (-1,3). May, (1976 and references cited there), has shown from linear stability analysis that for $R_o < 0$, the population goes to extinction and for $0 < R_o < 2$, the stable point $X^* = K$ is approached either exponentially or through damped oscillations depending on whether $R_o < 1$ or $R_o > 1$. For $R_o > 2$ the population exhibits stable cycles or chaotic dynamics.

I now turn to the situation when R_o and K are dependent on the genetic substructure of the population. Consider a diploid sexually mating population with n alleles $A_1,..,A_n$ at a single locus. Let N_{ij} be the population size for the genotype A_iA_j and let N_T be the size of the total population. Let R_{ij} be the intrinsic growth rate and K_{ij} the carrying capacity for the genotype A_iA_j. Then we can define the absolute selective value W_{ij} in terms of the individual's contribution to the population growth, (Roughgarden, 1979), as follows:

$$W_{ij} = F(R_{ij};K_{ij};N_T) \tag{2.6}$$

where F(.) is given by either (2.4) or (2.5).

In terms of the genotypic selective values W_{ij}, the evolution of the gene frequency and total population can be written as:

$$p_i(t) = p_i(t-1)\; W_{i.}(t-1) \;/\; < W(t-1) > \tag{2.7}$$

$$N_T(t) = <W(t-1)>\; N_T(t-1) \tag{2.8}$$

$$W_{i.}(t) = \sum_j W_{ij}(t)\; p_j(t) \tag{2.9}$$

$$<W(t)> = \sum_i W_{i.}(t)\; p_i(t) = \sum_i \sum_j p_i(t)\; p_j(t)\; W_{ij}(t) \tag{2.10}$$

where p_i is the frequency of the allele A_i and $<W>$ is the mean fitness averaged over the whole population.

Roughgarden (1979) has argued that R, the intrinsic growth rate, and K, the carrying capacity, have to be negatively correlated. His arguments can be summarized as follows. A large value of R implies that the organism rapidly allocates the energy it has acquired to the production of offsprings. On the other hand, individuals with high K would defer allocation of energy for production till later and thus utilise the energy for surviving under crowded conditions. Since energy and time in growth season are both limited, the organisms cannot have simultaneousl high R and K. There can either be a high value of R, and thus low valu of K, or a high value of K, and hence a low value of R, but not both. In the present model this is simulated through a simple linear relation ship between R and K, as follows:

$$K_{ij} = \{ \beta R_{ij} + 1.0\} K_o \tag{2.11}$$

where K_o is a constant corresponding to the carrying capacity when $R_{ij} = 0$. When $\beta = 0$, Equations (2.7) - (2.10) approach the neutral limit. For $\beta < 0$, R and K are negatively correlated and when $\beta > 0$, they are positively correlated. As we shall see later, the eventual population size converges to a value close to K_o. Thus, for convenience we refer to K_o, as the population size.

As mentioned in the introduction, at the molecular level, Kimura has estimated the neutral mutation rate per year to be around 10^{-7}. The actual mutation rate observed can exceed this value by an order of magnitude. Since generation time varies over two orders of magnitude, between 10^{-1} and 10 years for a majority of the organisms, we have chosen a value of 5×10^{-6} as the mutation rate per generation. A few calculations were done using values of 5×10^{-5} and 5×10^{-7}. The values of K_o chosen in these calculations vary between 100 and 500000. These values would thus cover most of the range of generation times and effective population sizes that possibly occur in nature.

Each population was allowed to evolve through 100000 generations. A total of 100 populations with different initial conditions and following different stochastic pathways was considered. All observables were calculated as the mean over the 100 populations and for the last 5000 generations in each population.

The following models were used in the calculations:

I. The uniform distribution model (UD model)

The values of R_{ij}, both for the homozygotes and for the heterozygotes were chosen from a uniform distribution of random numbers in the interval (-1,3) for F(.) given by (2.4) and (-1,8) for (2.5). The following submodels were considered:

(1.1) R-K correlated models (UDC model). In this submodel R and K are assumed to be correlated through Equation (2.11). Various values of β in the interval (-0.3,+0.1) were used in the calculations.

(1.2) R-K uncorrelated model (UDUC model). In this submodel the K_{ij}'s were chosen independently of R_{ij} from a uniform random distribution in the interval $(1,K_o)$.

(1.3) The β variable model (UDCβ model). In this model β_{ij} for genotype A_iA_j were chosen from a uniform random distribution in the interval (-0.3,0.0) and the K_{ij}'s were obtained through Equation (2.11).

II. The correlated heterozygote model (CH model)

In this model the values of R_{ii} for the homozygotes are chosen from a uniform random distribution as in the UD models. The R_{ij} values for the heterozygotes were obtained as follows:

$$R_{ij}(i = j) = \tfrac{1}{2}\{R_{ii}+R_{ij}\}\ (1 + \lambda(0,\sigma^2)) \tag{2.12}$$

where $\lambda(0,\sigma^2)$ is a normally distributed random variable with mean zero and variance σ^2. The values of σ^2 used in the calculations varied between 10^{-2} and 10^{-6}. All the submodels described in I also apply to the CH models.

Population sizes between 10^2 and 5×10^5 were used for K_o. Two types of calculations were done. In the constant environmental models, the values of R_{ij} and K_{ij}, once chosen, were kept fixed throughout the calculation. In the variable environmental models (VEM), the effects of fluctuations in the environment were assumed to be reflected in the values of K_{ij} through the following equation:

$$K_{ij}(t) = K_{ij}(0)\ \{1 + \eta_{ij}(0,\ \sigma_e^2)\} \tag{2.13}$$

where $\eta_{ij}(0,\ \sigma_e^2)$ is a normally distributed random variable with zero mean and variance σ_e^2 and $K_{ij}(0)$ is the initial choice of K_{ij} when the mutant first appeared. The subscripts (ij) are used in describing η to emphasize that the genotype fluctuations are uncorrelated, but all have the same variance σ_e^2, representing the fluctuations in the environment.

3. The Accuracy of the Numerical Calculations

As remarked in Section 2, our numerical method is much less accurate than the method used by Maruyama and Nei (1981) in their study of overdominant selection. The experimental data in the main is accurate to within 5 - 10%. Hence the philosophy followed in these calculations is to test a large number of models rapidly, without too high a cost in computing effort and with results that are within the accuracy of the data. As we shall see below, the present set of calculations are adequate to relate to a wide range of data.

In the first series of checks on the accuracy, the selection was turned off and the results compared with the exact predictions of the neutral model. In this case the comparisons were made with respect to the allel frequency distribution, the mean heterozygosity, the distribution of heterozygosity, the average number of alleles in the population and the substitution rate. The neutral model predicts the allele frequency distribution to be given by:

$$\Psi(x)\ dx = M(1 - x)^{M-1}\ x^{-1}\ dx \qquad (3.1)$$

where $M = 4N\mu$, N being the population size and μ the mutation rate per generation. The mean value of heterozygosity is obtained from the relation

$$H = M/(1 + M) \qquad (3.2)$$

The distribution of heterozygosity in the neutral model is more difficult to evaluate. Using a method due to Stewart, Fuerst et al. (1977) have computed the heterozygosity distribution for a few values of M and these are used in the comparison. The average number of alleles in the neutral model is given by:

$$n_a = \int_{1/2N}^{1} \Psi\ (x)\ dx \qquad (3.3)$$

where the integral is numerically evaluated. Finally, the substitution rate for genes is given in the neutral model by μ substitutions per generation, where μ is the mutation rate per generation.

In Figures 2 and 3, the results of the present calculations are compared with the predictions of the neutral for both the allele frequency distribution and for the heterozygosity distribution for values of $M = 0.01$, 0.1, 0.2 and 0.8. Our method is not useful for values of $M \lesssim 0.01$. Since the observed values of heterozygosity yield value of M within the range (0.01 - 0.30), the method used is sufficient to describe the data.

In the second series of checks, we study the convergence to neutrality as selection is made weaker. Consider the case when the selection coefficients, defined as $s_{ij} = 1 - W_{ij}$, are selected from a normal distribution of random numbers with mean zero and variance σ^2. As $\sigma^2 \to 0$, one converges towards the neutral model. The neutral model asserts that neutrality is obtained when the condition $2Ns \lesssim 1$, is satisfied, where s is the mean selective difference among the genotypes. In the present situation, since on the average the s_{ij}'s are distributed within the range σ about zero, the above condition is approximately

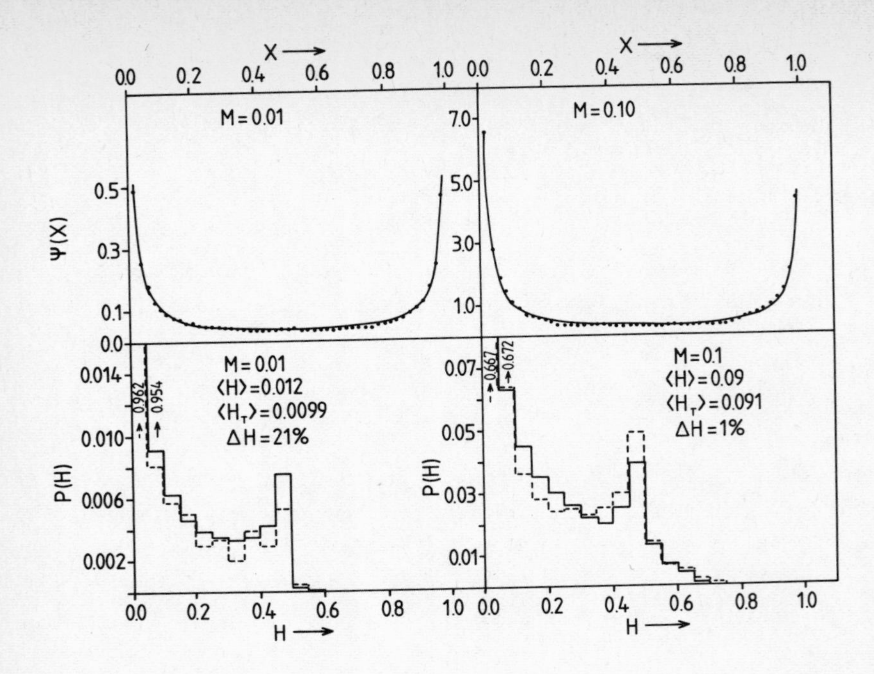

Figure 2. Comparison of the results of the ecological model calculations with selection turned off with the exact results from the neutral model. The top curve shows the allele frequency distribution and the bottom curve the heterozygosity distribution. In the top curve the full line represents the prediction from neutral model (Equation (3.1)) and the dots are the results from our numerical method. In the bottom curves, the full line are the results from the present calculations and the broken lines were obtained from the approximate calculations by Fuerst et al. (1977; see text for details). The mean heterozygosity $\langle H \rangle$, using our computer code, $\langle H_T \rangle$ using the exact neutral Equation (3.2) and the deviation of our values from the exact values, ΔH, are also given. In these figures $M = 4 N_e \mu$, where N_e is the effective population size and μ the mutation rate per generation.

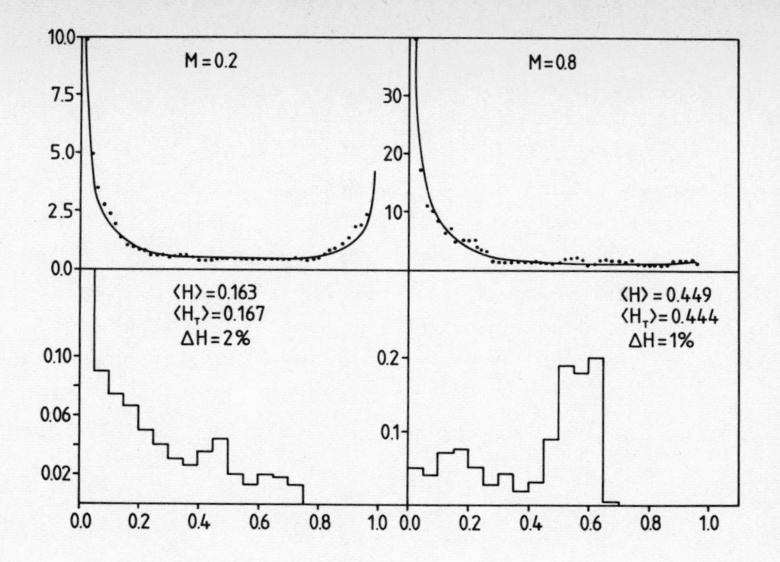

Figure 3. Same as Figure 2, with M = 0.2 and 0.8.

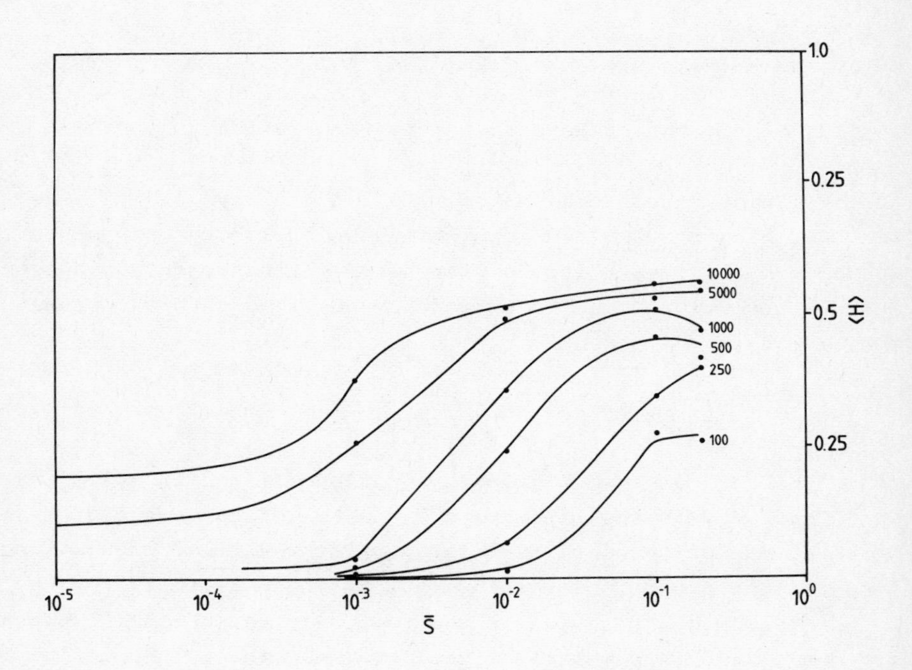

Figure 4. The variation of heterozygosity with mean selective strength $\bar{s}$ for various values of population sizes. See text (Section 3) for details.

given by $2N\sigma \lesssim 1$. Thus, for $\sigma = 10^{-3}$, population sizes below 500 would behave as if they contained only neutral alleles. The values of the mean heteroxygosity for various values of N and for $\sigma = 0.2, 0.1, 0.01$ were calculated and compared with the predictions of the neutral model. The results are shown in Figure 4. It is clearly evident from this figure that that for $\sigma = 10^{-3}$, N = 100 and 250 agree closely with the neutral model. For $N \geqslant 500$, the neutral model is approached much more slowly. Making a crude extrapolation of the results, it can be seen that the neutral model is obtained when $2N\sigma << 1$, for these values of N. Maruyama and Nei have calculated the case when one has overdominance and though their assumption of overdominance is different from the present model, their results are not at variance with the results of the present calculations.

The rest of the chapter is devoted to the results obtained using the ecological-genetic model. We first discuss the results pertaining to ecology. This is then followed by a discussion of the results for genetic diversity. In the last section I compare the present model with the neutral model and I also discuss some of the possible future developments of the model.

4. Why is Chaotic Motion Rare in Natural Populations?

From the available literature, Hassel, Lawton and May (1976), have analyzed the data for 24 field populations and 4 laboratory populations of seasonally breeding arthropods with discrete, non-overlapping generations. The population dynamics were analysed using the two parameter equation:

$$N_{t+1} = \lambda N_t \{1 + a N_t\}^{-\theta} \tag{4.1}$$

This is defined to be a two parameter equation, since the stability conditions depend only on the parameters λ and θ. Discussion of the stability criteria can be found in May (1976). For the field populations, they find that around 88% of the populations studied lie in the region of monotonic damping, around 8% in the region of damped oscillations and only one of the 24 populations yield values of λ and θ, indicating two-point stable limit cycle dynamics. There was no field population that showed dynamics with either oscillations higher than two-point cycle or chaotic motion. On the other hand, of the four laboratory populations studied, one indicated chaotic motion. Stubbs (1977), using a different form of density regulating function, arrived at the same conclusion, from a similar compilation of populations. More recently, Bellows (1981) had subsumed from the data used by Hassel _et al._ and by Stubbs, 14 best documented populations and analyzed them using a different density dependent form and arrived at very similar conclusions. Thomas, Pomerantz and Gilpin (1980) examined the populations of 27 species of _Drosophila_ reared at two different temperatures using the following two-parameter model:

$$N_{t+1} = N_t \, \mathrm{Exp}\{R(1 - \{N_t/K\}^{\theta})\} \tag{4.2}$$

They observed that _all_ the populations appear to have a stable fixed point. They argue that this tendency towards stability is a direct consequence of natural selection.

From the data presented above, it may be tempting to make the generalization that all natural populations exhibit stable behaviour, and that chaotic motion is extremely rare in natural populations. Considerable caution must be exercised in making such generalization, for the following reasons (May, 1976) : (i) There could exist inherent bias in the selection of the data. Populations that exhibit cycles of large order or are chaotic, fluctuate rapidly and hence may be very rare in most years. These then stand a higher chance of being excluded from the populations studied; (ii) No natural population can be regarded as isolated. The use of single species models with artificially built in density dependent regulation for describing the temporal pattern of the population is a gross over simplification of the complex interactions that occur within and between populations; (iii) The assumption of non-overlapping generations may be valid for arthropods, but is certainly not valid for describing the populations of many species, especially the vertebrates. In such cases, the complex interactions between populations and the age structure of the populations could, in theory, produce situations where chaotic dynamics occur (Guggenheimer _et al._, 1977). Detailed studies of natural populations with overlapping generations are sparse however, and hence it is not clear if most of them exhibit a high degree of stability in the sense used in this section.

In spite of all the reservations discussed above, one could still ask what role natural selection plays in populations that evolve away from chaotic behaviour. Such a study will not provide a unique explanation for the occurrence of a high degree of stability in natural populations, but would at least indicate one possible explanation for the rarity of chaotic conditions observed in the data discussed above.

Referring to Equations (2.4) - (2.10), the population dynamics is described in the model through

$$N(t+1) \;=\; N(t)\,\{1 + \langle R\rangle - \langle R/K\rangle\, N(t)\} \tag{4.3}$$

using the form (2.4). Similar equation would result if the form (2.5) had been used. In Equation (4.3), quantities $\langle x\rangle$ are averages of x_{ij} over all genotypes A_iA_j and are defined by:

$$\langle x\rangle \;=\; \sum_{ij} x_{ij}\, p_i\, p_j \tag{4.4}$$

where p_i are the gene frequencies. The stability criteria for (4.3) are determined by the values of $\langle R\rangle$ alone and are listed in Table 1 for both the forms (2.4) and (2.5).

Table I. Stability Criteria for Equation (3.6) when the Density-Dependent Regulation Forms (2.4) and (2.5) are Used

Equation (2.4)	Equation (2.5)	Stability Criteria
$R' < 0$	$R' < 0$	N = 0 : stable node
$0 < R' < 1$	$0 < R' < 1$	N = K': stable node
$1 < R' < 2$	$1 < R < 2$	N = K': stable focus
$2 < R' < 2.449$	$2 < R' < 2.526$	2-point cycle
$2.449 < R' < 2.570$	$2.526 < R' < 2.692$	2^k (k=2,3,4,..) point cycles
$2.570 < R' < 3.000$	$2.692 < R'$	Chaotic behaviour

NOTE: For the form (2.4) $R' = \langle R\rangle$ and $K' = \langle R\rangle/\langle R/K\rangle$ and though for (2 R' and K' are more complex, they can be well approximated by these expressions.

The distribution of $\langle R\rangle$ for population sizes K_o = 100 and 5000 and for the density-dependence (2.4) and (2.5) are shown in Figures 5 and 6 respectively. These distributions were calculated for the constant environment model (CEM). The distribution of $\langle R\rangle$ for the variable environment model (VEM) with 10% environmental fluctuation (σ_e = 0.1 ir

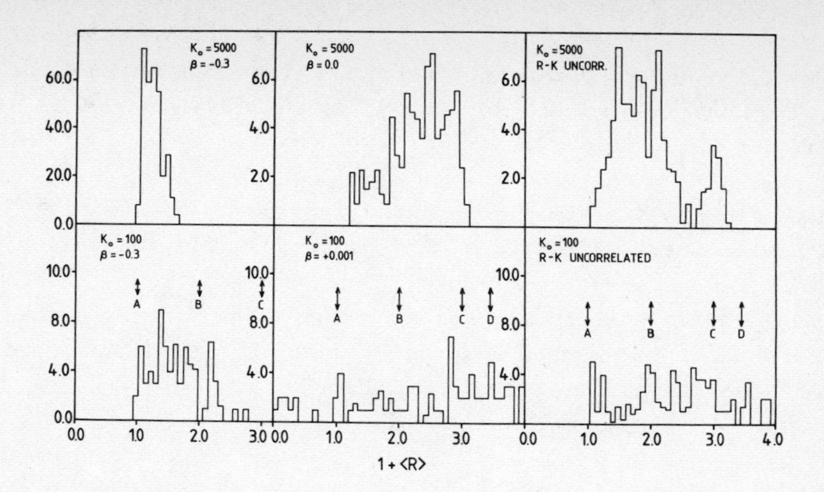

Figure 5. The distribution of $\langle R \rangle$ for $K_o = 100$ and 5000, when R and K are correlated ($\beta = -0.3$ and 0.0) and when R and K are uncorrelated, using the form (2.4). The arrows marked A - D represent the following:

$0.0 < \langle R \rangle < A$	:	Population goes to extinction.
$A < \langle R \rangle < B$	:	Monotonic rise to the stable point K^*.
$B < \langle R \rangle < C$	:	Stable point K^* reached through damped oscillations.
$C < \langle R \rangle < D$	:	Stable two-point cycle.
$\langle R \rangle < D$	:	Higher order cycles and chaos.

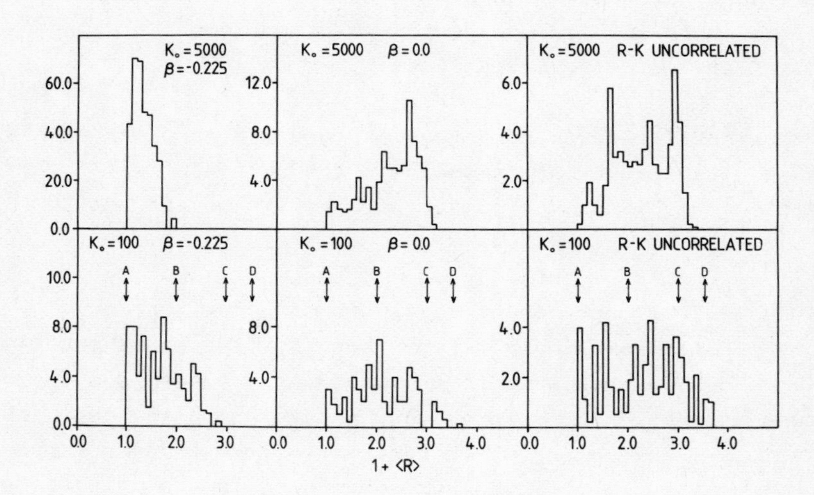

Figure 6. Same as Figure 5 for the form (2.5).

Equation (2.13)) is shown in Figure 7. In these figures the distributions are calculated for both the R-K correlated (see Equation (2.11)) and the R-K uncorrelated models. Finally, the variation of $<R>$ with β for CEM and for the form (2.4) is shown in Figure 8.

From these figures one arrives at the following conclusions:

(1) The exact form for the density dependent regulation is not important as can be seen from Figures 5 and 6, where both the forms (2.6) and (2.7) yield very similar results. Since the exact nature of the density dependence is not very critical, we have chosen for all later calculations the form (2.4).

(2) Small population sizes have a larger probability of entering into the chaotic regime, than large population sizes. When R and K are uncorrelated, there is a greater probability for the occurrence of chaotic dynamics.

(3) In the case when R and K are correlated, as β moves from negative values towards zero, the mean value of $<R>$ remains fairly constant till in the neighbourhood of zero, when it increases sharply, pushing the system more towards the chaotic region. As the value of β is increased further, the value of $<R>$ is decreased from its maximum value at $\beta = 0$. The decrease in the value of $<R>$ is sharper, the higher the population size. As we shall see in the next section, this is also the region where the population has a high probability of going to extinction.

(4) Environmental fluctuations tend to move the populations towards greater stability. As is shown in the next section, the price tha is paid for this is the higher probability of extinction.

Thus chaotic dynamics is not favoured by natural selection for large populations and for populations in fluctuating environments. Since the laboratory populations are in general very small and are usually maintained in controlled conditions, one would expect such populations to exhibit chaotic motion more often than natural populations which are generally large in size and which experience more or less rapid changes in the environment. What is demonstrated here is that, in the presence of a density regulating mechanism and environmental fluctuations, natural selection forces a high degree of stability on the populations. This then could provide a partial explanation for the data presented

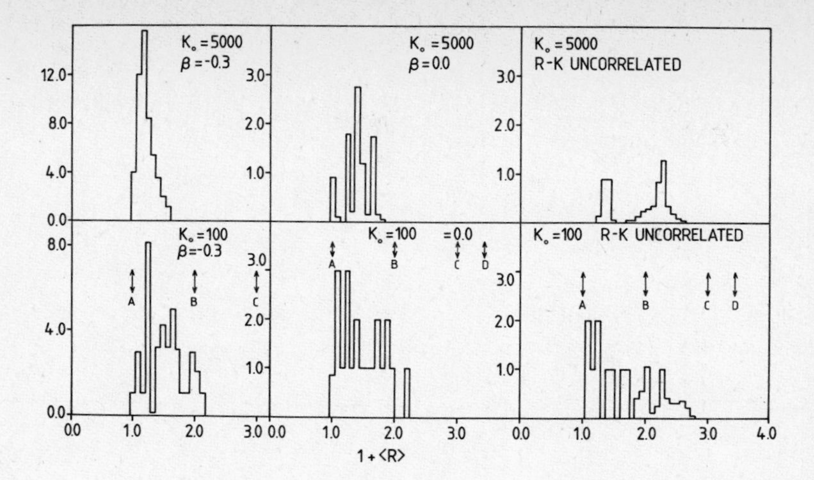

Figure 7. Same as Figure 5 for the form (2.4) when a 10% strength of fluctuation in the environment is included.

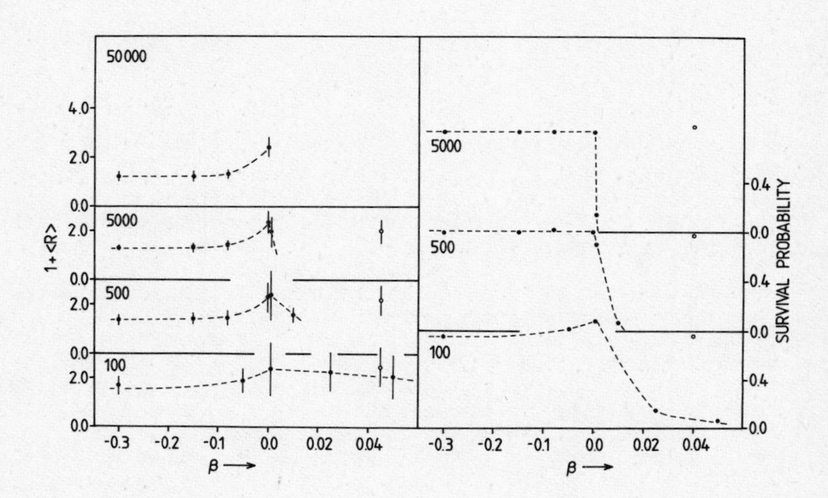

Figure 8. The set of curves on the left show the variation in the value of 1 + <R> as the value of β increases from -0.3 to +0.05, for various population sizes. The curves on the right show the variation of survival probability with β for various population sizes. The open circles in all the curves indicate the corresponding values when R and K are uncorrelated. The broken lines are smooth lines joining the points.

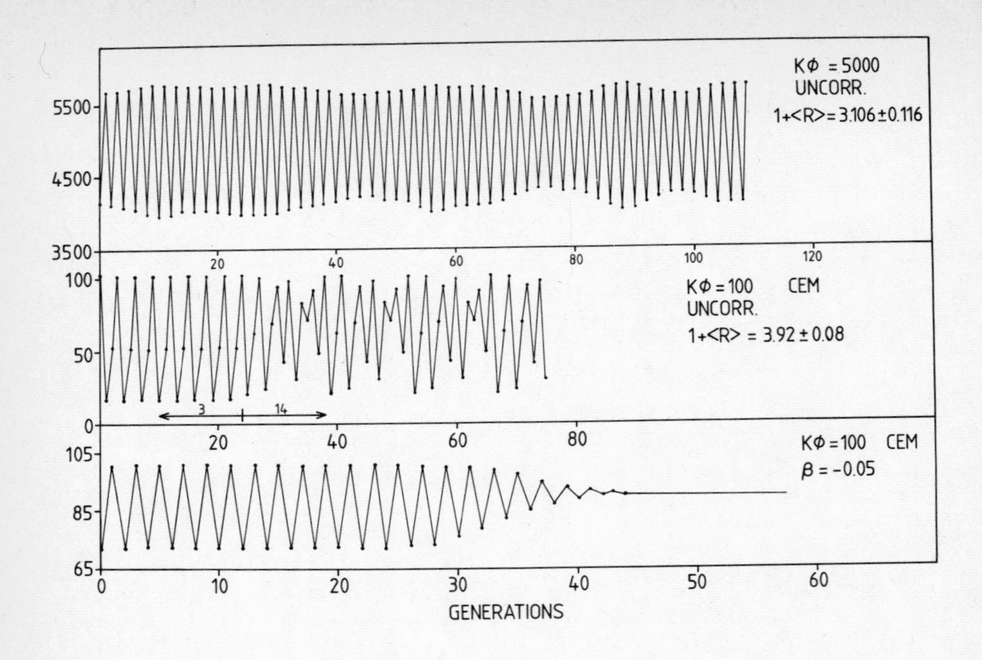

Figure 9. Three examples of populations exhibiting cyclic or chaotic behaviour. The top curve shows a two-point cycle for K_o = 5000 and R-K uncorrelated. The modulation of the curv is due to the continual input of mutants which slightly alter the value of the total population. The period of the modulation is around 20 generations which is the average number of generations for the introduction of a new mutant. Note that the zero in the figure corresponds to 99,500 generations. The middle curve is for K_o = 100 and R-k uncorrelated. There is an initial three point cycle which goe over to a 14-point cycle with the introduction of a new mutant. The three point cycle stretches backwards to a few 10,000 generations. This particular population went to extinction after 450,000 generations. Over half the populati that exhibited such chaotic behaviour, on the other hand, persisted for a million generations. The bottom curve shows an example of a two-point cycle that eventually goes over t a stable point behaviour with the introduction of a new mutant.

earlier.

The probability for the population to go into the unstable phase increases as the value of β approaches zero. As mentioned in Section 2, $\beta = 0$ corresponds very closely to the neutral model. Thus one would expect near neutral alleles to produce greater population instability in the presence of a density regulating mechanism. The maximum probability for chaos is obtained in the case when R and K are uncorrelated. In all these situations, increasing the population size or the amount of environmental fluctuation tends to dampen instability.

A few examples of the type of cyclic or chaotic motions that could be produced under conditions described above are shown in Figure 9.

5. Survival Probability for Populations

Another aspect of stability of populations is their persistence. Here we shall discuss the following questions:

(i) What is the probability of the population going to extinction in a constant environment as the value of β is changed?

(ii) For a given β how is the survival probability affected by environmental fluctuations?

There are primarily two reasons for a population to be driven to extinction. Firstly, the stochastic processes could move the population to the domain of R_{ij} for which $\langle R \rangle$ is less than zero. Secondly, the fluctuations, environmental or chaotic or both, could accidentally drive the population to extinction.

Since each case that is studied involves 100 populations with each population evolving for 100,000 generations, the survival probability is defined to be the fraction of the 100 populations that survive to 100,000 generations. The survival probability as a function of β for the form (2.4) is shown in Figure 8. The survival probability for $\beta \leq 0$ is around 80% and is very nearly independent of both the population size and β. This can be understood as follows. Since R_{ij}'s are chosen from the interval (-1,3), 25% of the values chosen would, on the average, lie below zero. If these were the values chosen in the first few mutations, then there is a good chance that they would yield an $\langle R \rangle$

that is less than zero. Thus the probability of the population going to extinction would be smaller than 25%. The choice in later generations would not seriously affect the survival probability, provided the population does not enter into the highly oscillatory or chaotic region. If the survival probability in this region of β is determined by the early choice of R_{ij}'s, then one would expect the populations that go to extinction to do so in very few generations. This is indeed what is observed. All the populations that are driven to extinction do so in 10 - 500 generations. In the case of the exponential form (2.5), since the R_{ij}'s are chosen from a much larger interval, the survival probability becomes correspondingly higher, as one would expect.

For positive values of β, the survival probability is very sensitive to the value of β and to the population size. The survival probability rapidly decreases with β, especially for large populations. For $\beta > 0$, from (2.11) it is seen that K_{ij} is an increasing function of R_{ij}. Consequently, K_{ij} becomes very small compared with N_T when R_{ij} goes below zero, increasing the selective values in this region. This forces the population to move towards the domain of negative R_{ij}'s with resultant extinction. The number of generations required is larger by one or two orders of magnitude than in the case $\beta \leqslant 0$. The survival probability can be increased by artificially allowing β to go negative when $R_{ij} < 0$. Such a procedure is arbitrary and has little justification. Also for such a model, even small environmental fluctuations would drive the population to extinction. Further, the genetic properties of the system becomes very peculiar and not in accord with observations.

I now turn to the effects of environmental changes on the survival probability. For $\beta > 0$, the survival probability is virtually zero even in the presence of very small fluctuations in the environment. This is not surprising, since we have already seen that the populations are very unstable even in a constant environment when $\beta > 0$. So I shall discuss only the situation when $\beta \leqslant 0$. The fluctuations are introduced through Equation (2.13) and we shall use $\Sigma_e \equiv \sigma_e$ as a measure of the fluctuation. The survival probability as a function of Σ_e for various population sizes and for various values of $\beta \leqslant 0$ are shown in Figure 10. The survival probability is seen to be fairly insensitive to the value of β except near $\beta = 0$. For $\beta < 0$, the larger the population the better it withstands severe environmental disturbances, in agreement with the widely held view.

When $\beta = 0$, the situation is the reverse. The survival probability for very small populations are very similar to the case when $\beta < 0$. On the other hand, for large populations the survival probability diminishes with Σ_e. This instability is a reflection of the rapid increase (or decrease) of $\langle R \rangle$ in the neighbourhood of $\beta = 0$, (see Figure 8). Evolutionary stability requires that the mean value of $\langle R \rangle$ be not far from unity. As remarked earlier, $\beta = 0$ corresponds to the neutralist's limit, especially for large populations. The above result is an indication of the structural instability of the neutral model. Eternal neutrality cannot survive evolutionary processes.

When $\beta > 0$ or when R and K are uncorrelated, the populations tend to extinction even for small disturbances in the environment. For $\beta > 0$, the instabilities appear even when β is arbitrarily set to be less than 0 for $R_{ij} < 0$. The fact that the R-K uncorrelated models produce unstable populations reinforces the assumption of negative correlation between R_{ij} and K_{ij}. The results quoted in this section and in Section 4 lead us to the following conclusions:

(1) Large populations are less likely than small populations to show fluctuating behaviour.

(2) Environmental fluctuations move populations towards the region of monotonic stability.

(3) Populations are highly unstable when R and K are positively correlated.

(4) The neutral limit, when K_{ij}'s are independent of genotypes and thus are constant, is structurally unstable. Small fluctuations drive the populations to extinction.

(5) The biological assumption that R and K are negatively correlated through the mode of sharing the finite resources between growth and reproduction is borne out by theory.

6. Predictions of the Model at the Genetic Level

6.a Allele frequency distribution

The allele frequency distribution, predicted by the neutral model through

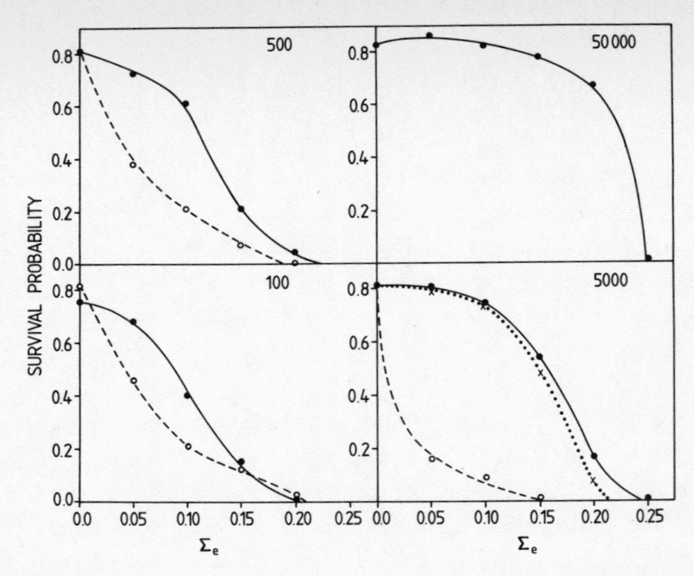

Figure 10. The variation of survival probability with the magnitude, Σ_e, of environmental fluctuation, for different population sizes and different values of β. Since the curves for $\beta = -0.08$ are very close to those with $\beta = -0.30$, only one case for $\beta = -0.08$ is shown.
_______ $\beta = -0.30$; ..... $\beta = -0.08$; - - - - $\beta = 0.0$.

Equation (3.1), is shown in Figure 11. In this figure the frequency distribution $\Psi(x)\,dx$, for the intervals $(2N)^{-1} - 0.01$, $0.01 - 0.05$, $0.05 - 0.1$, and the nine intervals between 0.1 and 1.0 having widths of 0.1, are shown. All subsequent frequency distribution curves also use the same intervals. The area of each bar, (product of height times the width), is proportional to the average number of alleles in that interval. All curves displayed in this figure were calculated with a mutation rate of 5×10^{-6}. For population sizes between 100 and 5000, the distributions are more or less U-shaped. The arm of the U near unity or zero is larger, depending on whether the population size is near 100 or near 5000, respectively. For large populations, with $N \lesssim 50000$, the distribution goes over to being L-shaped.

I shall use the term strong selection for the following model. The density dependence in (2.7) is switched off and the (relative) fitness coefficients W_{ij} are chosen from a uniform random distribution in the interval (0,1). Strong selection models yield probability density distributions for allele frequencies that are strongly peaked at intermediate frequencies, as seen in Figure 12.

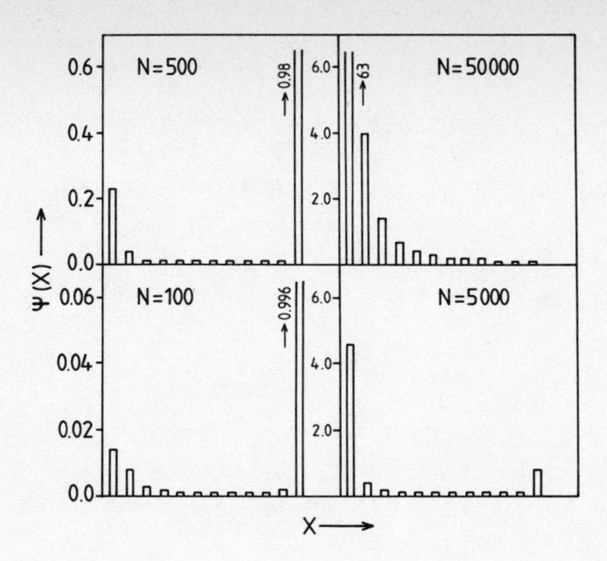

Figure 11. The allele frequency distribution predicted by the infinite allele neutral model. The first three boxes correspond to the frequency intervals (1/2N) - 0.01, 0.01 - 0.05 and 0.05 - 0.10 respectively. The rest of the boxes are for successive intervals of 0.1, from 0.1 to 1.0. These are the same intervals used by Fuerst _et al._ (1977).

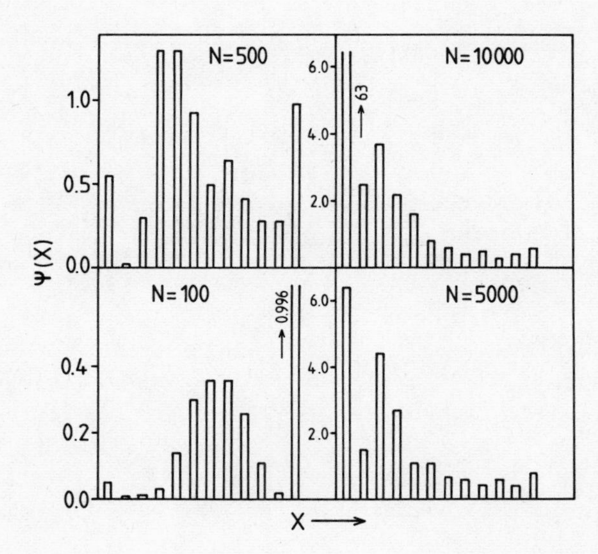

Figure 12. The allele frequency distribution predicted by the strong selection models. The frequency ranges are the same as in Figure 11. These results were calculated by Mani (unpublished).

The results of the uniform distribution model in a constant environment for the cases when R and K are correlated, with the value of $\beta = -0.3$ and 0.0 and when R and K are uncorrelated, are displayed in Figure 13. As seen from the figure, as β moves from a large negative value towards zero, the agreement with the neutral model becomes better. All the R-K correlated curves show a more or less U-shaped distribution. On the other hand, the uncorrelated cases approach the strong selection limit. This is not too surprising since in the presence of strong selection, the selection would dominate and the population regulation would tend to be weakened.

When environmental fluctuations are superimposed, selection operates with increased strength, particularly for small populations and for large negative β. This is understandable since for those populations that do survive, the genetic composition is such as to withstand externa disturbances. As seen earlier, small populations are less stable on the average than large populations in the presence of environmental fluctuations and consequently the selective effects are enhanced for small populations. The effects of environmental disturbance on population size and the value of β, are shown in Figure 14 and 15. For large fluctuations or as β approaches zero, the distributions become more "neutral

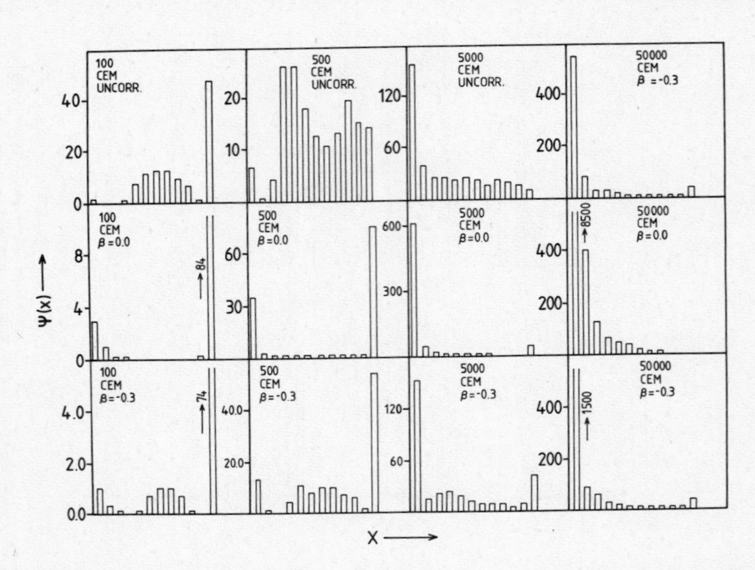

Figure 13. The allele frequency distribution predicted by the ecologica model with R and K correlated ($\beta = -0.3$ and 0.0) and uncorrelated. These results are for constant environment. The frequency ranges are same as in Figure 11.

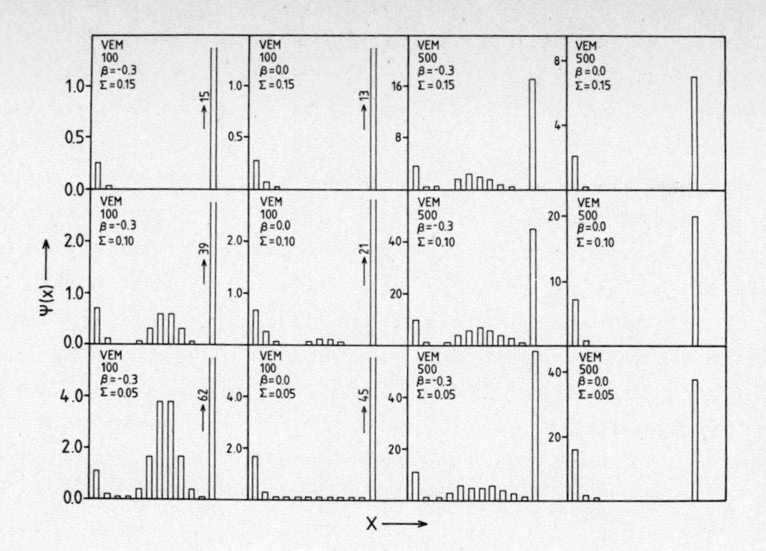

Figure 14. The allele frequency distribution predicted by the ecological model when environmental fluctuation is included. The frequency ranges are same as in Figure 11.

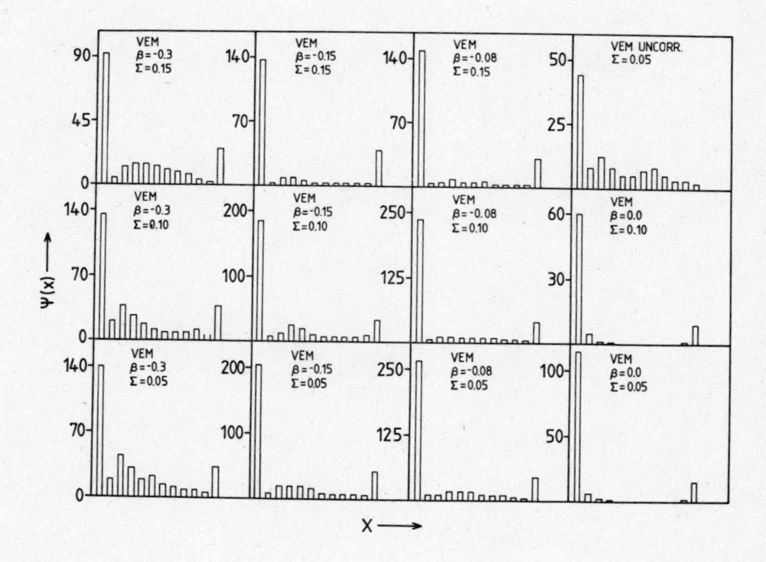

Figure 15. The allele frequency distribution predicted by the ecological model for $K_o = 5000$ and in the presence of fluctuating environment. The frequency ranges are same as in Figure 11.

Some calculations have been done using the correlated heterozygote model The results are shown in Figure 16. There have been a number of suggestions in the literature that the heterozygotes often have fitnesses that lie between the homozygotes. The correlated heterozygote models attempt to reproduce this situation. It is very evident from Figure 16, that as long as R and K are correlated (with $\beta \leqslant 0$) the results are in close agreement with the neutral model. The surprising result is that the variance σ^2, (see Equation 2.12), can change from 10^{-2} to 10^{-4} without producing significant change in the distribution. The R-K uncorrelated cases closely resumble the strong selection limit. In this case, the results are very dependent on the value of σ^2 used. As σ^2 is decreased, selective effects become weaker, as is evident from the rapid decrease in the peak at intermediate frequencies and corresponding increase in th terminal peaks, moving the distribution towards a U-shaped form. Finally, some calculations were performed with mutation rates 5×10^{-5} and 5×10^{-7}, and the distributions for these are also shown in Figure 16. For N = 5000, for example, the distribution for $\mu = 5 \times 10^{-5}$ is very similar to the distribution for N = 50000 and $\mu = 5 \times 10^{-6}$; similarly, the distribution with $\mu = 5 \times 10^{-7}$ agrees with that for N = 500 and $\mu = 5 \times 10^{-6}$. This also indicates the strong equivalence between the present model and the neutral model.

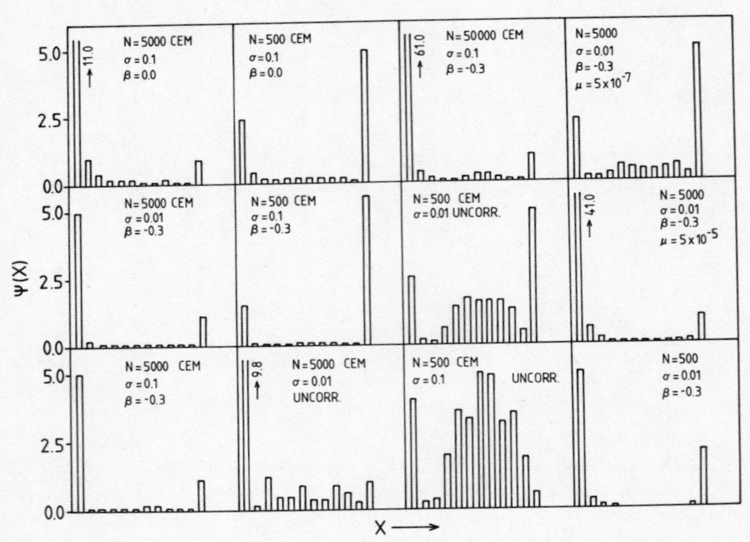

Figure 16. The allele frequency distribution predicted by the ecological model. These results are for the correlated heterozygote model in constant environment and with $\sigma = 0.1$ and 0.01. All the results, except those indicated, are for a mutation rate $\mu = 5 \times 10^{-6}$.

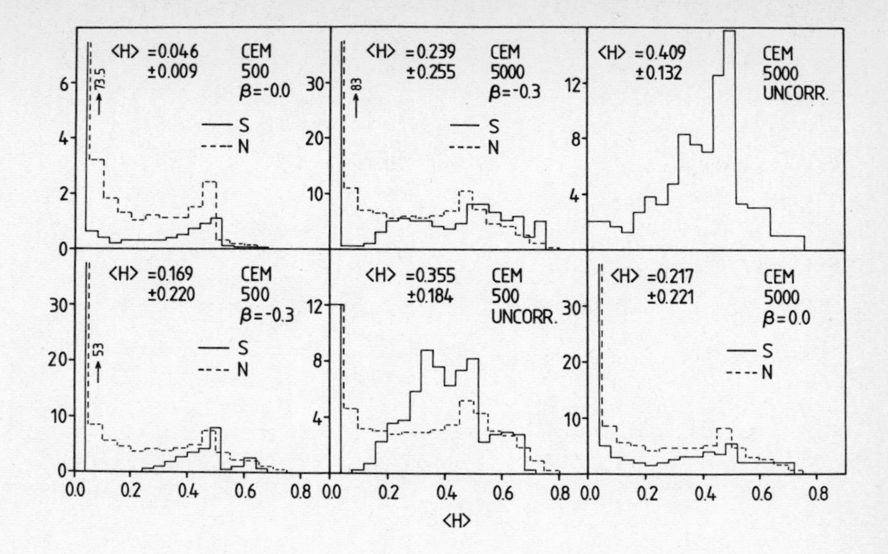

Figure 17. Comparison of the heterozygote distribution from the constant distribution model (full line) with the predictions of the neutral model (broken line; see text for details).

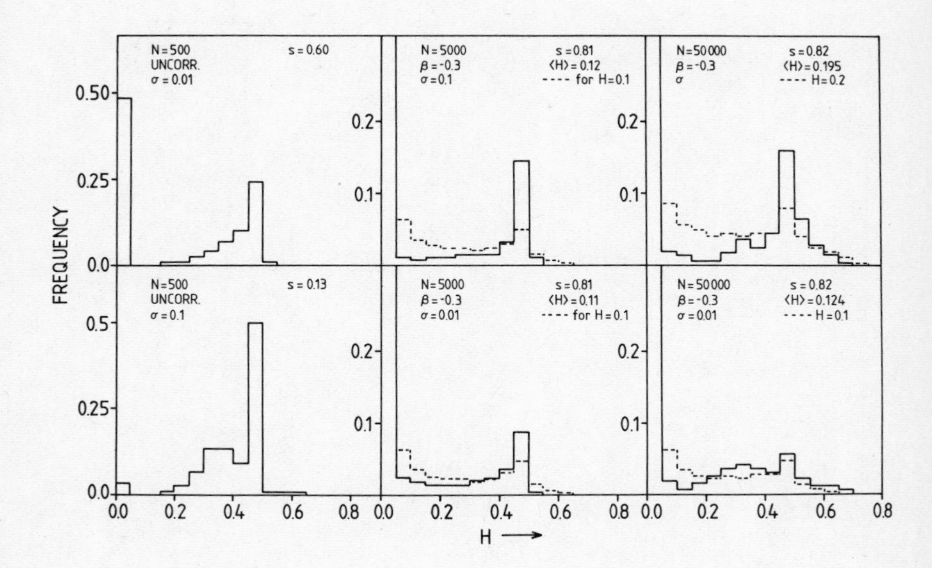

Figure 18. Comparison of the heterozygosity distribution predicted by the correlated heterozygote model (full line) with the predictions of the neutral model (broken line).

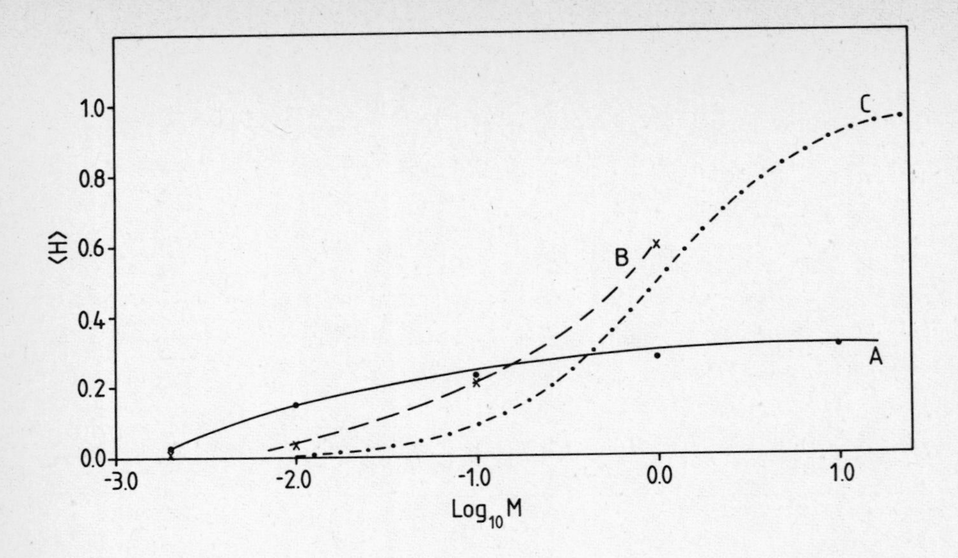

Figure 19. The variation of the mean heterozygosity with $M = 4\ N_e\mu$. Curve A : $\beta = -0.3$; Curve B : $\beta = 0.0$; Curve C : neutral model.

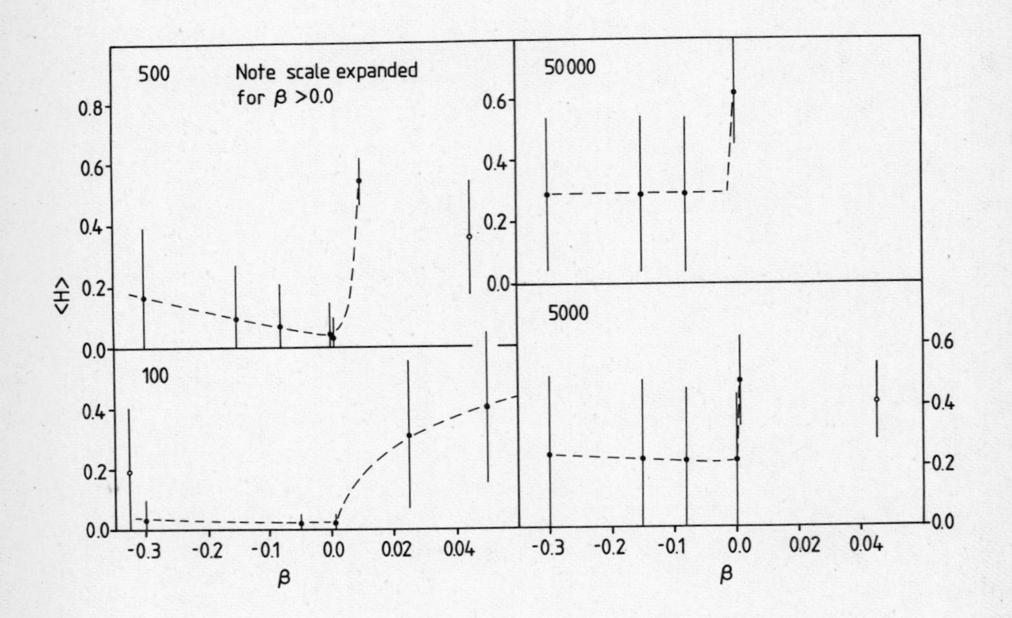

Figure 20. Variation of the mean value of heterozygosity, ⟨H⟩, with β for various population sizes.

6.b Single Locus Heterozygosity

We have already seen that the enzyme data yield a mean value of heterozygosity that is within the range of near zero to a maximum of 20%. The distribution of heterozygosity is very strongly peaked in the neighbourhood of zero and has small subsidiary peaks which are, in general, one or two orders of magnitude smaller in the interval between zero and unity. The most prominent of these subsidiary peaks occurs around $H = 0.5$. We have already discussed in the introduction some of the difficulties encountered by the neutral model in trying to explain consistently the relation between population sizes and heterozygosity. In the strong selection models, the heterozygosity reaches a limiting value of around 50-70% for very large populations. Also the heterozygosity is peaked at around 0.5 instead of at the origin.

Some typical examples of the distribution of heterozygosity for the uniform distribution model with constant environment is shown in Figure 17. In this figure these distributions are compared with the predictions of the neutral model. The neutral model calculations were taken from Fuerst _et al._ (1977). As is evident from the figure, except for the cases when R and K are uncorrelated, the agreements between the two models are reasonable. The ecological model shows a much more rapid decrease after the initial peak near zero, compared with the neutral model. The experimental data (see Fuerst _et al._, 1977) is not good enough to discriminate between the two models. In the case when R and K are uncorrelated, the distributions resemble those obtained with the strong selection models.

Some typical results of the correlated heterozygote model are displayed in Figure 18. For R and K correlated, as the variance σ (see (2.12)) is decreased, the distributions tend more closely to the neutral model. This trend is much less evident for the case when R and K are uncorrelated, though one can see the formation of a peak near zero and the reduction of the intermediate peak at 0.5 as the value of σ is decreased.

The variation of the mean heterozygosity $\langle H \rangle$ with $M = 4 N \mu$ for the present model with R and K correlated and for the neutral model are shown in Figure 19. Figure 20 shows the relation between β and $\langle H \rangle$. For small populations, $\langle H \rangle$ decreases as β increases from a negative value to zero. Beyond $\beta = 0$, $\langle H \rangle$ rapidly increases. For large populations ($N > 5000$) $\langle H \rangle$ is very insensitive to the value of β for $\beta < 0$; for

$\beta \geq 0$, <H> very rapidly increases with β. It is worthwhile remembering that the population is very prone to extinction for $\beta \geq 0$. These two figures indicate that the ecological model predicts a much weaker dependence of <H> on M than the neutral model. In the introduction we saw that the neutral model's explanation of the relation between population size and mean heterozygosity, based on the assumption of bottle necks, is very unsatisfactory. The very large compression of the population sizes for a given generation time, to yield theoretical effective population sizes that are almost the same, in order to reproduce the present observed heterozygosity cannot be reasonably accounted for through past "bottle-necks". This difficulty disappears in the ecological model where the heterozygosity is not strongly dependent on population sizes and converges to a value $\sim$ 30%. The dependence on β is also weak except in the neighbourhood of $\beta = 0$.

Another comparison between data and the neutral model used by Feurst, Chakraborty and Nei (1977) is the relationship between the mean value <H> of heterozygosity and its variance, Var(H). The variance of the single locus heterozygosity is related to the mean heterozygosity through (Stewart, 1976; Li and Nei, 1975):

$$\mathrm{Var}(H) = \frac{2M}{(M+1)^2\,(M+2)\,(M+3)} \tag{6.1}$$

In their paper, Fuerst, Chakraborty and Nei (1977) showed that within the considerable scatter in the data, the theory agreed with the experiments. Introducing the assumption of varying mutation rate with loci, they claimed a better agreement with data. The results of the present model are plotted in Figure 21 and compared with the data and with the neutral model relations given by Equations (3.2) and (6.1). The results of the present model were obtained by pooling all cases studied without imposing any bias except the omission of the R-K uncorrelated cases. It is easily seen from the figure that the ecological model represents the data to the same degree or even better than the neutral model. It is also as good as the modified neutral model with locus dependent mutation rate, as can be seen by referring to the Figures 1 and 2 in the paper by Fuerst *et al.* (1977). I suspect that the relation between the mean and the variance of the heterozygosity is not strongly model dependent.

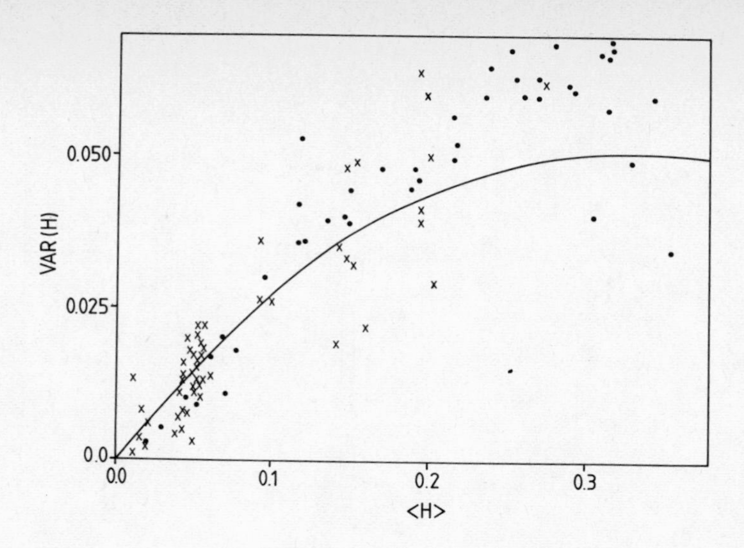

Figure 21. The relation between the mean and the variance of heterozygosity.
● Ecological model prediction; ——— neutral model prediction;
x Experimental data (from Fuerst <u>et al</u>. (1977)).

6.c The Fitness Distribution of the Homozygotes and the Heterozygotes

Figure 22 shows the distribution of the fitness parameters, W_{ij}, for the homozygotes (broken line) and for the heterozygotes (full line), for the uniform distribution model in a constant environment. The corresponding distributions for the correlated heterozygote model are displayed in Figure 23. The effects of fluctuations in the environment in fitness distributions can be seen in Figure 24.

The following conclusions can be drawn from these figures:

(1) For R and K correlated, small populations show reasonable heterozygote advantage, which for a given value of β, disappears as the population size is increased. In other words, the larger the population, the weaker the selection. This is opposite to the predictions of the neutral model, in which, for a given selection intensity, selection becomes weaker as the population size is reduced.

(2) For a given population size, as the value of β is increased from a large negative value to zero, the heterozygote advantage steadily decreases and eventually can even become slightly negative (hetero-

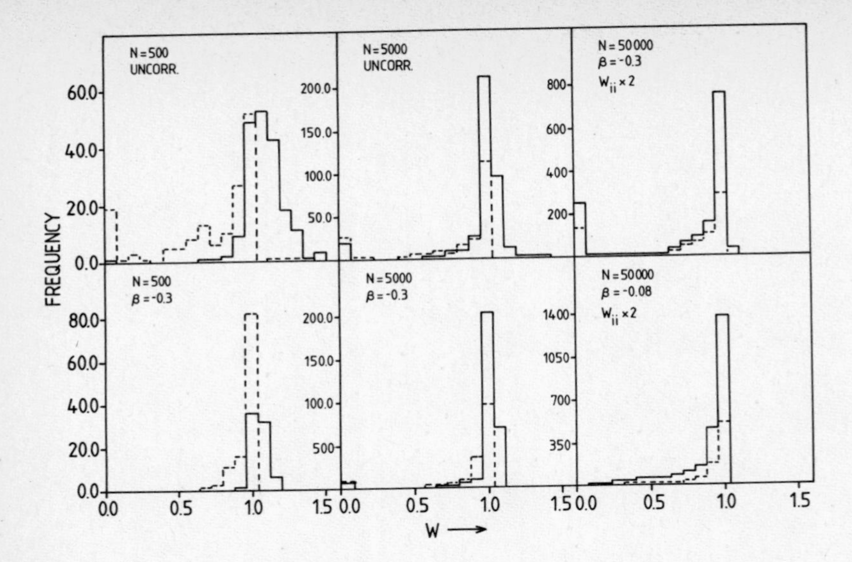

Figure 22. The distribution of the fitness parameter W_{ij} for the homozygotes (broken line) and for the heterozygotes (full line). All the cases shown are for the uniform distribution model in constant environment.

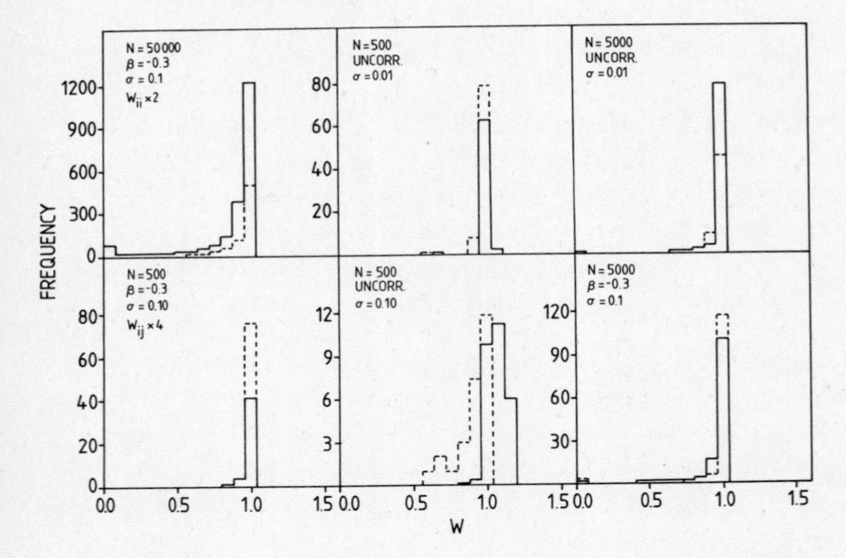

Figure 23. The distribution of the fitness parameters W_{ij} for the homozygotes (broken line) and for the heterozygotes (full line) for the correlated heterozygote model in constant environmen

zygote disadvantage), an effect due to the density dependence. When $\beta = 0$, the heterozygote advantage is exactly zero. For positive values of β, the heterozygote advantage increases very rapidly. These results are shown in Figure 25 where the heterozygote advantage ΔW, defined by

$$\Delta W = \{\langle W_{ij}(i \neq j)\rangle - \langle W_{ii}\rangle\} / \langle W_{ii}\rangle \tag{6.2}$$

is plotted as a function of β. In Equation (6.2) $\langle W_{ij}\rangle$ represents the average over the homozygotes or the heterozygotes, depending on whether $i = j$ or $i \neq j$, respectively.

(3) For R and K correlated, the population is seen to support slightly deleterious alleles. For small populations, the deleterious alleles are mainly contained in the homozygotes. As the population size increases, the deleterious heterozygotes correspondingly increase. As the value of β is increased from a negative value towards zero, the deleterious heterozygotes overtake the corresponding deleterious homozygotes, yielding the slight heterozygote disadvantage mentioned earlier. For this case, the fraction of alleles that are deleterious tend to decrease so that when $\beta = 0$, there is no resultant heterozygote advantage or disadvantage, ($\Delta W = 0$). For large negative values of β and large population sizes, a growing proportion of lethal alleles appears. As the value of β is increased towards zero, the lethal alleles are removed through purifying selection. It is interesting to note that the present model predicts slightly deleterious alleles in the population, reminiscent of the Ohta model (Ohta, 1972, 1973).

(4) When R and K are uncorrelated, there is a very significant heterozygote advantage in the uniform distribution model. This is not surprising, since, as discussed earlier, the R-K uncorrelated case tends towards the strong selection models. In the correlated heterozygote models, this advantage is reduced as the value of σ is reduced as the value of σ is decreased, as expected. Also, in the correlated heterozygote models, the population supports a fraction of deleterious alleles, especially for large populations and when R and K are correlated.

(5) Comparing Figures 22 and 24, it is seen that the effect of environmental fluctuations is to move the populations towards neutrality.

The price that is paid for this is population extinction, as discussed in an earlier section.

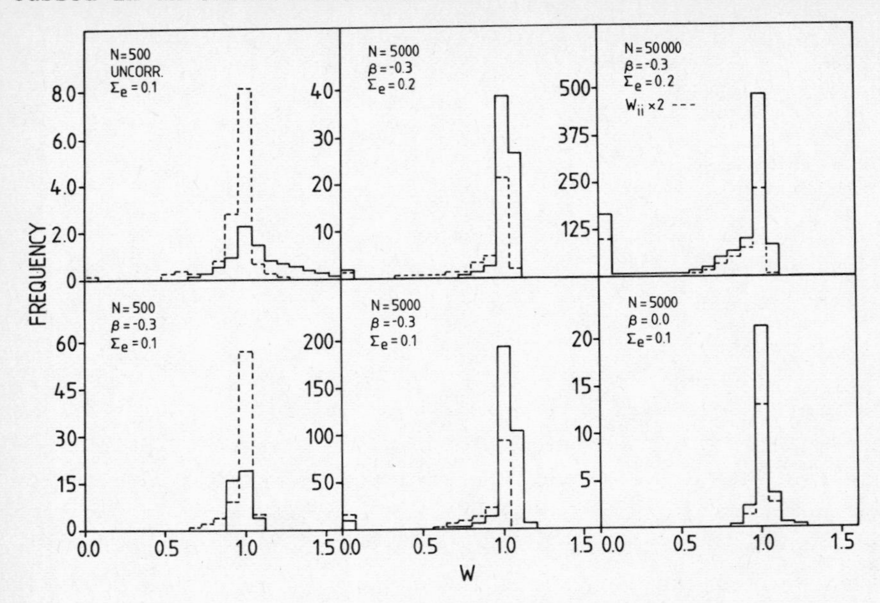

Figure 24. The distribution of the fitness parameter W_{ij} for the homozygotes (broken line) and for the heterozygotes (full line) for the constant distribution model in the presence of environmental fluctuation.

I now turn to a discussion of the distributions of R_{ij} and K_{ij}. Referring to Equations (2.4) - (2.11), when $\beta = 0$, $K_{ij} = K_o$ for all genotypes. As discussed in an earlier section, evolutionarily stable populations would tend towards a total population size of $N_T = K_o$. Thus the stability criteria provide in this case a weak selective force. For $K_{ij} = K_o$ W_{ij}'s are independent of genotypes and hence one would expect a uniform distribution of R_{ij}'s. This is seen in Figure 26, where the distribution of R_{ij}'s for the homozygotes and the heterozygotes are plotted for the case $K_o = 5000$ and $\beta = 0$. The values of the expected mean and the value corresponding to one standard deviation are also shown in the figure. The slight tendency of the heterozygotes to increase with R_{ij} is due to the selective effects (or constraints) imposed through the stability criteria. This can be seen by referring to Figure 5, where the mean value of <R> is seen to be around 1.34 for $K_o = 5000$ and $\beta = 0$. This value of <R> is within the stable point region (see Table 1). As seen from Figure 26, most of the values of R_{ij} fall within one standard deviation of the mean.

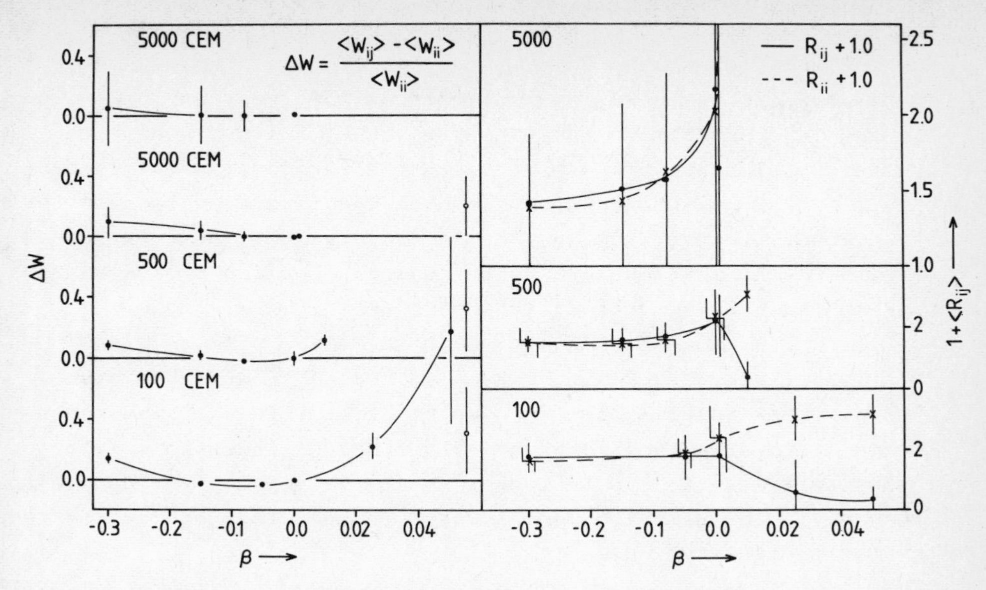

Figure 25. The curves on the left show the variation of the heterozygote advantage (or disadvantage) with β, for various values of population sizes. The heterozygote advantage (or disadvantage) is defined by

$$\Delta W = \{ \langle W_{ij}(i \neq j) \rangle - \langle W_{ii} \rangle \} / \langle W_{ii} \rangle$$

When $\Delta W > 0$, one has heterozygote advantage and when $\Delta W < 0$ there is disadvantage. The curves on the right show the variation of $\langle R_{ij} \rangle$ for the heterozygotes (full line) and for homozygotes (broken line) with β. All the results are for the uniform distribution model in constant environment.

Some typical examples of the distribution of R_{ij} and K_{ij} for the homozygotes and the heterozygotes, are given in Figures 27 and 28. The following points are worth noting:

(1) For negative values of β (R-K negatively correlated), the homozygote distributions are wider than those for the heterozygotes, but the mean value of K_{ij} is the same in both cases; the mean value of R_{ij} is higher for the heterozygotes than for the homozygotes.

(2) As the value of β is increased, but still negative, the distributions of R_{ij} become broader and of K_{ij}, narrower. The width does not increase by a large amount even when one goes from $\beta = -0.3$ to -0.08. Since the distribution becomes almost uniform for $\beta = 0$, the width of the R_{ij} distribution has to increase very rapidly in the neighbourhood of $\beta = 0$. This confirms the statement made earlier, that except in the neighbourhood of zero, the results are not very sensitive to the value of β.

(3) In the case when R and K are uncorrelated, the distributions show the existence of a small broad peak between -1.0 and 0.0, a minimum around 1.0 and the curve rising to a maximum at around 3.0. The homozygote distribution, on the other hand, follows the heterozygote distribution up to 1.0 and thereafter remains fairly constant. The K_{ij} distributions for both the heterozygotes and the homozygotes are very similar, having a very sharp peak at K just below K_o. The large heterozygote advantage arises from a sharp rise in the heterozygote distribution for $R_{ij} > 1.0$.

(4) As K_o increases to 50000, the R_{ij} and the K_{ij} distribution show a narrow peak superimposed on a constant background. The peaks become narrower as environmental fluctuations increase to 20%. These contribute to the fact that for large populations and in the presence of large environmental disturbances, the populations tend towards neutrality, (see Figure 13-15).

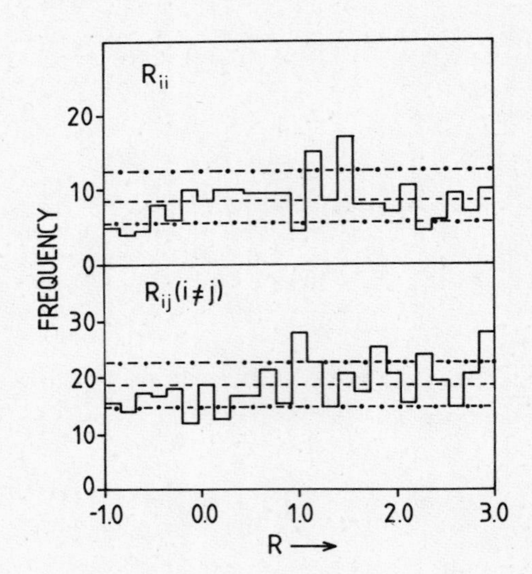

Figure 26. The distribution of the intrinsic growth parameter R_{ij} for $\beta = 0$ and for the uniform distribution model in constant environment. The broken line is the expected uniform distribution and the two lines on either side are the limit of one standard deviation, (see text for details).

The variation of the mean values of the intrinsic growth rate $\langle R_{ij} \rangle$ $(i \neq j)$ and $\langle R_{ii} \rangle$ with β is shown in Figure 25. It is seen that for $\beta < 0$, $\langle R_{ij} \rangle$ and $\langle R_{ii} \rangle$ are nearly equal while for $\beta > 0$, they rapidly diverge. The high probability for extinction when $\beta > 0$ is due to this

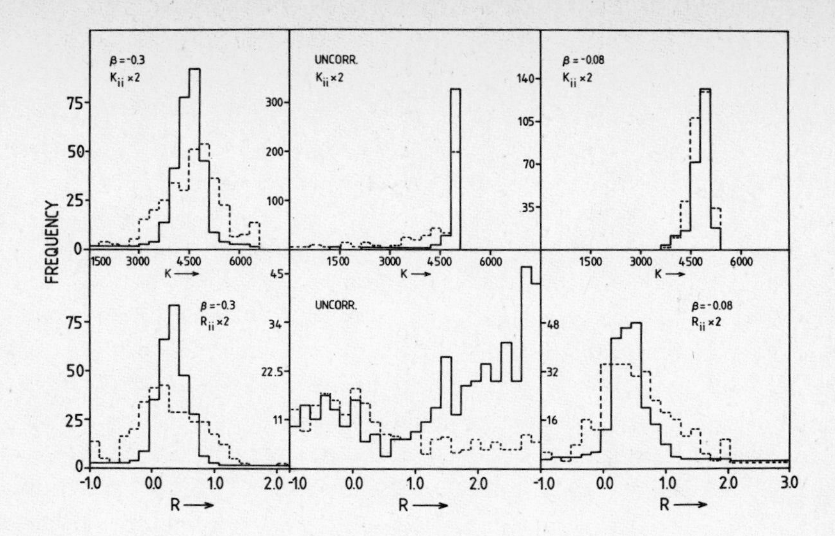

Figure 27. The distribution of R_{ij} and K_{ij} for the heterozygotes (full line) and for the homozygotes (broken line) for the uniform distribution model in constant environment. All the results are for K_o = 5000.

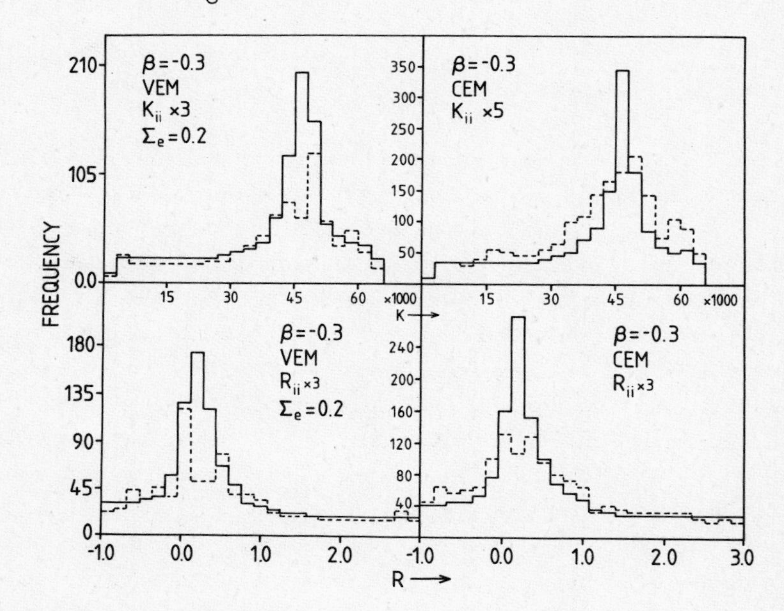

Figure 28. The distribution of R_{ij} and K_{ij} for the heterozygotes (full line) and for the homozygotes (broken line) for K_o = 50000. The lower curves are for R_{ij} and the upper curves are for K_{ij}. The effects of environmental fluctuation are shown in the curves on the left.

divergence.

6.d Average Number of Polymorphic alleles at a Single Locus

In the neutral model the average number of polymorphic alleles is given by Equation (3.3). The average number of alleles in the present calculations are obtained by counting the number of alleles that are present in the population with frequencies $> 1/2N$ and averaging them over the last 5000 generations and over the 100 populations studied. These two numbers would then be comparable. The relation between the average number of alleles and the population size N is shown in Figure 29. Here the solid dots are for $\beta = -0.3$ and the crosses for $\beta = 0$. Also included are the standard deviations for these points. The solid line represents the prediction of the neutral model. As expected, $\beta = 0$ is in good agreement with the neutral model. When $\beta < 0$ the variation of the number of alleles with N rises much less steeply than the neutral predictions. In general, the experimental data indicate that there are between 1 - 5 alleles at any locus for populations with widely differing sizes. Thus the present model is in closer accord with experiment than the neutral model.

6.e What is the Optimum value for β?

Throughout this section I have pointed out that the results are not very sensitive to the value of β, as long as $\beta < 0$. But, as can be seen from Figures 8, 20 and 25, β does have a small influence on the results when $\beta < 0$. Further, as β is increased towards zero, the populations come closer to the predictions of the neutral model. For $\beta > 0$, the populations are very unstable and the results are crucially dependent on the value of β. It is interesting to ask if there is an optimum value of β that can be obtained through natural selection. As discussed in an earlier section, if this optimum value turns out to be at or very close to zero, then the population would be structurally unstable and would not survive through the evolutionary time.

The optimum value of β has been investigated using the variable β model described in Section 2. The distributions of β_{ij} for the homozygotes and the heterozygotes and the mean value $\langle\beta\rangle$ defined by

$$\langle\beta\rangle = \sum_{ij} \beta_{ij}\, p_i\, p_j \tag{6.3}$$

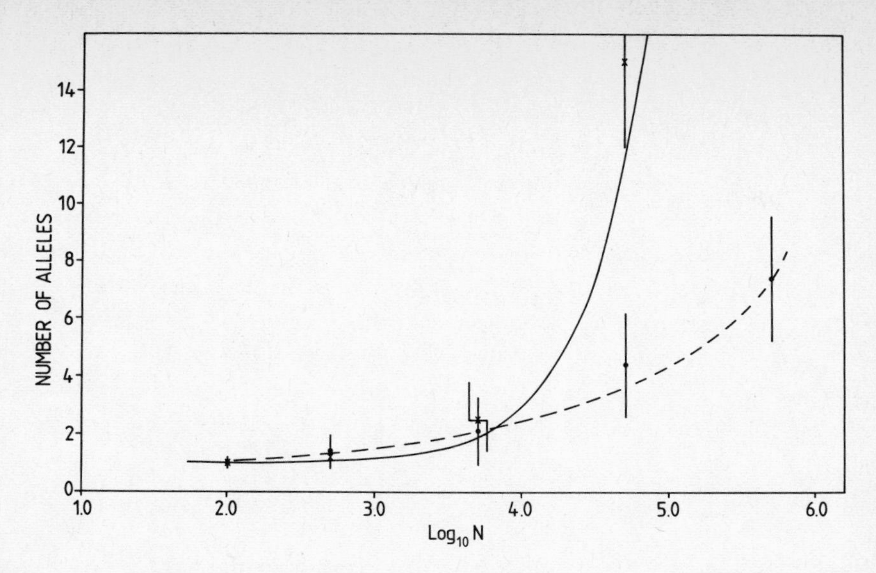

Figure 29. The average number of alleles as a function of population size N predicted by the ecological model and by the neutral model.

——— neutral model; – – – – ecological model, β = - 0.3

X ecological model, β = 0.0

are shown in Figure 30. The distribution of the heterozygotes is sharply peaked around zero and has a mean value of - 0.063; the homozygotes are more uniformly distributed, with a suggestion of peaking around 0.06 and the distribution has a mean value of - 0.137. Interestingly, the distribution of <β> indicates a broad peak, with maximum at - 0.05 and a mean at - 0.083 ± 0.056. Thus natural selection produces an optimum value of β that is below zero so that the population is stable and sufficiently near zero, ensuring near neutral conditions. These results are all for K_o = 5000. The results for K_o = 500 and 5000 using the uniform distribution model and the correlated heterozygote model are tabulated in Table II The range of β_{ij} used in these calculations was (- 0.3, 0.0). When the range was extended to include positive values of β, the populations showed a high probability to go extinct; the populations that survived, on the other hand, had an optimum value of β very close to the values give in Table II.

From Table II it is seen that <β> for the uniform distribution model is almost independent of population size, being around - 0.08. For the correlated heterozygote model, its value slightly increases (moves

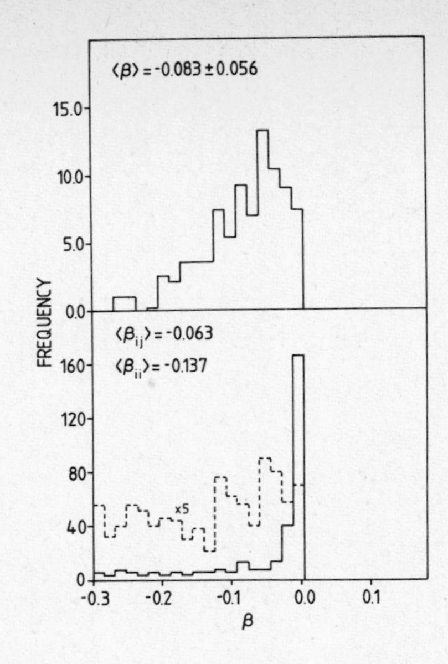

Figure 30. The bottom curve shows the distribution of β_{ij} for the heterozygotes (full line) and for the homozygotes (broken line) for the variable β sub-model (see text for details). The top curve is the value of β_{ij} averaged over the whole population. The mean values of the various distributions are also shown. These curves are for the constant distribution model and for K_o = 5000.

Table II. Mean Values of the Distributions of β_{ij} $(i \neq j)$, and β for the β Variable Model

K_o	S*	Model**	β_{ij}	β_{ii}	$\langle\beta\rangle$
500	0.82	UDC	-0.028	-0.112	-0.082 ± 0.061
500	0.85	CH(σ=0.1)	-0.066	-0.064	-0.057 ± 0.060
500	0.85	CH(σ=0.01)	-0.070	-0.068	-0.065 ± 0.062
5000	0.87	UDC	-0.063	-0.137	-0.083 ± 0.056
5000	0.87	CH(σ=0.10)	-0.075	-0.072	-0.039 ± 0.046
5000	0.87	CH(σ=0.01)	-0.073	-0.064	-0.036 ± 0.042

*S denotes the survival probability.

** UDC denotes uniform distribution model; CH is for correlated heterozygote model with variance σ^2.

towards zero) as the population size is increased. As the value of σ is decreased, from 0.1 to 0.01, the value of ⟨β⟩ is hardly changed. As σ is decreased, the values for the homozygotes and the heterozygotes become equal, this being achieved by the homozygotes moving towards zero

The calculations in Table II can be criticised on many points. It is naive to assume that genes at a single locus would control both the intrinsic growth rate R_{ij} and the correlation parameter β_{ij} which represents the mode and efficiency of resource allocation. Further, chosing β_{ij} values from a uniform random distribution may bear little relation to real populations. The description of population regulation through the two parameters R_{ij} and β_{ij} is a gross distortion of the real world. Despite all these shortcomings, it is gratifying to note that the process of natural selection tends to produce an optimum value of $\langle\beta\rangle$ that yields results in accord with observation while maintaining the stability of the population. Also, the fact that the reasonable variation in the value of $\langle\beta\rangle$ does not significantly change either of the above requirements indicates the robustness of the model.

6.f Rate of Gene Substitutions

One of the arguments that led to the neutral theory is concerned with the equality of substitution rates in phylogenetic trees constructed from contemporary protein sequences from various species. Wilson, Carlson and White (1977) have compared the protein sequence evolution in mammals whose generations are long (whales, elephants and humans) with that in mammals whose generations are short (rodents, tree shrews and rabbits). They compared pairs of mammals with common ancestors and with short and long generation time. Such comparison avoids uncertainties in divergence time. In spite of considerable scatter of the results, their analysis clearly demonstrates that the number of nucleotide substitutions depend on chronological rather than generation time. Langley and Fitch (1973, 1974) have compared the number of nucleotide substitutions with the estimated divergence time, based on amino acid sequences of cytochrome c, myoglobin, hemoglobin α- and β- chains, two fibrinopeptides and the insulin C-peptides in seventeen species of mammals. Their analysis indicates that there exists significant deviation from a strictly stochastic molecular clock. Fitch (1976) maintains that in spite of this, longer term averages over a much larger number of proteins than the seven used above would produce estimates that are linear with respect to paleontological time. The data collected by Langley and Fitch indicate that the rate of substitution for the primates is significantly lower than for the other mammals considered. This could be either due to errors in divergence time as argued by Wilson, Sarich and others or due to a reduced rate of molecular evolution in the primates, a view held by many primate paleontologists. In spite of such conflicting data sets and arguments, the concensus of opinion is in favour of molecules evolving at a fairly constant rate with some suggestion of saturation. Though the examples quoted above are all for mammals, some nonmammalian data do exist and the reader can find the details and further references in Wilson, Carlson and White (1977).

The difficulty of relating the above results at the nucleotide level wit the present model arises from our rather <u>ad hoc</u> definition of the gene. If the gene is defined to be the region (or regions) of the chromosome producing a single functional protein, then, since the number of amino acid molecules varies considerably between various proteins, one has no direct method of ascribing to the gene the number of substitutions to be expected. Using this definition of the gene one can ask what is the prediction of the neutral model and compare the predictions of the present model with the neutral model. As pointed out by Ewens (1979), such predictions cannot be used as a test for neutrality or, on the same basi for the present model.

For selectively neutral alleles, the probability for fixation of new mutants is $(2N)^{-1}$ and there are $2N\mu$ new mutants per generation, where μ is the mutation rate per generation. It then follows that the rate of substitution, being equal to the number of new mutants that eventually reach fixation, is given by μ. If η be the number of substituions of nucleotides per year and if there are A amino acid sequences in the protein associated with the gene, then clearly $\mu = 3\,A\,g\,\eta$, where g is the generation time in years. The value of 10^{-7} arrived at by Kimura and Ohta (1971) and mentioned in the introduction was obtained by assuming A = 300. The fixation rate of 1/2N quoted above is only applicable for strictly neutral alleles; if one includes deleterious mutants, this woul no longer be true and the substitution rate would not be equal to the mutation rate, and would not be independent of the effective population size N. In this context it is worth remarking that the effective population size is related to the generation time and this is evident in the plot for $N_{e(T)}$ in Figure 1. The broken line shown in the figure was calculated using the following relation between N and g:

$$N = 1.78 \times 10^{5}\, g^{-1.25} \tag{6.4}$$

Nei has disputed the deleterious mutation model of Ohta on the ground that the substitution rate is not independent of N. But with the type of relation (6.4) between generation time and N and from a fixation time

Table III. Gene Substitution Rates per Generation

K_o	β	G^* x 10^5	μ x 10^{-5}	S^{**} x 10^{-5}	N^+ x 10^{-5}
500	-0.08	1.0	5.00	1.20 ± 1.00	5.00
500	-0.08	3.0	5.00	0.80 ± 1.10	5.00
500	-0.08	1.0	0.05	0.16 ± 0.30	0.05
500	-0.08	3.0	0.05	0.12 ± 0.12	0.05
500	0.00	1.0	5.00	4.10 ± 1.70	5.00
500	0.00	3.0	5.00	4.60 ± 1.50	5.00
500	0.00	1.0	0.05	0.15 ± 0.36	0.05
500	0.00	3.0	0.05	0.12 ± 0.16	0.05
5000	-0.08	1.0	5.00	1.80 ± 0.81	5.00
5000	0.00	1.0	5.00	1.40 ± 0.70	5.00
5000	-0.08	1.0	0.05	0.60 ± 0.49	0.05
5000	0.00	1.0	0.05	0.30 ± 0.58	0.05
50000	-0.08	1.0	5.00	1.40 ± 0.41	5.00
50000	0.00	1.0	0.05	0.60 ± 0.40	0.05

* Number of generations.
** Substitution rate from present model.
\+ Substitution rate predicted by the neutral model.

that differs from 1/2N, it is quite conceivable that the Ohta model could reproduce the observed data, bearing in mind the large scatter in the data and other uncertainties that are involved.

In Table III some results on the substitution rate per generation using the present model are given. These were averages over 25 populations only. All the results are for the uniform distribution model in constant environment. The correlated heterozygote model and variable environment yield very similar results. For comparison the predictions of the strictly neutral model are also given. It is seen that for $\beta < 0$, the substitution rates are not strictly dependent on population size; this is true for $\beta = 0$ except for small populations. The substitution rate is not equal to the mutation rate with the result that apart from the case when $\beta = 0$ and $\mu = 5 \times 10^{-5}$, it is smaller than the neutral model prediction by a factor between 2 and 2.5 when $\mu = 5 \times 10^{-5}$; for $\mu = 0.05 \times 10^{-5}$, it is larger than the neutral result by a similar factor. It should be noted that our values have large errors associated with them.

As discussed in an earlier subsection, the present model shows that the population has a small fraction of deleterious alleles. The discrepancy between the neutral model and the present model noted above arises from this. From the discussions given in this section it should be clear that the ecological model is not at variance with the data on gene sub-substitutions given the various uncertainties both in the data and in the interpretation. Gene substitution rates do not provide an unambiguous test of the models but only indicate whether the models produce reasonable and consistent results.

7. Conclusions

The primary aim of this work has been to demonstrate that the data from allozyme studies need not necessarily indicate a non-Darwinian process. At present there exists a dichotomy in the understanding of biological phenomena. I have on my bookshelf a number of books on diverse biological topics, mostly concerned with macroscopic (phenotypic) properties which are all discussed more or less within the framework of Darwinian selection. These topics include various aspects of sexual evolution (Maynard Smith, 1978; G. C. Williams, 1975; P. O'Donald, 1980; E. L. Charnov, 1982), coevolution (Futuyma and Slatkin, 1983), evolutio ary ecology (Pianka, 1974; Roughgarden, 1979), animal conflicts (Maynar Smith, 1982), sociobiology (E. O. Wilson, 1975), pre biotic evolution (Kuppers, 1983) and many others too numerous to list here. On the othe hand, the evolution of protein molecules and the evolution of the DNA i the post biotic era, are considered in terms of the non-Darwinian, neutral theory (Kimura and Ohta, 1971; Nei, 1975). Either this dichoto is an intrinsic and in some ways a unique property of biological system or, like the representation of the elephant by the seven blind men of Hindustan, they are just convenient models to describe the particular set of topics or observations that one is interested in. My own training and background make me accept the second of the two explanations. If so many aspects of macroscopic features, such as behaviour, sex, lif history, competition etc. are influenced by selective mechanisms to a greater or lesser extent, then it is legitimate to investigate the reductionist relationship between such macroscopic phenomena and the unde lying microscopic or genetic world. The present ecological model is a crude attempt at this. I must emphasize that this does not imply the extreme view of the absolute selfish gene or that every tiny feature at the phenotypic level is necessarily influenced by natural selection or possesses adaptive significance. Many aspects of the microscopic world do resemble the predictions based on neutrality and thus what is needed is a synthesis between these two opposing types of observations.

The neutral model, in its original form, is a one parameter model, the parameter being $M = 4 N_e \mu$. That many features of the enzyme data, at least qualitatively, agree with the predictions of such a simple model cannot be easily disregarded. It indicates that, at least at the molecular level, mutation and drift are of fundamental importance in producing diversity. But as we have seen in the introduction, when quantitative tests are made, one observes deviation from the theoretical prediction. This is highlighted in the relationship between population size and heterozygosity. The neutral theory then has to invoke assumptions relating to bottle necks or relating to slightly deleterious genes to provide an explanation. In criticising Ayala and Valentine's hypothesis for adaptive strategy, Nei and Graur (1983) make the point that there is no theoretical basis for this hypothesis and that it is difficult to include such a concept within the framework of a mathematical theory. The same type of objections could equally well apply to "bottle necks". Bottle necks are convenient hypothesis to explain some of the shortcomings of the theory but they have no predictive content in them. In the same paper (Nei and Graur, 1983) the authors admit that ".. of course, some proportion of non-deleterious mutations must be advantageous; otherwise no adaptive evolution can occur. However, if a few percent of non-deleterious mutations are advantageous, as postulated by the neutral mutation theory, adaptive change of organisms can easily be explained". If alleles at each locus have a small probability of being advantageous, then it follows that during the course of the evolutionary period a good fraction of the loci behave as predicted by selective models. Enzyme polymorphisms, though admittedly scanning a specific subset of possible enzymes, do not show patterns with recognisable selective functions. It is possible to think of mechanisms involving environmental fluctuations to explain this, but I feel such explanations are difficult to quantify. They have little predictive value and are not easy to justify. The present model, on the other hand, shows that near-neutrality is the norm because of ecological constraints but that the populations have an underlying diversity that can, in specific circumstances, be acted upon by selective forces; these circumstances depending on the complex interactions within the population, between populations and with the environment.

Ohta predicts that the rate of evolution is more rapid in small populations than in large populations. Her observation is based on the fact that the selection coefficient of a mutant gene is variable over both spatial and temporal dimensions due to environmental conditions. Thus, for large populations, spanning large spatial dimensions, a mutant, to have an overall advantage, must be able to survive varied environmental conditions, while such spatial variation is not present for small populations. Also small populations have a higher chance of fixation of slightly deleterious mutations. Thus gene substitution in small populations could be higher than in large populations. The present model is in agreement with Ohta's view. We have already seen that small populations show more selective differences than large populations. Thus one could expect the rate of evolution to be more rapid in small populations. Referring to Table III, one also sees that near $\beta = 0$, the substitution rate is much higher in small populations than in large populations. For $\beta < 0$, this happens when the population size is much smaller than 500. Whether this would promote a higher rate of speciation is an open question. Since small populations are more prone to extinction than large populations, there might perhaps be an optimum range of population sizes that balance extinction and evolutionary rates in such a manner as to provide the most efficient mechanism for speciation.

There could be many objections to the present model. Some of these are discussed in the appendix. The model, as any theoretical model, is a very poor representation of the real world. There are three criteria for the building of models. Firstly, the model assumptions should at least reflect biological reality. Secondly, the model should be able to explain a much wider class of phenomena than previous models. Thirdly, the model should be predictive, at least in a statistical sense. The model presented in this chapter satisfies, more or less, all these requirements. The real biological world is so complex that it would be foolish to claim that any particular model encompasses all observations. What the ecological model presented here shows is that it is possible to reproduce neutrality within the framework of Darwinian theory. It also shows that many of theories of coevolution (Roughgarden, 1983; May and Anderson, 1983) have a much wider significance.

There is still much work to be done. The extension of the model to include multi locus systems, migration etc. is needed. Also future studies should involve ecological problems such as evolution of competition, coevolution of species, evolution of sex and stability of ecological systems.

References

Bellows (Jr.), T. S. (1981)
The descriptive properties of some models for density dependence.
J. Anim. Ecol. 50, 139 - 156.

Chakraborty, R., Fuerst, P. A. and Nei, M. (1978)
Statistical studies on protein polymorphism in natural populations.
II. Gene differentiation between populations.
Genetics 88, 367 - 390.

Chakraborty, R., Fuerst, P. A. and Nei, M. (1980)
Statistical studies on protein polymorphism in natural populations.
III. Distribution of allele frequencies and the number of alleles per locus.
Genetics 94, 1039 - 1063.

Charnov, E. L. (1982)
The theory of sex allocation
Princeton University Press, Princeton, New Jersey.

Ewens, W. J. (1979)
Mathematical population genetics
Springer-Verlag, Heidelberg.

Fitch, W. M. (1976)
"Molecular evolutionary clocks" in Molecular evolution
(Ed.) Ayala, F. J., Sinauer Associates, Sunderland, Massachusetts.

Fuerst, P. A., Chakraborty, R. and Nei, M. (1977)
Statistical studies on protein polymorphism in natural populations.
I. Distribution of single locus heterozygosity.
Genetics 86, 455 - 483.

Futuyama, D. J. and Slatkin, M. (1983)
Coevolution
Sinauer Associates, Sunderland, Massachusetts.

Goodman, M. (1981)
Decoding the pattern of protein evolution.
Prog. Biophys. and Mol. Biol. 37, 105 - 164.

Guckenheimer, J., Oster, G. and Ipaktchi, A. (1977)
The dynamics of density dependent population models.
J. Math. Biol. 4, 101 - 147.

Hamilton, W. D. (1980)
Sex versus non-sex versus parasites.
Oikos 35, 282 - 290.

Hassel, M. P., Lawton, J. H. and May, R. M. (1976)
Patterns of dynamical behaviour of single-species populations.
J. Anim. Ecol. 45, 471 - 486.

Itoh, Y. (1979)
Random collision processes of oriented graphs.
Institute of statistical mathematics (Japan), Research memorandum No. 154, 1 - 20.

Kimura, M. and Ohta, T. (1971)
Protein polymorphism as a phase of molecular evolution.
Nature, 229, 467 - 469.

Kimura, M. and Ohta, T. (1971)
Theoretical aspects of population genetics.
Princeton University Press, Princeton, New Jersey.

Kuppers, B-O (1983)
Molecular theory of evolution.
Springer-Verlag, Heidelberg.

Langley, C. H. and Fitch, W. M. (1973)
"The constancy of evolution: A statistical analysis of the α and β hemoglobins, cytochrome c and fibrinopeptide A"
in Genetic structure of populations (Ed.) Morton, N. E.
University Press of Hawaii, Honolulu.

Langley, C. H. and Fitch, W. M. (1974)
An examination of the constancy of the rate of molecular evolution.
J. Mol. Evol. 3, 161 - 177.

Li, W. -H. and Nei, M. (1975)
Drift variance of heterozygosity and genetic distance in transient states.
Genet. Res. 25, 229 - 248.

Maruyama, T. and Nei, M. (1981)
Genetic variability maintained by mutation and overdominant selection in finite populations.
Genetics 98, 441 - 459.

May, R. M. (1978)
Theoretical Ecology: Principles and Applications. (Ed.) May, R. M.
Blackwell, Oxford.

May, R. M. (1981)
"Non linear problems in ecology and resource management"
Les Houches Summer School on Chaotic Behaviour of deterministic Systems, July, 1981.

May, R. M. and Anderson, R. M. (1983)
"Parasite - Host Coevolution" in Coevolution (Eds.) Futuyma, D. J. and Slatkin, M. Sinauer Associates, Sunderland, Massachusetts.

Mayo, O. (1970)
Fixation of new mutants.
Nature 227, 860.

Maynard Smith, J. (1978)
The Evolution of Sex.
Cambridge University Press, Cambridge.

Maynard Smith, J. (1982)
Evolution and the Theory of Games.
Cambridge University Press, Cambridge.

Nei, M. (1975)
Molecular Population Genetics and Evolution.
North-Holland, Amsterdam.

Nei, M., Fuerst, P. A. and Chakraborty, R. (1976)
Testing the neutral mutation hypothesis by distribution of single locus heterozygosity.
Nature 262, 491 - 493

Nei, M., Fuerst, P. A. and Chakraborty, R. (1978)
Subunit molecular weight and genetic variability of proteins in natural populations.
Proc. Nat. Acad. Sci. (USA). 75, 3359 - 3362.

Nei, M. (1980)
"Stochastic theory of population genetics and evolution"
in Proceedings of Vito Volterra Symposium on Mathematical Biology
Lecture Notes in Biomathematics, Vol. 39., Springer-Verlag, Heidelberg.

Nei, M. (1983)
Chapter 3 in this volume.

Nei, M. and Graur, D. (1983)
Extent of protein polymorphism and the neutral mutation theory.
Evolutionary Biology (in press)

O'Donald, P. (1980)
Genetic models of sexual selection.
Cambridge University Press, Cambridge.

Ohta, T. and Kimura, M. (1971)
On the constancy of the evolutionary rates of cistrons.
J. Mol. Evol. 1, 18 - 25.

Ohta, T. (1972)
Population size and rate of evolution
J. Mol. Evol. 1, 305 - 314.

Ohta, T. (1973)
Slightly deleterious mutant substitutions in evolution.
Nature 246, 96 - 98

Pianka, E. R. (1974)
Evolutionary Ecology.
Harper & Row, New York.

Pimm, S. L. (1982)
Food Webs.
Chapman and Hall, London.

Propping, P. (1972)
Comparison of point mutation rates in different species with human mutation rates.
Humangenetik 16, 43 - 48.

Roughgarden, J. (1979)
Theory of population genetics and evolutionary ecology: An introduction.
Macmillan, New York.

Stewart, F. M. (1976)
Variability in the amount of heterozygosity maintained by neutral mutations.
Theor. Pop. Biol. 9, 188 - 201.

Stubbs, M. (1977)
Density dependence in the life cycle of animals and its importance in K- and r- strategies.
J. Anim. Ecol. 46, 677 - 688.

Tanner, J. T. (1966)
Effects of population density on the growth rates of animal populations.
Ecology 47, 733 - 745.

Thomas, W. R., Pomerantz, M. J. and Gilpin, M. E. (1980)
Chaos, asymmetric growth and group selection for dynamic stability.
Ecology 61, 1312 - 1320.

Vogel, F. and Rathenberg, R. (1975)
Spontaneous mutations in Man.
Adv. Hum. Gen. 5, 223 - 318.

Williams, G. C. (1975)
Sex and Evolution.
Princeton University Press, Princeton, New Jersey.

Wilson, E. O. (1975)
Sociobiology: The New Synthesis.
Harvard University Press, Cambrdige, Massachusetts.

Wilson, A. C., Carlson, S. S. and White, T. J. (1977)
Biochemical evolution.
Ann. Rev. Biochem. 46, 573 - 639.

Wright, S. (1938)
Size of population and breeding structure in relation to evolution.
Science 87, 430 - 431.

APPENDIX

The final draft of my manuscript was sent to Prof. M. Nei to enable him to include in his contribution, his criticism of the ecological genetic (EG) model. Professor Nei agreed that I should give a rebuttal to his objections and this I do in this Appendix.

One of the objections to the EG model that is raised is concerned with the use of discrete generations with a specific single species model based on the two parameters, r, the intrinsic growth rate and K, the carrying capacity. In particular, Nei remarks that the fact that N, the population size, can exceed K, makes K a biologically meaningless parameter. It should be noted that N exceeding K can also occur in equivalent overlapping generation models where there is a delay in response between resource utilization and population growth. In discrete generation models this delay is already built-in implicitly. This feature has been extensively discussed by May (1976). Most insect populations studied have non-overlapping generations and though the use of such simple single species models is extremely inadequate to describe real populations, what the EG model shows is that weak selection can result from the constraints imposed on the population through ecological interactions. To consider K, the carrying capacity, to be an absolute upper bound for the population is a very naive interpretation. When energy that goes into reproductive effort is small, (small values of r), K does describe the limit to population size. On the other hand, when resources are being rapidly consumed in reproductive effort (high value of r), then the population can rise above the value of K but would eventually crash to a value below K and this is a consequence of the time delay in response. Thus K is a biologically meaningful parameter, though in a real field situation it would be an averaged parameter describing the effects of various types of interactions that are neglected in the single species models.

Chaotic dynamics is the norm in a large class of non-linear equations. Such dynamics can also occur in the case of overlapping generations under suitable conditions. Chaotic behaviour may reflect on our inadequacy in formulating the dynamical equations of multi-species systems. No theoretical model can ever claim to describe observations absolutely, especially for such complex systems as occurs in biology, with their highly non-linear character and large stochastic components. But the ecological models do describe more or less the population properties and enable one to understand and inter-relate their behaviour of different populations. Using such models, surprisingly no field populations exhibit chaotic features while at least one laboratory population can be classified as being in the chaotic regime. Though caution should be exercised in interpreting these results, what is shown by the EG model is that natural selection imposes severe conditions on the range of r and K such that the population that survives is stable. As remarked elsewhere (Mani, 1983), ecological stability can be understood only if one includes genetic variability in the ecological models.

The results of the EG model do tend more towards the neutral limit than the strong selection limit, provided there exists a negative correlation between the intrinsic growth rate r and the carrying capacity K. The aim of the model is not to reproduce the results of the infinite allele neutral model exactly. If it did so, then the EG model would be very uninteresting. It is reasonably well established that the single parameter infinite allele model does not describe the enzyme data in detail. What is so remarkable and attractive concerning the single parameter neutral model is that it does give a good qualitative description of the data. For example, the allele frequency distribution in the case of enzyme polymorphism U-shaped as predicted by the neutral model but when the data are considered in greater detail, one observes an excess of alleles at small frequencies and a small but noticeable peak in the distribution at intermediate frequencies, (Ginzburg, 1983). On the other hand, the spectra derived from the neutral theory have a minimum in the range of intermediate frequencies. The increase in occurrence of rare alleles could be explained in terms of slightly deleterious mutations as proposed by Ohta (1973). Ohta's model assumes no dominance and, as remarked by Ginzberg (1983), it cannot explain the occurrence of the small increase in the allele frequency spectra at intermediate frequencies. Ginzburg proposes a compromise theory which allows for the possibility of selective dominance within a "neutralistic" system and which describes the experimental data better. The EG model is a "neutralistic model with dominance present and this occurs naturally in the theory.

Another objection to the EG model is that it has too many parameters and hence its success is probably due to chance. The advocates of the neutral model point out that the selection theories embedded in population genetics do not yield predictions that agree with the data on enzyme polymorphism. Consider then the number of parameters that occur in a selection model for finite populations with mutations included. These are, the effective population size N_e, the mutation rate μ, and the fitness coefficients W_{ij} for the genotypes A_iA_j. If W_{ij} values are chosen from some random distribution, then strictly the parameter set is much smaller than the complete set of $\{W_{ij}\}$. In the EG model the parameters involved are N_e, μ, R_{ij} and K_{ij} where R_{ij} and K_{ij} are the genotypic values of intrinsic growth rate and carrying capacity, respectively. Since R_{ij} and K_{ij} are negatively correlated through the correlation parameter β, we have essentially one extra parameter, namely, β. The results of the model are very insensitive to the value of $\beta < 0$. Thus, in effect, the classical selection models and the EG model have the same number of parameters. It is then unfair on the one hand to claim that classical selection models cannot explain the enzyme data and on the other hand to object that the EG model, which has effectively no extra parameters and has the added constraint of population regulation, as having too many parameters. Further, the extreme one-parameter neutral model is inadequate to yield details of the experimental data and one has to implicitly or explicitly impose additional parameters. Nature is not built on a single parameter and if a model such as the EG model, with more parameters that are derived through biological reasoning not only yields a better description at the enzyme level but also is capable of describing many macroscopic biological features, then surely one should seriously consider such models. Nei further states that "his study does not give much insight into actual selection mechanism". This depends on our concept of selection. Within the Darwinian framework, selection is a phenotypic property which relates to the success of the individual through breeding in the presence of competition for resources, mates etc. From this view point, the EG model does indeed define selection through growth rate and resource utilization, even though the description could be improved by including inter-species interactions.

Another major comment made by Nei relates to the observation that the substitution rate in the third codon is much higher than the rate in the first two codons. Since in the first two positions most of the changes of nucleotides are expected to lead to amino acid changes while only a quarter of those in the third codon yield corresponding changes in amino acid, the purifying selection on deleterious genes would result in a larger rate of substitution for the third codon. Similarly, functional genes are observed to have a much lower substitution rate compared to pseudogenes and for reasons similar to the one given above. These facts, according to Nei, are only consistent with the neutral theory. Nei goes on to say, "On the other hand, selectionists believe that most nucleotide substitutions are caused by positive Darwinian selection, in which case the rate of nucleotide substitution in functionally unimportant genes or parts of genes is expected to be relatively lower because mutations in these regions of DNA would not produce any significant selective advantage". The use of the term "positive Darwinian selection" is misleading. Though some "selectionists" may use Darwinian selection in this sense, it is not what is contained in the original formulation by Darwin. The EG model is based on Darwinian selection and it clearly indicates that the substitution rate would be larger by a factor of five or so for functionally unimportant genes when compared with structural genes. A pseudogene is described in the model by $\beta = 0$. Referring to Table III in chapter 4, one sees that for $\beta < 0$ the substitution rate is reasonably constant but that it increases rapidly when $\beta = 0$. Thus the larger substitution rate in non-functional genes is consistent with the EG model.

REFERENCES

Ginzburg, L. R.,(1983) Theory of Natural Selection and Population Growth, Benjamin/Cummings, Menlo Park, California.

Mani, G. S., Genetic Diversity and Ecological Stability in Evolutionary Ecology, (Ed. B. Shorrocks) Blackwell, Oxford.

May, R. M.,(1976) Models for Single Species Populations in Theoretical Ecology: Principles and Applications, (Ed. R. M. May), Blackwell, Oxford.

Ohta, T., (1973) Slightly deleterious mutant substitutions in evolution, Nature, 246, 96-98.

THE POSSIBILITY OF STRESS-TRIGGERED EVOLUTION

Christopher Wills
Department of Biology, C-016
University of California, San Diego
La Jolla, CA 92093

ABSTRACT

A consideration of recent data from the molecular biology of pro- and eukaryotes has led Harrison Echols and myself independently to very similar ideas about the way in which evolution might proceed discontinuously, a way which he calls "inducible evolution" and which I call "stress-triggered evolution." Echols deals primarily with important observations on the stress-induced "SOS" system in *E. coli*, which leads to an increase in point mutations. Such systems may play an important role in evolution. The essence of the idea of stress-triggered evolution, on the other hand, is that there exist specific genetic architectures in organisms that have co-evolved with *internal, cooperative* mutagenic agents such as viruses, plasmids and transposable elements such that environmental stress will tend to trigger bursts of genomic change. This article explores ways in which such cooperative interactions could have evolved through both individual and group selection, and suggests the kinds of experimental tests that would be necessary to attempt to verify the existence of such a process.

I present here a possible mode of discontinuous evolution. Similar suggestions have recently been made by Harrison Echols (1981, 1982), and by Belyaev and Borodin (1982). Since all of us arrived at these ideas independently, there are considerable differences in approach and emphasis, which I will explore here along with (as is the author's prerogative) a discussion of my own viewpoint. These ideas are being published elsewhere in more extended form (Wills, 1983).

The essence of what I have called stress-triggered evolution and what Echols calls inducible evolution is that there is a far more complex feedback process between the environment and the genome of the organism than has commonly been supposed. This feedback means that the rate of mutational change -- and possibly the rate of evolution itself, though the connection between these two processes is by no means self-evident -- may be altered by environmental changes that are not normally considered mutagenic. Such environmental changes might include various types of physiological stress.

This viewpoint should not be considered Lamarckian, or Lysenkoist. The Lamarckian view of evolution suggests that a specific environmentally-altered aspect of the phenotype is somehow transferred back to the germ line genotype and that this specific change is passed on to the next generation. To extend an analogy recently suggested by Dawkins (1982), it is as if the addition of icing to a cake should somehow alter the recipe of the cake, so that subsequent cakes baked with that recipe would end up with that particular type of icing. Only one rather specialized instance of Lamarckian inheritance has been suggested recently, that of the possible transfer of modified genetic information coding for specific antibodies from differentiated lymphocytes back to the germ line, perhaps through the agency of viruses (Gorczynski and Steele, 1981). While this is not a biologically impossible or even unlikely process, the evidence for it has been strongly criticized (Brent _et al_., 1981).

Echols makes the point that the bacterial SOS response, which is induced by environmental factors that interfere with DNA replication, acts to increase both general and site-specific recombination and the rate of point mutations. The reaction of the genome to a particular environmental stress is therefore not the induction of a specific set of mutations, but rather a generalized increase in mutation rate. There should therefore be, as with most mutagenic agents, only a coincidental relationship between the agent and the types of mutations and regions of the genome which the agent affects. A simple increase in the mutation rate due to environmental effects is not Lamarckian.

As will be developed in the course of this paper, I part company with Echols somewhat in this matter. My prediction is that the SOS system is only one of a number of systems that allow complex interactions between genomes and the environment to occur. Some of these interactions may be the result of co-evolution of the genome and the mutagenic agents that affect it. It may therefore not be surprising that some environmental stimuli might result in specific genomic changes. The probability of occurrence of a specific mutational change might be very high if the mutagenic agent (say a virus or a plasmid) which can cause the change is triggered into activity by a specific environmental alteration. But there need not be -- and indeed is unlikely to be -- any obvious connection between the mutational change and the environmental alteration which triggers it. Even if such a connection is more than coincidental -- if, for example, an increase in the temperature can trigger an increase in the rate at which temperature-resistant mutants appear -- it should be possible to explain such an effect on the basis of a combination of group and individual selection, rather than by resorting to a Lamarckian explanation. I will spend some time at the end of this paper doing so.

Echols' concept of inducible evolution is based chiefly on the observation that the mutation rate to point mutations in bacteria can be increased by influences that inhibit DNA synthesis. The simplest explanation for this increase is straightforward; factors which interfere with the reactions catalyzed by DNA polymerase may decrease the fidelity of replication of the enzyme. I would like to extend and elaborate this basic concept by pointing out that the entire definition of the mutation process is undergoing dramatic alteration at the present time, as a result of recent discoveries in molecular biology. Mutagenic agents are both more numerous and more complex in their interaction with the genome than had previously been imagined.

It is now possible to classify mutagenic agents in many different ways, and I will offer a classification which may be useful in view of the arguments to be developed later. Mutagenic agents can be classified along a continuum from external, autonomous agents to internal, cooperative agents. Some examples follow.

External, _autonomous_ agents are agents that act from outside the cell, and that show little interaction with the genome. That is, all regions of the genome and all DNA sequences are equally likely to be affected by a truly autonomous agent. Of course, there are no completely autonomous mutagenic agents. Certain regions of the chromosomes are more susceptible to breakage by X-rays than are others. Ultraviolet light will be more likely to damage thymine-rich regions of the DNA. Nonetheless, these agents do act from outside the cell. There are few ways an evolutionary lineage can respond to such external agents except by evolving more efficient repair mechanisms. The fact that the mutation rate per generation is approximately constant for organisms with a large range of generation times suggests that this evolution has occurred. Further, repair mechanisms can be induced in mammalian cells by the process of irradiation itself (Tuschl et al., 1980). These responses may be highly complex, but they may be likened to a fire department in a big city. Fires may break out anywhere in the city, and not all can be put out in time.

Internal, cooperative agents are far more sophisticated, and may be much more far-reaching in their evolutionary consequences. Such agents work from inside the cell, and they are often confined in their effects to specific regions of the genome. Thus they act cooperatively with the genome in producing their effects. They are likely to be much more susceptible to environmental alterations than are external agents. Thus, internal cooperative agents are likely to be subject to coevolutionary pressures. Consider bacteriophage Mu (Taylor, 1963). This bacteriophage uses replicative transposition to insert itself into many sites in the genome. The results of many of these insertions, particularly if they disrupt a structural or regulatory gene, are mutations. This source of mutations is absolutely dependent on the invasion of *E. coli* by Mu, and it seems certain that the ability of Mu to integrate into the genome and make the bacterium lysogenic has been the result of a long period of co-evolution.

Much more rapid co-evolution can be seen in the rapidity with which bacterial populations have accumulated resistance transfer factors. These factors, in which one or more genes conferring resistance to antibiotics have been incorporated into a variety of plasmids, have made a recent appearance in bacterial populations, but the plasmids have been there for much longer. A high frequency of plasmids -- which however lack these genes -- has been found in a great range of bacteria collected and preserved from the 1930's (Hughes and Datta, 1983). The whole evolution of resistance transfer factors has taken place since the introduction of antibiotics, but the machinery which facilitates this evolution was in place and able to respond to the new environmental stress (the introduction of these new chemicals into the environment) with tremendous speed. It seems likely that this machinery was the result of a long history of co-evolution, and that many genes have been transferred in the past between closely and even remotely related types of bacteria by these transfer factors.

Thus, the evolution of resistance transfer factors is a paradigm of evolutionary processes which until recently have not been considered by theorists or indeed by most thinkers about the evolutionary process. Plasmids have co-evolved with the genomes of a number of host bacteria, in such a way as to facilitate the transfer of genetic information between them. The introduction of a new environmental stress has resulted in the rapid transfer of newly-advantageous mutants among these strains. Such rapid evolution could not have occurred without the previous co-evolution of what are in effect mutator or mutation-triggering sequences.

The alteration of genomic sequences may often be intimately related to environmental changes that themselves would not be considered mutagenic by the usually standards applied to an external autonomous mutagen. Internal cooperative agents may act to magnify the effects of the genome of relatively slight environmental alterations. For example, there are transposons of E. coli which both facilitate (Kleckner, 1981) and repress (Biek and Roth, 1980, 1981) the insertion of further copies of the transposon at other points in the bacterial genome. Further, insertion sequences are found to insert more commonly at low than at elevated temperatures (Cornelis, 1980; Arber et al., 1981).

A great deal of work has been done on the conditions that define lysogeny and the initiation of the lytic cycle in lambda phage, and once again the indications are that the environment can play a very large role. Herskowitz and Hagen (1980) have reviewed the influence of the environment on the probability that a given infection will be lysogenic rather than lytic, and list a number of treatments such as a short period of starvation that greatly increase the probability of lysogenization. Once a lysogenic strain has been established, a number of quite mild regimens such as treatment with hydrogen peroxide or organic peroxides (Hayes, 1970) can initiate the lytic cycle. It may be argued that this is not a mutagenic process, but if we define mutation as any heritable alteration in the genome (exclusive of regular crossing-over or uncomplicated gene conversion) then the integration and release of lambda phage from the bacterial genome must count as mutational events. Synthesis of the cII protein which is important in the switch between the lysogenic and lytic states has recently been shown to depend on a number of host genes (Hoyt et al., 1982).

While bacteria are the organisms of choice for the study of environmental effects on the rate of point mutation and rearrangement of the genome, there are some indications that higher organisms may be subject to similar influences. A great variety of environmental effects influence the rate of recombination in higher organisms (for review, see Catcheside, 1977). If these also have an effect on such processes as plasmid insertion and unequal crossing-over, they could directly influence the production of new genes and the recombination of non-alleles into new mosaic genes by these processes. Environmental effects may be strong in the production of somatic mutations in certain plant tissues. In *Zea*, *Nicotiana*, *Portulaca* and *Antirrhinum*, lowering the temperature increases the degree of mosaicism in leaf or endosperm tissue (Fincham, 1973). The genetic basis of this mosaicism is not fully understood, but the role of movable elements in generating mosaicism in *Zea* (McClintock, 1953) is now being substantiated at the molecular level (Burr and Burr, 1981).

In eukaryotes in general, it has always been a puzzle that many changes in the genome that can be detected at the chromosomal and now at the molecular level are not of the sort that can easily be generated by external autonomous mutagenic agents. There appears to be a remarkable regularity to some genomic changes, almost a directionality, which has not as yet been tied in to selective pressures but which strongly suggests that such pressures are operating. To give just one particularly puzzling example, consider the pattern found by Keyl (1965) in the amounts of DNA of the polytene chromosomes of *Chironomus thummi thummi* and *Chironomus thummi piger*. When hybrids between the two species were examined and the amount of DNA in bands compared directly in the hybrids, it was found that the amount of *C. t. thummi* was always either the same as that of *C. t. piger*, or two-, four-, eight- or sixteen-fold greater -- never less. Two things are surprising about these observations; first, the regular nature of the increase, and second, the fact that it is *always* an increase. If some random process such as molecular drive (Dover, 1982) were producing this increase, one would expect an increase to be found in the bands of the other species at least some of the time. Further, there is no obvious external autonomous agent that could have produced these particular genomic changes, any more than there are obvious external agents that can lead to the shifts in amount and distribution of highly redundant DNA which is such a hallmark of the differences between closely-related species (Peacock, et al., 1977). In the latter case, one must rely on unequal crossing-over, perhaps coupled with molecular drive or strong selection. In the former case, there appears to be a regular mechanism for duplicating the large segments of the genome which are bounded by polytene interband regions. How often do these processes occur, and by what precise mechanism? What can trigger these changes? It is difficult to imagine processes more unlike those with which we are familiar as being caused by external, autonomous mutagenic agents. I would predict that the ever-growing role of viruses and plasmids in our understanding of prokaryotic mutational changes will be demonstrated to have parallels with the eukaryotic situation. A further prediction is that the environment will be found to play a large role in triggering these changes.

Evolutionary consequences of stress-triggered changes in the genome:

Echols very properly points out that if bursts of mutations can be induced by environmental alterations, standard evolutionary theory would suggest that this should have a very profound effect on the rate of evolution. The same point has been made by Belyaev and Borodin (1982). I suspect that this approach may be a bit too simplistic. One must ask: what kinds of mutations are being induced, and is it possible that different categories of mutational change may have different degrees of impact on the evolutionary process? Standard evolutionary theory suggests that evolutionary change occurs through the appearance of random mutations, most of which are lost but a small fraction of which spread through the population, displacing the old alleles and the loci at which they arise. They may do so through some combination of selection, drift, migration, selection on nearby linked loci and (possibly) molecular drive. Certainly, it is straightforward to attribute to these processes much evolution at the molecular level and at the DNA level, although endless arguments have risen about the relative importance of the various factors involved in the spread of new alleles. The same processes can be assumed to apply even to such dramatic changes as alteration of chromosome number, although extreme conditions may have to be postulated such as severe restriction in population size. Of the many difficulties which arise to complicate this simple picture, let me mention two particularly relevant ones.

1. There is no evidence whatsoever that an increase in the external, autonomous mutation rate results in a concomitant increase in the rate of the evolutionary process (Maynard Smith, 1977). There are enormous differences in evolutionary rate among different organisms. Why do living fossils (the bowfin fish, the coelacanth, the ginkgo and so on) show little morphological change while other lineages exhibit periods of very rapid evolution? Why do some lineages, such as the cetaceans, undergo explosive evolutionary change followed by long periods of relative morphological stasis? It seems unlikely that all the periods marked by tremendous evolutionary divergence (the Cambrian, Permian and Paleocene for animals, the Cretaceous for flowering plants) were also marked by a great increase in external, autonomous mutagenic agents. Otherwise, the adaptive radiation of the Hawaiian Drosophila or the African cichlid fish would have to be attributed to localized increases in the externally-caused mutation rate. It seems on the contrary to be far more likely that there are many different classes of mutational process, some of which may have a

larger impact on morphological evolutionary change than others. Just as it is productive to classify mutant alleles into many different categories in order to understand the impact on a population of different classes of genetic variant (Wills, 1981), it should also be productive to classify mutagenic agents according to the impact they are likely to have on the evolutionary process. While I cannot predict the ultimate shape of such a classification, I am confident that it will be far more complex than the classification utilized by evolutionary theorists at the present time.

2. Standard evolutionary theory has not fully considered the implications of the coevolution of mutagenic agents and the genomes of the organisms they affect. There appear to me to be two general ways in which this coevolution could occur.

First, there could be an immediate benefit to the organism in which the mutation occurs. In the course of the evolution of the immune system, for example, it may be that the first instances of translocation and joining of genes, resulting in the "two or more genes-one polypeptide" situation which is so characteristic of this system (Tonegawa, 1983) may have been mediated by viruses or transposable elements. They would therefore not have occurred initially with any regularity, but they might have conferred an immediate advantage on the organism in which they occurred because they greatly extended the capabilities of the immune system. Strong selection for coevolution of the virus or transposable element and the genome of the organism would eventually have resulted in a higher frequency and greater specificity of this particular translocation -- specificity in the sense that a particular well-defined region of the DNA would be translocated rather than a region in which the breakpoints might vary from one translocation event to another. It seems unlikely in the extreme that this could happen without coevolution of both the mutagenic agent and the region which it affects. Indeed, the mutagenic agent appears now to have been incorporated into the genome, producing its "mutational" change with complete regularity.

There is a second and more subtle way in which the mutagenic agent and the genome could evolve, however. This route has a direct bearing on the central argument of this paper, that stress may trigger mutational changes in the genome and that these changes may be more likely to be derived from the action of internal, cooperative mutagenic agents than of external, autonomous ones.

To begin with, I assume that there is no particular advantage, and there may be a decided disadvantage, to an increase in the frequency of mutations and genomic rearrangements during the normal interactions of a species with its environment. A good deal of theoretical speculation has been expended on examining the evolution of the mutation rate itself (e.g. Leigh, 1973), but the attention of theorists has not yet been turned to the possibility that the mutation rate may have evolved to be very low under some conditions and higher under others. Such evolution would be unlikely to occur if the source of most mutations is due to external, autonomous agents, since the impact on the genome of such agents is so widespread that the only selective response might be expected to be an increase or decrease in sensitivity of a large part or all of the genome simultaneously. If there has been coevolution of the genome and internal, cooperative agents, it might be expected to take the form of a low mutation rate caused by these agents during normal environmental conditions and a much higher rate during abnormal conditions when an increase in the genetic variance of the population might be of advantage to the survival of the population as a whole.

This last argument is a group selection one, and since group selection arguments are not popular among many evolutionists, let me specify exactly what sort of group selection I have in mind.

Group selection has always suffered from two difficulties in the minds of evolutionists. The first is that, since the unit of selection is the group or the species instead of the individual, it is likely to be extremely slow. This is particularly the case if the group selection is acting at the species level, since the lifetime of a species is very long compared with that of an individual. The second problem is that group selection and individual selection may often be in conflict. The classic example of group selection suggested by the work of Wynne-Edwards (1965), is of a gene which when fixed will limit the size of a population, thus preventing it from over-exploiting its environment and being driven to extinction. The effects of such a gene would be to decrease the Darwinian fitness of an individual carrying it, and so such a gene would be unlikely to become fixed in a population unless the population underwent a severe size bottleneck. On the other hand, if it did become fixed, its effects could immediately be counteracted by any subsequent mutation which increased the Darwinian fitness of an individual carrying it.

These difficulties with group selection are very hard to surmount theoretically, and they stem from the fact that both group and individual selection affect exactly the same properties of the population -- namely the components of fitness of the individuals making up the population. It is not surprising under these circumstances that individual selection tends to win out over group selection (Wilson, 1975).

I suspect, however, that these difficulties with group selection can be surmounted if the genetic characters which enhance the survival of the group are very different from those which affect the survival of the individual, so that they are not in direct conflict with each other. One suggestion along these lines has been made by Steven Stanley (1979). He suggests that sexual reproduction itself, for which theorists have yet to find a short-term selective advantage, may only confer an advantage during times of rapid environmental change. Then, unusual recombinants of existing genetic variability may be more likely to survive. The rest of the time, he suggests (rather unpersuasively in my opinion) that the whole elaborate machinery of sexual recombination is essentially neutral in its impact on fitness. Those lineages which retain the ability to reproduce sexually are those most likely to survive in evolution.

The same argument can be used with respect to stress-triggered evolution, with the advantage that a more persuasive case can be made for the selective neutrality or near-neutrality of the nuclear and extra-nuclear genetic mechanisms that initiate stress-triggered changes. They are more likely perhaps to be selectively neutral for long periods of time than the highly elaborate mechanisms of sexual reproduction. This is particularly the case if it is recalled that sexual reproduction is in constant danger of being supplanted by partheogenetic modes of reproduction, while little will happen to a lineage that loses its stress-triggered mutational mechanisms. But imagine a lineage in which various sequences capable of triggering genomic changes are present. These sequences would have no effect on the individual except for the presumably small cost of the replication of some extra DNA. They would only exert their effect at times of environmental change, at which point they would aid in the generation of phenotypic variability on which selection could act. Those lineages which contained such "triggering" sequences (perhaps viruses, retroviruses, plasmids, transposable elements) would be most likely to evolve sufficiently in order to be able to survive, either by phyletic evolution or by giving rise to new distinct species. Thus, lineages harboring these sequences would be more likely to survive. Those that do not would be more likely to go extinct -- and the members that _did_ survive would tend to undergo little phenotypic change. Further, and to make things even more complex, the structure of the genomes of these successful lineages and the structures of their triggering sequences would have co-evolved in such a way as to make particular genetic changes leading to particular phenotypic alterations more likely. The dice might be loaded in such a way that a particular environmental stress is likely to give rise to a suite of genetic changes that includes a high proportion of changes that will prove adaptive in the new circumstances. This would not be Lamarckian, although it sounds like it. Rather, it would be the result of group selection and would be very slow and perhaps rather ineffective. Nonetheless, it would be possible because the characters being selected for are not normally in conflict with individual fitness.

Some interesting possibilities for the interpretation of future observations are worth thinking about now. Consider the highly organized segments of the genome that are now becoming accessible to the new techniques of molecular biology. I fully expect that the organization of these segments will be found to be such that mutational rearrangements giving rise to specific phenotypic alterations have a high probability of occurrence. This is already becoming apparent in the arrangement of the immune system genes. While it may be argued that this organization is the result of individual selection, a role for group selection can be suggested in the evolution of morphogenetic supergenes in mimetic butterflies. Evolution which results in a genetic architecture increasing the likelihood of appearance of a new mimetic morph would greatly enhance the survival of those mimetic species carrying that particular genetic architecture.

While it may be argued that individual selection will play the predominant role in the evolution of such architectures, it also seems likely that their evolution will be accelerated during periods of environmental stress -- defined broadly as an environment or set of environments to which the organisms have not previously been exposed. Further, reticulate evolution may play a role in accelerating this process. If, as is becoming apparent, mechanisms exist for the transfer of genetic information across species barriers in eukaryotes, then a supergene or other genetic architecture capable of rapid stress-triggered evolution could be passed by such a process to other organisms. I might even suggest, rather daringly, that reticulate evolution mechanisms may facilitate the rapid evolution seen in large groups of congeneric species -- African cichlid fish, or Hawaiian Drosophila, for example. And it is the lack of such potential which may help to prevent the further evolution of living fossils lacking nearby congeners with which such genetic information can easily be exchanged.

Experimental tests: It seems to me that experimental tests of the possibility of stress-triggered evolution fall into four categories. First, as was pointed out earlier, simply demonstrating an increase in the mutation rate as a result of stress has no direct bearing on whether such an increase affects the speed of evolution.

What will have to be done is first to determine whether such an increase occurs. If it can be shown to be the case, then, the kinds of genetic changes triggered by stress will have to be investigated at the DNA level. Third, evidence that similar changes occur in the process of evolution will have to be found. And finally -- and this is the hardest step of all -- evidence will have to be accumulated that such changes actually result in phenotypic alterations with an important impact on the fitness of the organisms in which they have occurred. All these steps will be necessary before the circle of argument is fully closed.

We are making some tentative investigations of the first of these four steps, using the easily-quantified parameter of growth rate in the yeast *Saccharomyces cerevisiae* (Blatherwick and Wills, 1971). It will obviously be a very long time, and may require the attentions of many investigators, before all four levels of argument are fully investigated and the ideas set forth here in outline form are confirmed or rejected. Nonetheless, I feel sanguine about the possibility that stress-triggered evolution actually does occur, if only because it is a biologically possible mode of evolution. If there is one thing we have learned about the evolutionary process, it is that it is so opportunistic that every avenue open to it seems to have been taken.

ACKNOWLEDGEMENTS

I would like to thank many colleagues who have, sometimes unwittingly, helped in the development of these admittedly far-fetched ideas. I would especially like to thank Ted Case, Jim Crow, Harrison Echols, Peter Geiduschek, Mae-Wan Ho, Steve Howell and Dan Lindsley.

REFERENCES

Arber, W., Humbelin, M., Caspers, P., Rief, H.J., Iida, S. and Meyer, J. 1981. Spontaneous mutations in the *E. coli* prophage P1 and IS-mediated processes. Cold Spr. Harb. Symp. Quant. Biol. 45:38-39.

Belyaev, D.K. and Borodin, P.M. 1982. The influence of stress on variation and its role in evolution. Biol. Zbl. 101:705-714.

Biek, D. and Roth, J.R. 1980. Regulation of Tn5 transposition in Salmonella typhimurium. Proc. Natl. Acad. Sci. USA 77:6047-6051.

Biek, D. and Roth, J.R. 1981. Regulation of Tn5 transposition. Cold Spr. Harb. Symp. Quant. Biol. 45:189-192.

Blatherwick, C. and Wills, C., 1971. The accumulation of genetic variability in yeast. Genetics 68:547-557.

Brent, L., Rayfield, L.S., Chandler, P., Fierz, W., Medawar, P.B. and Simpson, E. 1981. Supposed Lamarckian inheritance of immunological tolerance. Nature 290:508-512.

Burr, B. and Burr, F.A., 1981. Detection of changes in maize DNA at the shrunken locus due to the intervention of Ds elements. Cold Spr. Harb. Symp. Quant. Biol. 45:463-465.

Catcheside, D.G., 1977. The Genetics of Recombination. Edward Arnold, London.

Cornelis, G. 1980. Transposition of Tn951 Tnlac and cointegrate formation are thermosensitive processes. J. Gen. Microbiol. 117:243-247.

Dawkins, R., 1982. The Extended Phenotype. Freeman, San Francisco.

Dover, G. 1982. Molecular drive: a cohesive mode of species evolution. Nature 299:111-117.

Echols, H., 1981. SOS functions, cancer and inducible evolution. Cell 25:1-2.

Echols, H., 1982. Mutation rate: some biological and biochemical considerations. Biochimie 64:571-575.

Fincham, J.R.S. 1973. Localized instabilities in plants--a review and some speculations. Genetics (suppl). 73:195-205.

Gorczynski, R.M. and Steele, E.J. 1980. Inheritance of acquired immunological tolerance to foreign histocompatibility antigens in mice. Proc. Natl. Acad. Sci. USA 77:2871-2875.

Hayes, W. 1970. The Genetics of Bacteria and Their Viruses. 2nd Ed., Wiley, New York.

Herskowitz, I. and Hagen, D. 1980. The lysis-lysogeny decision of phage lamda: explicit programming and responsiveness. Ann. Rev. Genet. 14:399-445.

Hoyt, M.A., Knight, D., Das, A., Miller, H. and Echols, H. 1982. Control of phage lamda development by stability and synthesis of cII protein: role of the viral cIII and host hflA, himA and himD genes. Cell 31:565-573.

Hughes, V.M. and Datta, N. 1983. Conjugative plasmids in bacteria of the 'pre-antibiotic' era. Nature 302:725-726.

Keyl, H.-G. 1965. A demonstrable local and geometric increase in chromosomal DNA of Chironomus. Experientia 21, 191-193.

Kleckner, N. 1981. Transposable elements in prokaryotes. Ann. Rev. Genet. 15:341-404.

Leigh, E. 1973. The evolution of mutation rates. Genetics 73:1-18.

Maynard Smith, J. 1977. The limitations of evolutionary theory. Pages 235-242 in R. Duncan and M. Weston-Smith, eds. The Encyclopedia of Ignorance. Pergamon Press.

McClintock, B. 1953. Mutation in maize. Carnegie Inst. Wash. Year Book 52:227-237.

Peacock, W.J., Appels, R., Dunsmuir, P., Lohe, A.R. and Gerlach, W.L. 1977. Highly repeated DNA sequences: chromosomal localization and evolutionary conservatism. In International Cell Biology 1976-77, ed. by B.R. Brinkley and K.R. Porter, pp. 494-506, Rockefeller, New York.

Stanley, S.M. 1979. Macroevolution, Pattern and Process. Freeman, San Francisco.

Taylor, A.L. 1963. Bacteriophage-induced mutation in Escherichia coli. Proc. Natl. Acad. Sci. USA 50:1043-1051.

Tonegawa, S. 1983. Somatic generation of antibody diversity. Nature 302:575-581.

Tuschl, H., Altmann, H., Kovack, R., Topaloglou, A., Egg, D. and Günther, R. 1980. Effects of low dose radiation on repair processes in human lymphocytes. Radiat. Res. 81:1-9.
Wills, C. 1981. Genetic Variability. Oxford Univ. Press, London.
Wills, C. 1983. Stress-triggered evolution (submitted).
Wilson, E.O. 1975. Sociobiology. Harvard, Belknap Press, Cambridge, Mass..
Wynne-Edwards, V.C. 1965. Self-regulating systems in population of animals. Science 147:1543-1548.

Bio-mathematics

Managing Editor: S. A. Levin

Volume 8

A. T. Winfree

The Geometry of Biological Time

1979. 290 figures. XIV, 530 pages
ISBN 3-540-09373-7

The widespread appearance of periodic patterns in nature reveals that many living organisms are communities of biological clocks. This landmark text investigates, and explains in mathematical terms, periodic processes in living systems and in their non-living analogues. Its lively presentation (including many drawings), timely perspective and unique bibliography will make it rewarding reading for students and researchers in many disciplines.

Volume 9

W. J. Ewens

Mathematical Population Genetics

1979. 4 figures, 17 tables. XII, 325 pages
ISBN 3-540-09577-2

This graduate level monograph considers the mathematical theory of population genetics, emphasizing aspects relevant to evolutionary studies. It contains a definitive and comprehensive discussion of relevant areas with references to the essential literature. The sound presentation and excellent exposition make this book a standard for population geneticists interested in the mathematical foundations of their subject as well as for mathematicians involved with genetic evolutionary processes.

Volume 10

A. Okubo

Diffusion and Ecological Problems: Mathematical Models

1980. 114 figures, 6 tables. XIII, 254 pages
ISBN 3-540-09620-5

This is the first comprehensive book on mathematical models of diffusion in an ecological context. Directed towards applied mathematicians, physicists and biologists, it gives a sound, biologically oriented treatment of the mathematics and physics of diffusion.

Springer-Verlag
Berlin
Heidelberg
New York